D0855076

A CENTURY OF DARWIN

By courtesy of the Hulton Picture Library

CHARLES DARWIN AT THE AGE OF 40
From an engraving by T. H. Maguire

A CENTURY OF DARWIN

Edited by

SAMUEL ANTHONY BARNETT

Essay Index Reprint Series

BOOKS FOR LIBRARIES PRESS
FREEPORT, NEW YORK

Reprinted 1969 by arrangement

STANDARD BOOK NUMBER:
8369-1019-2

LIBRARY OF CONGRESS CATALOG CARD NUMBER:
71-76891

PRINTED IN THE UNITED STATES OF AMERICA

CONTENTS

CHAP.		PAGE
	Preface	xi
1	THEORIES OF EVOLUTION by C. H. Waddington	1
2	SPECIES AFTER DARWIN by Theodosius Dobzhansky	19
3	THE THIRD STAGE IN GENETICS by Donald Michie	56
4	DARWIN AND ANIMAL BREEDING by John Hammond	85
5	DARWIN AND CLASSIFICATION by R. A. Crowson	102
6	DARWIN AND THE FOSSIL RECORD by A. S. Romer	130
7	DARWIN AND EMBRYOLOGY by Gavin de Beer	153
8	THE STUDY OF MAN'S DESCENT by Wilfrid Le Gros Clark	173
9	THE "EXPRESSION OF THE EMOTIONS" by S. A. Barnett	206
10	SEXUAL SELECTION by J. Maynard Smith	231
11	DARWIN AND CORAL REEFS by C. M. Yonge	245
12	DARWIN AS A BOTANIST by J. Heslop-Harrison	267
13	DARWINISM AND THE SOCIAL SCIENCES by Donald G. MacRae	296
14	NATURAL SELECTION AND BIOLOGICAL PROGRESS by J. M. Thoday	313
15	DARWINISM AND ETHICS by D. Daiches Raphael	334
	REFERENCES	360
	INDEX	367

CONTRIBUTORS

S. A. Barnett, Senior Lecturer in Zoology, University of Glasgow.

Sir Gavin de Beer, F.R.S., Director, British Museum (Natural History).

Sir Wilfrid Le Gros Clark, F.R.S., Dr. Lee's Professor of Anatomy, University of Oxford.

R. A. Crowson, Lecturer in Taxonomy, University of Glasgow.

Theodosius Dobzhansky, Professor of Zoölogy, Columbia University, New York.

John Hammond, C.B.E., F.R.S., Reader Emeritus in Agricultural Physiology, University of Cambridge.

J. Heslop-Harrison, Professor of Botany, The Queen's University, Belfast.

Donald G. MacRae, Reader in Sociology, University of London.

Donald Michie, Honorary Research Assistant, Royal Veterinary College, University of London.

D. Daiches Raphael, Senior Lecturer in Moral Philosophy, University of Glasgow.

A. S. Romer, Director, Museum of Comparative Zoölogy, Harvard College.

J. Maynard Smith, Lecturer in Zoology, University College, University of London.

J. M. Thoday, Head of Genetics Department, University of Sheffield.

C. H. Waddington, F.R.S., Professor of Genetics, University of Edinburgh, and Director, Agricultural Research Council Institute of Animal Genetics.

C. M. Yonge, C.B.E., F.R.S., Regius Professor of Zoology, University of Glasgow.

TEXT-FIGURES

PAGE

1. Heads of four of Darwin's "finches" from the Galapagos Islands, showing the shapes of the beaks 3
2. The effect of an environmental stress on the fruit-fly *Drosophila* 16
3. Three stages of race- and species-formation 29
4. The races of the salamander *Ensatina eschscholtzi* in California 47
5. The process of species-formation in time 50
6. The "two-way" and "one-way" theories of heredity 57
7. Some mammalian cells 58
8. The "chromosome theory" of heredity 60
9. The chromosomes of the fruit-fly *Drosophila melanogaster* (female) 61
10. Types of antigen produced by *Paramecium* in a typical experiment 64
11. The final state of the "one-way" theory of heredity: "dualism" 68
12. Transplantation of cell-nuclei from embryos into eggs in the frog 73
13. Survival rates after nuclear transplantation 74
14. Glavinic's experiment with two varieties of tomato 81
15. The effect on egg-production of the selective breeding of hens 90

PAGE

16. Diagram showing how the different tissues of the body compete for food from the blood stream 93

17. Plan of McMeekan's experiment on the growth of pigs 94

18. Differences in the body-proportions of well-fed and poorly-fed pigs at 16 weeks 94

19. Changes in the body-proportions of pigs, brought about by selection under high-plane nutrition 95

20. Pigs of the same live weight (200 lb.) but reared along differently-shaped growth-curves 96

21. Changes in the milk-yield of Indian herds after adoption of modern feeding and selection techniques 100

22. The phylogenetic diagram from *On the Origin of Species* 104

23. Amphibians, illustrating pædomorphosis 108

24. Comparison of the Australian lungfish, the modern cœlacanth, a fossil cœlacanth and the carp 112

25. A "living fossil" among the spiders, *Liphistius malayanus* 116

26. A fossil lamp-shell, *Lingula quadrata* 116

27. Crustacea with bivalve shells 124

28. Some of the grotesque Pleistocene mammalian inhabitants of South America 133

29. Map of South America, showing where Darwin collected fossil vertebrates 135

30. Pædomorphosis in Ammonites 161

31. The pædomorphic origin of Chordates 165

PAGE

32. The pædomorphic origin of man 167

33. The evolutionary relationships of men and apes 175

34. The skulls of a male gorilla and modern man compared 188

35. The cultures of the old stone age 189

36. The skulls of a female gorilla and a South African " man-ape " compared 191

37. The skulls of Java man and modern man compared 193

38. Palates and teeth of male gorilla, " man-ape " and Australian aborigine compared 195

39. Pelvic bones of " man-ape ", gorilla, chimpanzee, orang utan and modern man compared 197

40. The canine tooth of modern man 203

41. Baboon " in placid condition " and " pleased by being caressed " 209

42. Dog " approaching another dog with hostile intent " 213

43. Cat " terrified at a dog " 214

44. Dog " in a humble and affectionate frame of mind " 219

45. Cat " in an affectionate frame of mind " 225

46. Chimpanzee " disappointed and sulky " 227

47. The analogies between advertisements and the competitive signals of animals, and between signboards and non-competitive signals 237

48. The courtship dance of the fruit-fly *Drosophila* 241

49. Part of a young colony of coral (*Porites haddoni*) 256

PAGE

50. Two of Darwin's diagrams of coral reef formation 259

51. Map showing present distribution of dwarf birch (*Betula nana*) in the British Isles 272

52. Diagrams of orchid flower by Darwin and Vermeulen 277

53. Darwin's diagram illustrating heterostyly in the primrose 285

54. Darwin's diagram of the circumnutation of a seedling of cabbage 290

55. Changes in the average hair-number resulting from artificial selection in populations of flies 325

PLATES

Charles Darwin *Frontispiece*

FACING PAGE

I. " Industrial melanism " in moths 60

II. Stockard's bulldog-bassethound cross 61

III. Seaward edge of growing coral reef at low tide 246

IV. Skeleton of a reef-building coral 246

V. Ifaluk atoll, Western Carolines 247

PREFACE

It was once said of Darwin's ideas that they must be false, because they were dangerous. Nowadays Darwinism is respectable, as this volume shows. A few writers still believe that Darwinism is dead, but this is due to ignorance both of Darwin's work and of its position in twentieth century biology.

Most of the chapters that follow show Darwin as a biologist of extraordinary penetration, industry and scope. His work, and that of his contemporaries, inspired the researches of (among others) embryologists, botanists, geologists, classifiers and even social anthropologists—often working on specific themes which had already been developed by Darwin himself. Nevertheless, Darwin as a scientist remains to many an enigma, even after, or perhaps because of, a detailed study of his works. The probable explanation is the one implied by Crowson in chapter 5, when he writes that Darwin was not a modern, professional biologist at all (as T. H. Huxley was), but the last prominent representative of an older tradition: he was a prosperous amateur, primarily a naturalist. Although, as Heslop-Harrison shows in his remarkable chapter on Darwin as a botanist, Darwin was an experimenter too, his principal rôle was in description and in drawing inferences from large masses of ordered data. His skill in inference is well illustrated by Yonge in his account of the work on corals (chapter 11).

By far the most important generalization propounded by Darwin was the theory of natural selection. This volume marks the centenary of his first announcement of this theory in a joint paper with Wallace. Like Darwin among biologists, the theory of natural selection occupies a dominating and yet a peculiar place in biology. Darwin himself never formulated it in a logically valid way: he adopted Herbert Spencer's expression, " the survival of the fittest ", without waiting to ask: what is the definition of fitness? Darwinian fitness, however, is measured by the capacity to produce offspring, and

Spencer's phrase, in a biological context, is hardly more than a tautology: the survival of those that survive. A rigorous expression of the theory of natural selection was given by C. H. Waddington, when he wrote: "Natural selection is an inevitable consequence of genetical variation in fitness". [6][1] If a population of a species includes individuals which contribute to different extents to later generations; and if these differences are, at least partly, genetically determined; then there will be a corresponding change, with time, in the genetical make-up of the population.

Biological fitness, then, is nothing to do with athletic prowess or general physical health, *unless* these are correlated with superior achievement in leaving descendants. And natural selection is not an *agent*, like a farmer choosing seed or bulls: it is the name for a process which arises from the nature of living things, in particular, from their inheritable variability.

By a paradox the process of natural selection has, over short periods, a conserving effect rather than a modifying one. In a constant environment the great majority of the individuals of a species are very precisely fitted to their habitat, and almost any change from the typical will be a disadvantage: consequently, many variant forms are eliminated before they can reproduce, or at least leave fewer descendants than the "wild type". Over long periods a species may display "trend evolution", that is, a tendency to alter slowly and consistently in a specific direction. J. B. S. Haldane has described this as an almost unimaginably slow process: for characters such as the lengths of bones he shows that "no measurable evolutionary changes are to be expected in most species in 10,000 years" [2]. This is the type of evolutionary change that can be followed in detail in the fossil record of certain groups, including the South American mammals studied by Darwin during his famous voyage, and described here by Romer in chapter 6.

The crucial question of *divergence*, the formation of new and separate forms from a common stock, has to be studied by different methods; and these, with the results so far obtained, are brilliantly summarized by Dobzhansky in chapter 2. As

[1] Throughout this book references are indicated by numbers in square brackets. The bibliographies of the different chapters are on pages 360–366.

he shows, despite the slowness of organic evolution, it is possible to identify stages in species-formation in existing forms.

The theory of natural selection implies that the changes, either of trend evolution or of divergence, are related to the demands or opportunities offered by the environment: in other words, that the changes are adaptive. It is easy to talk about " adaptation " in a general way, and to leave the content of this notion exceedingly vague. This is avoided if we specify what an organism is adapted for in any given instance, and how it differs from others which are less well adapted. An example may be taken from the appearance, in Britain, in large numbers, during the past half-century, of dark forms of many species of moths. This phenomenon, first rigorously studied by J. B. S. Haldane in 1924, has recently been shown by H. B. D. Kettlewell [3] to be a particularly elegant example of the effect of a change in the environment on the genetical structure of populations. In areas subject to atmospheric pollution by the filth from chimneys, the dark forms are at an advantage compared with the light, because they are less easily seen by predatory birds (plate 1). In this case, a man-made change of environment has led to adaptive change in animal populations which can be demonstrated by observation and experiment.

Evolution by natural selection depends, then, on the adaptability of organisms, that is, on the presence in each population of genetical variation. While natural selection, in most circumstances, tends to reduce variation, the mechanism of heredity—the chromosomes and genes—is a means of conserving it. The importance of maintaining variability is discussed especially by Thoday in chapter 14. This was a problem which defeated Darwin, and led him to fall back on Lamarckism, or the " inheritance of acquired characters ", as a source of variability. In the past fifty years genetics—the most notable post-Darwinian achievement in general biology—has gone far to dispose of this question, as Waddington illustrates in chapter 1, though Michie strikingly shows in chapter 3 that the present situation in biology is not quite as simple as it is sometimes made out to be.

Despite all this, for many laymen " Darwin " means, not

natural selection, but "the monkey theory". The present situation regarding the study of man's descent is expounded by Le Gros Clark in chapter 8, but the lay response reflects much more than a novel development in human biology. For Darwinism represents a major revolution in human thought—one that quite transcends any single scientific question. It has helped to extend our concept of the past, from a belief in a creation a few thousands of years ago to an attitude in which it is possible to doubt whether the notion of a beginning has any meaning. Darwinism is consequently incompatible with literal acceptance of any of the myths of creation; it gives man a place in nature, in which he resembles all other species in being a product of evolution; and evolution itself is subject to matter-of-fact investigation. Even the mysteries of animal and human behaviour (chapters 9 and 10) can be objectively studied, and must in any case be regarded as consequences of the quite unmythical process of natural selection.

Darwin was not himself responsible for all this. It would be an absurd lapse into the cult of the individual to say that he was. Darwin's work and ideas were, on the contrary, to a material extent a *product* of the intellectual climate of his time. It has often been pointed out that his emphasis on conflict and struggle reflected the nature of the acquisitive society of which he was a privileged member. Similarly, the interpretation of natural selection in terms of Tennyson's "Nature, red in tooth and claw" was agreeable to those who were doing well out of private enterprise. Meanwhile, as MacRae points out in chapter 13, the other camp, that of socialism, welcomed Darwinism both for its rationality and for its emphasis on the ever-changing nature of living things—including man. This welcome was however not agreeable to Darwin: the autographed copy of Marx's *Capital* remains uncut at Downe House.

The representative character of Darwin's thought within biology is still more easily shown. The parallel and entirely independent proposal of the theory of natural selection, by A. R. Wallace, is an outstanding example of the very common phenomenon of simultaneous discovery in science. The idea of evolution itself was already familiar, if conventionally rejected, and was certainly implied in the geology of Darwin's time, as well as the biology.

The noisiest impact of Darwinism on our ideas of man as a social animal was not political, in the narrow sense, but ethical. The subject of Darwinism and ethics has been thoroughly confused by two misunderstandings. One of these is fully exposed by Raphael in chapter 15: it is the belief that we should accept something as good because it is "natural"; for instance, that if organic evolution is a product, in Darwin's words, of " the war of nature, . . . famine and death " then social evolution *ought* to be so too. This type of error has often been exposed, but has not yet been routed.

Secondly, there is the notion that, because the theory of evolution by natural selection conflicts (whether in fact or in appearance) with religious doctrine or belief, it follows that Darwinism leads to immorality. Perhaps this view is now rarely held. It is based on the assumption that there is a high positive correlation between adherence to theism and moral behaviour. This is hardly supported by the lives of men such as Darwin himself or T. H. Huxley, and it is contraverted by the researches, for example, of W. A. Bonger. [1]

A more difficult question has been very judicially discussed by D. Lack. He concludes that " the theory that man's moral behaviour has been evolved . . . by natural selection fails to account for the essential aspects of moral experience " [4]. An opposite point of view has been expressed by many others, notably M. F. A. Montagu [5], but equally within the framework of the highest ethical principles.

It is obvious that this is not the sort of question that can be answered to everyone's satisfaction. Nor is it of a kind with which scientists, as such, are usually concerned. Most scientific work, like most of Darwin's, consists of painstakingly assembling facts (or apparatus), and of drawing conclusions, usually tentative, on specific points of detail. For most scientists today, moreover, the details on which they work come within applied science. Occasionally, the facts uncovered, or the conclusions reached, are disturbing to conventional beliefs. A century ago this was conspicuously the case with the theory of evolution. Today, Darwinism may be respectable but, like all science, it remains dangerous to fixed habits of thinking.

March 1958 S. A. BARNETT

1

Theories of Evolution

By

C. H. Waddington

For everyone, except perhaps a few pedantic historians, the " Theory of Evolution " means " Darwin " as unmistakably as " Relativity " means " Einstein ". This is rather surprising, because the pedants are actually quite right, as their habit is. The theory of evolution—which means the theory that living things do not remain unchangeably the same from generation to generation, but gradually alter in character until they eventually become significantly different from their early ancestors—that had been thought of by many people before Charles Darwin. In fact his own grandfather, Erasmus Darwin, wrote an elaborate exposition of the idea some fifteen years before Charles was born, in the form of a long and unfortunately remarkably boring poem entitled *Zoonomia*; while a little later, within a year or two of Darwin's birth in 1809, there appeared the fully worked-out evolutionary theory of a man who is still Darwin's most important rival, the French biologist Lamarck. In vaguer and more general forms one can find the idea even in much earlier writers. The Greek philosophers, who thought of almost every notion which mankind has been able to conceive, certainly thought of this one too; but in their usual way they tasted it, and sniffed at it, but never really tested it by experiment or detailed observation.

The reason why it is Darwin's name which we associate with the idea of evolution, rather than that of Lamarck or one of the other earlier writers, is to be found partly in the fact that his contribution to the subject was pre-eminently one which could be, and was, submitted to extensive and detailed testing. There were certainly other factors at work, perhaps more important ones. In particular, Darwin wrote at a time when the intellectual world was just becoming ready to consider, and eventually to accept, the revolutionary change in philosophical outlook which a belief in evolution rather than special creation inevitably implied. But it was in the first place its commonsense, down-to-earth quality which gave Darwin's theory its impetus towards winning agreement. Intelligent and open-minded men felt that its arguments were incontrovertible. " How very silly we were," they said, " not to have thought of that ourselves " [1].

SPECIES

As everyone knows, the title of Darwin's most famous book is *On the Origin of Species*; and the context in which his evolutionary theory was first presented was that of the doctrine of the constancy of species. Animals and plants can, for most practical purposes, be classified into different kinds, each of which man has been in the habit, apparently since the days of Adam, of identifying by a characteristic name. When this classification is done carefully by professional biologists, the resulting groups are known as species.[1] The question immediately suggests itself: Are the species really quite distinct from one another or do they shade off into each other, so that one can draw only rather artificial dividing lines for purposes of convenience? The answer which was given in Darwin's day, and which still seems on the whole an adequate account of the situation, is that among living animals and plants the groups are usually distinct from one another, although in some types of organism they are clearer than in others, and in a few they are extremely uncertain and vague.

In so far as species are, and remain, distinctly different from one another, it is possible to accept the story of Creation given in the book of Genesis as a reasonably adequate account of how living things have come into being. That is to say, one can believe that in some sense a Creator has made each species as it exists today. If, on the other hand, species are found not to be truly distinct from one another, and in particular if it can be shown that one species becomes converted into another by mundane processes occurring before our own eyes, then one can scarcely escape the conclusion that the story of Genesis omits some points of considerable interest and importance. The pre-Darwinian writers on evolution had shown that

[1] [See also chapter 2.—ED.]

the fact that most species today are fairly well defined does not finally dispose of the matter. There remains the possibility that they gradually change as time passes, so that what was once species A eventually becomes converted into species B. It had already been recognized in Darwin's time that the question whether this is so or not is a fundamental problem for biology. Darwin began by believing, like most of his contemporaries, in

1.—Heads of four of Darwin's "finches" from the Galapagos Islands, showing the different shapes of the beaks. Darwin found different, but closely related, species of this group on different islands, and concluded that they had each diverged from a common ancestor, resembling the mainland form, as a result of long-continued geographical isolation. (From the *Journal of Researches.*)

the fixity of species. During his five-year voyage round the world on the surveying and exploring expedition in the *Beagle* he gradually accumulated a huge body of observations which convinced him that he had been wrong and that species do become modified with the lapse of time.

It is unnecessary to relate the facts which led Darwin to this conclusion. They were quite complicated and extremely detailed. In general the point was that species—even clearly distinct species—fall into groups which closely resemble each other, the whole of one group of species sharing certain common features in which it differs from other groups. Many facts about such species-groups, particularly their geographical distribution, suggested to Darwin that they could be most easily explained by the hypothesis that the different species comprising a group had originated from a common ancestor.

For instance, in the isolated archipelago of the Galapagos Islands, Darwin found a number of species of a kind of bird usually referred to as a " finch ", although it was not identical with the British bird of that name. The Galapagos finch-species are not found anywhere else, and although they are quite well-defined, separate species, these share many features with each other. This is just what one would expect if they had all originated within the archipelago by divergence from a single ancestral species which had in some way once succeeded in reaching that isolated locality. Looking at a large number of such cases, Darwin argued that an explanation in terms of evolution is more convincing than the alternative, which supposes that the existing species have been specially created to exhibit geographical distributions and morphological relationships of some apparent regularity which are, however, quite meaningless. It is, of course, difficult to guess what the ways of a Creator would be, but it would be nothing less than blasphemous to accuse him of building into his creation apparent clues to an underlying pattern which was in reality non-existent.

NATURAL SELECTION

In the *Origin* Darwin brought together vastly more factual material of this kind than had been previously assembled, and thus put the case for the mutability of species very much more convincingly than any earlier author ; but above all, his contribution was to suggest a plausible mechanism by which the changes might be produced. He found the clue in the work of Malthus, who had argued that man—the human species—would always tend to increase in numbers to the limit set by the amount of foodstuffs which he could produce from the region he inhabited. Darwin pointed out that most species of animals and plants can, potentially, reproduce much faster than man; they are often extremely prolific of eggs and seeds. Between the conception and the maturity of a new generation, many deaths occur, and in most species the proportion which die before being able to reproduce themselves is enormous. If individuals of any one species vary among themselves—and Darwin gave voluminous evidence that they do—and if the variations are inherited—and again Darwin demonstrated many examples in which they are—then, he argued, it is inevitable that some variant types that are weaker than their fellows will more often die before becoming parents ; while others, the more efficient ones, will more frequently survive to pass their qualities on to the next generation. This is the principle of what Darwin called natural selection. It is a simple and unassailable argument. Given that more offspring are conceived than are necessary to preserve the numbers of the population, and that there are inherited variations between individuals, natural selection *must* occur [2].

Darwin had produced convincing evidence that species do change, and he had shown that there is a biological mechanism which must inevitably produce changes. The two points together were enough to convince nearly all biologists that evolution does in fact occur. This conviction has turned out to be one of the major germinal notions which colours the whole of man's idea of the kind of world he lives in. Accepting it, we can no longer regard any part of the living world as immutable, existing simply as its unchanging self. We are forced to adopt one of the old Greek views—that everything is flux and process; everything has a history; it is its history which has moulded its character, and it is in terms of its history that its nature is to be understood. In its effect on man's general mode of thought about the world, the theory of evolution is at least as far-reaching as the most important earlier contribution of science, namely, Newton's laws of motion and concepts of mass and force.

The theory as Darwin advanced it was not as complete and self-sufficient a system as Newton's. In the latter, there were no obvious loose ends; everything concerning the motion of all entities, from grains of sand to the whole earth itself, found a place within the system, and a place which appeared quite adequate until, after some two hundred years, far more precise methods of measurements and the study of the infinitesimally small and unimaginably huge worlds of the atom and the stars enabled Einstein to formulate a still more inclusive scheme. In Darwin's theory there was from the beginning a major gap.

Granted that natural selection will cause changes of some kind in living organisms, can we be sure that these changes will be adequate to account for the evolutionary alterations which would be necessary to convert one species into another? The answer to this must obviously depend on the kinds of hereditary variation which occur in natural populations and are thus offered as raw materials for natural selection to work on.

LAMARCKISM

Darwin was very well aware that he did not fully understand variation. He knew that in any large population of animals or plants no two individuals are quite alike, but he was very uncertain how these variations arose; and, although he knew that to some extent at least some of them are inherited, he did not know how biological heredity operates. His ideas on these subjects changed during the course of his life, and it is doubtful if he ever felt them to be quite satisfactory. In the main, he adopted the theory of heredity which was usual in his time, namely that both mother and father contribute hereditary qualities to their offspring, which appears as a synthesis or blend of its two parents. This theory is perhaps the simplest, and what would occur to one first when considering the matter, but it presents several difficulties. For instance, it provides no obvious reason why brothers should ever differ. Again, there is a point which particularly troubled Darwin: if variations blend in inheritance, then crossing between the variant types in a population will produce later generations which come closer and closer to a uniform intermediate averageness. On a system of blending inheritance, in fact, variation will gradually disappear, and evolution will come to a standstill for lack of anything for natural selection to work on, unless there is some agency which makes new variation as fast as the blending effects of cross-breeding gets rid of it. Darwin searched assiduously for something which could be supposed to produce variation. He was never quite certain that he had found anything satisfactory, but, sometimes at least, he was tempted to fall back on the earlier suggestion of Lamarck, which he had originally derided, that the conditions under which an organism lives produce in it effects which are inherited, and which make up for the individuality lost through blending.

The fate of Lamarck's theory has been very different from that of Darwin's. Whereas Darwin became the high priest of orthodoxy in biology, Lamarck has been for the most part regarded as a cranky and light-weight heretic. His supporters in recent times have been in the main old-fashioned biologists who have opposed the whole modern advance in the study of heredity. The only group who have tried to make his doctrines into a part of the orthodox core of biological theory are a school of Russian biologists associated with Lysenko, who have argued that because in some of his writings Darwin supported Lamarck's ideas we should do so too; but almost all biologists outside their own coterie remain quite unconvinced.

One of the reasons why Lamarck's theories have been so much less attractive to biologists than Darwin's is to be found in the type of concept which he employed. Darwin's theory of natural selection deals in hard facts which could be counted and entered in a ledger: numbers of eggs fertilized, of offspring which reach maturity and so on; all notions which the most hard-headed materialist can find congenial. Lamarck on the other hand started from the concept of the Will, an idea which gives most biologists a rather queasy feeling, since any theory which can find a place for it must involve notions beyond those necessary to cover the facts of the inorganic world. Lamarck's theory boldly starts from the consideration that animals choose, by an exertion of the will, to conduct their lives in a certain manner. This, he goes on, involves an effort to use some of their organs in appropriate ways. And the kernel of his theory is that the effects of such use or disuse of particular organs—the enlargement of muscles which is the result of exercising them, the keenness of sight which is produced by continual practice in visual discrimination, and so on—are inherited and passed on to the next generation.

Apart altogether from the awkwardness for science of a concept such as that of the will, this theory has many obvious imperfections. For instance, it hardly applies, in any full sense, to plants; and, although there is no obvious reason why evolution should occur in just the same manner in the plant as in the animal kingdom, a theory which applies to only one of these two realms is clearly incomplete. But the crucial point is, of course, the last step in the argument given above.

Is it, or is it not, the case that the effects of use and disuse are inherited? It should be possible to decide by experiment.

In the years around the end of the nineteenth and beginning of the twentieth centuries, a great deal both of thought and experiment were devoted to the question. In most of this work, the problem was stated in rather wider terms than those implied by the phrase " use and disuse ". The expression " an acquired character " was used to describe any feature in which an individual member of species differed from the normal type owing to influences which had impinged on it during its lifetime, whether those influences involved excessive use or disuse of certain organs or operated in some other manner. And the question was asked, are acquired characters inherited? A considerable amount of experimental work failed to reveal any examples which quite convincingly demonstrated such inheritance. Some experiments were undoubtedly badly planned. For instance, it was shown, after considerable effort, that the surgical removal of a part of an animal does not lead to the production of offspring in which that part is lacking, even when the operation has been carried out for several generations; a conclusion which had in fact already been tested, and much more thoroughly than in the laboratory, by the centuries-old practice of circumcision amongst Jews and others. But even much more sensible experiments never yielded unequivocal evidence that acquired characters were transmitted to the offspring.

Reinforcing the experimental work, theoretical consideration of the problem failed to discover any means by which an external agent impinging on the body of a developing plant or animal could influence its germ cells in such a way as to produce an effect on the next generation similar to that caused in the organism immediately affected. Darwin indeed invented a theory of heredity which might have accounted for such an effect. He supposed that all the cells of the body give off small particles (which he called gemmules) which become accumulated in the germ cells and transmit through them to the next generation the exact condition of the cells from which they rose. But this was a purely hypothetical suggestion, advanced at a time when there was not yet an adequate body of facts on which to found any firmly based theory of heredity;

and as such facts become available, Darwin's theory of " pangenesis ", as he called it, was found to be quite untenable.[1]

MENDELISM

The resolution of Darwin's difficulty about the disappearance of variation under a system of blending inheritance was in fact found, not by calling in older theories such as those of Lamarck, but by experiments which revealed the true mechanism of biological inheritance. Mendel's work should, indeed, have been available to Darwin, since it was published in 1865, but, as has often been recounted, it suffered one of the most remarkable fates which has ever befallen a major scientific advance, and remained quiet unappreciated and forgotten until rediscovered in 1900. The essential point of Mendel's discovery was that a variation occurs by an alteration in some discrete particular hereditary unit; during reproduction, these units become reassorted into new combinations, but each retains its identity and does not blend with any other similar unit. It is true that any given hereditary unit may, in some of the individuals in which it is carried, have little or no obvious effect on the character of the adult organism; but even when this is the case, experiment shows that the unit is still present unchanged and may be passed on to later generations in which, under various circumstances, conditions may be more favourable for its effects to be noticeable. Thus Darwin's problem does not arise. Biological inheritance does not involve any true blending of the qualities of the two parents, and there is no tendency for variation to be diminished by cross-breeding.

The rise of modern genetics, built on the foundations laid by Mendel, added considerably more to Darwinism than the solution of this difficulty about the preservation of variety. Within a few decades after 1900, biologists gained a thorough understanding of the mechanisms by which hereditary qualities are passed on from generation to generation. It is obviously in terms of this knowledge that Darwinism must be expressed. Natural selection is effective only in so far as it brings about changes in the hereditary qualities of a population. The essential of its action is that certain organisms, bearing certain hereditary qualities, leave more offspring than others with

[1] [For a further discussion, see chapter 3.—ED.]

other qualities. The hereditary units are nowadays usually spoken of as " genes "; the fundamental effect of natural selection is to alter the frequency with which particular genes are present in the population of animals or plants which is being considered. Darwin's theory of natural selection, in the form in which it becomes rephrased when it is translated into terms of gene frequencies, is often known as neo-Darwinism.

NEO-DARWINISM

This has been the dominant type of evolutionary thought for the last quarter of a century. It has opened the way to very considerable advances in our knowledge and understanding of evolutionary processes as they occur in nature. Of all its successes, two are perhaps to be regarded as of outstanding importance. The first was the demonstration that the inheritance of minor variations such as differences in size or strength, where there may be a continuous gradation from smaller to larger with no clearly distinct categories, follows the same basic rules which apply to such sharply marked characters as the coat colours of cattle or the contrasting flower colours of the sweet-pea. The only difference is that where the hereditary characters are clearly distinct, they are usually influenced by only one or a few pairs of genes, whereas if we find a continuous series of intermediate types, there are likely to be a larger number of factors involved. The point is important, because among wild animals and plants it is much more usual to find continuous variation, and it would therefore seem probable that it is this type which provides the main raw material for evolution. If, as some people thought at first, the discovery of definite hereditary units applies only to cases where the variations are clearly distinct, then Darwin's difficulty about the disappearance of variation would not have been solved after all; but this fortunately turned out to be a false alarm.

The second major contribution of neo-Darwinism was even more important. It was mentioned above that an animal may carry a hereditary factor without showing any sign of its presence. This can occur because each organism actually contains two of each type of gene, one received from its father and one from its mother. The two members of such a pair of genes may not be absolutely identical, though they belong to the same type; they are often symbolized by the same letter printed in different ways, for instance **Aa.** In general, one detects the existence of the genes by the fact that an animal with two **A**'s looks different, in some way, from one with two **a**'s. And one would expect an animal with one **A** and one **a** to be intermediate. So it may be; but very often it looks identical with the **AA** type, showing no sign that one of the **A**'s has been replaced by an **a**. In such a situation one speaks of the **A** gene as " dominant " and the **a** one as " recessive ".

The fact that genes may be recessive means that mere inspection of a population of animals or plants is not adequate to disclose all the hereditary variation it may contain. There may be many recessive genes, but in so far as they are carried by organisms which also contain the corresponding dominant, they will remain invisible; it is only when two of them come together, to produce an individual of **aa** type, that their presence will be detected. When wild populations were investigated with this possibility in mind, it was found that they do indeed usually contain many recessive genes, and therefore a much wider range of hereditary variation than would have been thought at first sight. This means that there is more for natural selection to work on; if conditions change, so that some new type of adaptation becomes desirable, the population may be able, as it were, to pull something new out of the hat.

The flexibility of a species is still further increased by the fact that the effect of any one pair of genes depends to quite a large extent on what other genes the organism also contains. One may, for example, have a pair of genes, say **B** and **b**, which affect the deposition of dark pigment in the hair, with **B** dominant and producing dark hair and **b** recessive and producing paler hair; but it would probably be found that the exact shade of darkness produced by **B** and the degree to which it is dominant over **b** (that is, whether **Bb** is exactly like **BB,** or is somewhat paler) are both dependent on what other genes are present. Again, when this theoretical possibility was investigated, it was found that it is actually realized in nature. Wild populations contain not only many recessive genes with strongly marked effects, but also a large number of genes which cause minor modifications and adjustments.

In recent years, more and more importance has been attributed to genes with such slight, relatively inconspicuous effects. During the earlier years of neo-Darwinism, it was usual to picture a wild population as consisting of individuals, all of which contained, on the whole, the same set of so-called " wild-type " genes, although each might have one or two aberrant recessives. This is still the accepted view as far as genes with major effects are concerned, but nowadays many biologists think that evolutionary advances depend much more on the genes which control continuously varying characters, and find reason to suppose that wild animals and plants are less uniform in their content of genes of this type. Instead of both members of a pair being similar, **AA** or **aa,** they are more often unlike, of the **Aa** pattern. In fact, in most populations there are more than two representatives of each type of gene, so that some animals may be $\mathbf{a_1a_2}$, others $\mathbf{a_1a_3}$, $\mathbf{a_2a_3}$, or $\mathbf{a_3a_4}$, etc.

One speaks of all the genes which occur in some individual or other of the population as its " gene pool ". The most recent developments of neo-Darwinist theory emphasize the large number of different representatives of the genes which are found in gene pools, and are very much concerned with the problem of how this heterogeneity is controlled. Analysis of wild populations suggests that all the genes in any particular interbreeding population show a certain concordance with one another, in the sense that, whatever particular set of these genes an individual contains, the resulting animal will be reasonably fit and efficient; whereas if, by cross-breeding for instance, genes from one gene-pool become mingled with those from another, the result is likely to be less favourable. Such observations are the basis for what is known as the concept of the " co-adapted gene-pool ", that is, the notion that natural selection gradually restricts the collection of genes in a population to those which can " go along together " in any combination. This idea is the most advanced point to which neo-Darwinist theory has yet arrived.

MUTATION

Although the development of the science of genetics, from Mendel to the present day, has provided us with a very much more concrete, detailed and precise understanding of the operations of natural selection, it is difficult to feel that it has quite satisfactorily completed Darwin's theory by an adequate account of the nature and origin of variation. In the discussion above, hereditary variation between animals and plants of the same species has been attributed to the fact that in different individuals a given type of gene is sometimes represented by one form, sometimes by a slightly different one. These variant forms of one and the same basic type of gene are known as "alleles". A full theory of evolution must contain some account of how new alleles arise, and what sort of variation they may produce in the organisms in which they are exhibited.

Darwin, as has been mentioned, at some periods of his life was tempted towards Lamarck's idea that the use and disuse of various organs tended to produce hereditary variations—that is, what would nowadays be spoken of as "mutations" of one allele into another, new type; but more recent work has failed altogether to find evidence for this. Geneticists have, however, been able to ascertain certain facts about the occurence of such mutations. In the first place, they happen in all animals and plants in quite normal conditions, under circumstances in which it has proved impossible to implicate any particular factor in the situation as a cause of their occurrence. Again, when allele-changes occur in this way, we have as yet no means of foretelling in which gene-type they will happen or what the nature of the change will be; that is, if there are gene-types **A, B, C, D,** etc., one cannot predict when change will affect **A** or **B** or **C**; nor whether, if **B** mutates, it will alter to $\mathbf{B_1}$, $\mathbf{B_2}$, $\mathbf{B_3}$ or some other possible allele.

These limitations on our knowledge apply even to the experimental methods which have been discovered for influencing mutation. Muller, in 1927, found that X-rays and similar ionizing radiations increase the frequency with which mutations occur; and more recently certain substances have been shown to act in a similar manner when fed to animals. But still we cannot choose beforehand which gene-type will be caused to undergo alteration, nor specify what the change will be.

Geneticists have had to conclude that gene mutation—the origin of new alleles, which gives rise to the hereditary variation on which evolutionary progress is based—is essentially a "random" process. The gene is pictured as being unstable; perhaps simply because it is so large. It behaves rather like a house of cards; sometimes it falls down for no ascertainable reason, and it is still more likely to topple if someone jogs the table—which is an analogy for the effects of X-rays or the mutation-stimulating chemicals. If one conceives of a gene in terms of that sort, one has an explanation of our inability to predict either the time at which a given gene will alter or the exact nature of the change it will undergo.

To a tidy mind, this is not a very satisfying way of filling the gap in Darwin's theory concerning the origin of variation. One would like to have a coherent, logical connection between new variation and something else in the organism's world—a connection as closely knit as Lamarck provided with his theory of the origin of variation through the exercise of will and the effects of use and disuse. Instead, we are offered two signs saying "Road Closed". It is argued that one cannot hope to discover any specific cause for the occurrence of a new allele, nor any limitation on the kind of variation it may produce.

It is unlikely that either of these roads will remain permanently barred to the advance of scientific knowledge. Possibly, even probably, it is true that genes are unstable entities, which may sometimes alter spontaneously for unascertainable reasons. But this can hardly be the whole story. The genes must be involved in several ways in the complicated chemical processes going on in the cells containing them. For instance, each time the cell divides, a new gene must be manufactured out of something, to make a new copy for the new cell. Again, genes certainly affect the character of the cells containing them—that is, after all, how their presence is discovered—and they can do so only by taking some part in the operations on which the cell's life depends. It seems most probable that some, at least, of the mutations of a gene are connected with its activities of these kinds. We still have almost no detailed knowledge of what such connections may be, but one can look forward to an eventual understanding which will make the occurrence of a new mutation less blatantly incomprehensible than it is at present.

THE EFFECTS OF MUTATION

We can already see a little further up the other closed road. To say that allele-changes can produce new variation of any and every type does not cover everything we know about the subject. In the first place, a new gene mutation can cause an alteration only to a character which the organism had had in previous generations. It could not produce a lobster's claw on a cat; it could only alter the cat in some way, still leaving it essentially a cat. Secondly, there are reasons for believing that the structure of an animal or plant may be conditioned, by natural selection, so as to be more easily altered in certain ways than in others. An adult organism is brought into being by a whole succession of developmental processes which gradually mould the fertilized egg into the final form. A new gene operates by influencing these processes in some unusual way. If it is equally easy to divert development in any direction, then (if the gene changes are random in nature) the resulting variation will be of all possible kinds. But if, for some reason, the developmental processes are easy to shift in certain directions and less easy in others, the changes produced by mutation will often be exhibited as one of the patterns which the system has a tendency to assume.

There are reasons, both theoretical and experimental, for believing that natural selection would in fact operate to produce developmental systems which show such characteristics. An organism during its lifetime has to meet and deal with a whole host of circumstances, which are collectively referred to as its environment; for instance, the climate, the need to catch prey, to escape its enemies and so on. Quite often it is found that the response of an organism to such demands includes modifications of its structure. If muscles are used vigorously and often they usually become larger and stronger; an animal in an unusually cold climate may grow longer fur; or a plant subjected to a mountain climate become shorter and less luxuriant. One could multiply almost indefinitely examples of such adaptations, as they are usually called. They are, of course, " acquired characters " of the type which Lamarck thought to be inherited, but which we nowadays think not to be. But although the changes themselves are not inherited, one would expect the *capacity* to undergo such appropriate modifications of the development system to be a hereditary character. If this were so, natural

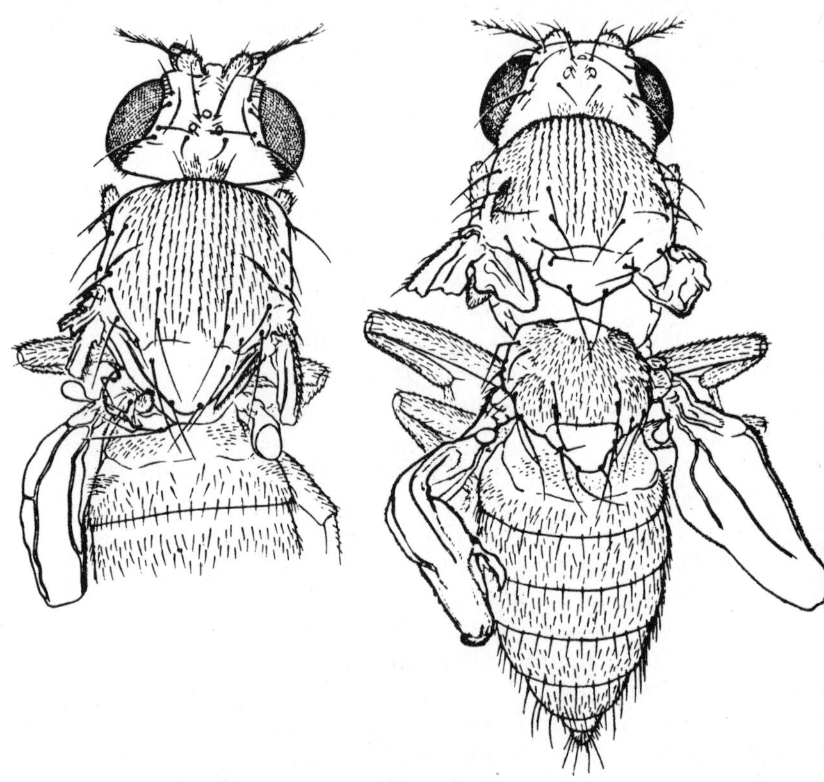

2.—On the left is a drawing of the fruit-fly *Drosophila*. Its wings have been removed to show a small lump and rudimentary extra wing appearing between the thorax and abdomen; this has been developed as a response to a particular environmental stress (treatment of the eggs with ether vapour). After many generations of selection from a stock in which initially about half the flies reacted like this, a strain was obtained which was made up mainly of individuals such as that shown on the right; there is now a complete secondary thorax, with a quite good extra pair of wings. In this selected strain, most individuals show this extreme modification even if they are not subjected to the ether vapour stress; and practically all do if the stress is applied. (From Waddington, *Evolution*, 1956, **10**.)

selection would favour those organisms which had a high capacity to become adapted to an abnormal situation, and would operate against those which either failed to become modified or were affected in useless or harmful way. It would, in fact, build in to the developmental system a tendency to be easily modified in directions which are useful in dealing with environmental stresses and to be more difficult to divert into useless or harmful paths [3].

I have recently made some experiments to test this line of thought. They showed quite clearly that, as expected, individuals show inherited differences in the way in which they react to environmental stresses, and that it is possible by selection to build up a hereditary constitution which is very easily modified in the desired direction. In fact, it was often found that, after a large number of generations of selection for those animals which were changed in a certain way, one had arrived at a constitution which spontaneously developed into the chosen form without the need for any particular environmental stress to push it in that direction. Presumably the alleles which make an individual liable to become modified in a particular manner under stress also tend to produce that type even without stress; and when enough of such alleles have been brought together into a single animal by selection of its ancestors for many generations, the original " modification " appears as its " normal " form. In this rather roundabout manner a character which was in earlier generations an acquired one and developed only under special conditions of stress may, in later descendants, come to be inherited in the absence of any environmental peculiarity. The result could be something rather similar to the Lamarckian " inheritance of acquired characteristics ", although it needs many generations of selection to bring it about, and its mechanism is quite different from, and less direct than, anything which he contemplated.

These experiments also give some grounds for confidence in the general line of argument given above concerning the origin of new variation in general. Selection for the capacity to react to the environment does indeed, as was suggested, build up genetical constitutions which have certain characteristic modifications which can be easily produced and others which can be formed only with difficulty; and chance alterations in the nature of the alleles will therefore have some tendency to produce one or other of these preselected types of adult. Thus even if we do not succeed in discovering any rationality or orderliness in the actual processes by which new alleles originate, the kinds of variation which will appear in the animals and plants on which natural selection operates is not quite senseless and unrelated to anything else; to some extent, at least, it is dependent on the kinds of stresses to which the animals have been responding in the past. And it seems likely that it is just those variations whose nature is most closely determined by the built-in instabilities of the developmental system which will be most favourable in relation to natural selection, and thus of most importance for evolutionary progress.

This conclusion goes some way—though by no means the whole way—towards filling the major gap in Darwin's theory of evolution. It also leads our attention back to some of the factors which Lamarck had originally stressed. The environmental stresses, which we have considered to result ultimately in the formation of preferred modes of modification of the developmental system, are dependent to a large extent on the kind of environment in which an animal chooses to live, and the manner in which it chooses to exploit its surroundings. What we are suggesting is, to give a simple example, that if a population of animals chooses to live by burrowing in the soil after worms, it will be selected for its capacity to respond to the stresses of such a life by developing large and powerful forearms, provided with thick claws and hands suitable for digging; and that once its hereditary constitution has been selected so that these modifications are easily produced, some new allele is likely to appear which will produce them. It is not entirely meaningless to use Lamarck's terminology, and say that the animal's will is involved in the choice it makes as to how it will sustain its life. And the effects of use and disuse of its organs, although not inherited directly, are acted on by natural selection in such a way that they set a stage on which new variations will appear; and set it so that it is probable that some of these variations will be appropriate for the demands which life is making. Darwin may not have been so wrong as many have since thought him in feeling that there was something—he was never quite sure what—in Lamarck's views.

2

Species After Darwin

By

THEODOSIUS DOBZHANSKY

A CENTURY ago, in one of the greatest revolutions in the history of human thought, Darwin demonstrated beyond reasonable doubt that man is a part of nature and kin to all life. Existing species, including man, have evolved from very different ancestors, and they continue to evolve.

Darwin's thesis that species evolve was not a new one. A. R. Wallace reached a similar conclusion simultaneously with Darwin, and Lamarck and Erasmus Darwin were evolutionists two generations before him. There are reasons to think that Descartes and Buffon made the same discovery more than a century earlier, but they chose to withhold it from publication because evolutionism was about as popular in seventeenth- and eighteenth-century France as Communism is now in the U.S.A. or Capitalism in the U.S.S.R. Darwin obviously had predecessors who anticipated some of his discoveries. But Darwin did what none of his predecessors had done; he adduced in favour of his evolutionary views a store of facts which biologists could interpret in no way other than that in which Darwin interpreted them. Carlyle called genius the " transcendent capacity of taking trouble ", and Darwin did possess such a capacity (which Carlyle then proved unable to recognize). To establish the fact of evolution a tremendous mass of intellectual rubble had to be swept away. Science is cumulative knowledge, and verification of a scientific idea is often worth more than the idea itself.

One of the greatest among seventeenth-century philosophers, John Locke, thought that " the boundaries of species, whereby men sort them, are made by men ". But Linnaeus, the father of systematic biology, would have none of Locke's man-made species. To Linnaeus, species were God-created entities, which man can apprehend and describe, but can neither change nor observe changing. Darwin made the belief in the artificiality as well as in the absolute fixity of species equally obsolete. Species evolve from varieties and races, and give rise to genera, families, and other categories. And yet, species are somehow more natural and tangible entities than races, genera and other groupings of the natural classification of organisms.

However, as Darwin himself quite clearly realized, he had far from solved the problem of the nature of biological species. Just what are the attributes which make species fundamental biological entities? Darwin wrote: "In determining whether a form should be ranked as a species or a variety, the opinion of naturalists having sound judgment and wide experience seems the only guide to follow." The wisdom of this advice cannot be gainsaid, but it poses some awkward problems. Just how do naturalists of sound judgment and wide experience recognize their species? How do we know, for example, that mankind is a single species divided into races, and not a group of several species? Or how do we know that the lion and the tiger are distinct species, although hybrids between them, so-called tiglons, are among the attractions in some zoological gardens?

A satisfactory understanding of the nature of species was simply unobtainable in Darwin's day. It had to wait for the development of a then non-existent biological science, namely, genetics. And although the foundation of genetics was laid in 1866 by Mendel, Darwin was unfamiliar with Mendel's work, as were most other biologists until 1900. More than half a century had to elapse before genetics reached a point when it was able to begin to analyse the differences between species. Even now, a century after Darwin, the species problem remains the subject of much active study and of lively controversy [2].

The existence of species is perceived intuitively even by people without formal training in biology. In most languages, the names of conspicuous animals and plants refer usually to what biologists call species. The writer had a striking demonstration of this fact in the equatorial rain-forests of Brazil. Hundreds of species of trees may grow on each square mile in these forests, but trained botanists are often at a loss to identify some of the species without examination of the fruits and flowers, which are not always available. Yet, native woodsmen are usually able to supply vernacular names of the trees after examination of the appearance, smell, and taste of the bark and the foliage. Experience has shown that at least nine times out of ten there is one-to-one correspondence between the vernacular and the scientific names of the species.

Species are facts of everyday experience to systematic zoologists and botanists, from Linnaeus down to our own time. The species concept is an operational tool of prime importance in systematic biology, and systematic biology in turn is clearly an indispensable auxiliary in all other biological studies. One must know which organisms one has observed or experimented with. On the other hand, genetics is concerned mainly with varieties, races, or subspecies (these terms are practically synonymous, and we shall use them interchangeably) of a rather limited number of species. Nevertheless, it appears that genetics may help to make more communicable the idea of species which systematic biologists, and even people at large, have been using for centuries, if not for millennia.

In what follows, it will be expedient to consider first the bases of the biological classification, as they appear from the vantage point of genetics. The results of this consideration will then be compared with those arrived at by systematists working with the aid of their time-honoured methods. It will be shown that the species established by either genetical or by systematic methods are, with relatively few interesting exceptions, exactly the same things. One could hardly imagine a better attestation of the reality of the natural phenomena called " species ".

GENETICAL BASIS OF ORGANIC DIVERSITY

Every organism builds its body from materials drawn in from the environment, that is, from the food which it consumes. However, the food materials are assimilated—modified and transmuted into a likeness of the assimilating body and of its ancestors. A similar directed transmutation is the essence of heredity. Heredity is, in the last analysis, self-reproduction. The units of heredity, and hence of self-reproduction, are corpuscles of macromolecular dimensions, called genes. The chief, if not the only, function of every gene is to build a copy of itself out of the food materials; the organism, in a sense, is a by-product of this process of gene self-synthesis.

Heredity is a conservative force. Self-reproduction tends to go on endlessly, at least as long as the supply of the susceptible environment is not depleted. Evolution occurs because the conservatism of heredity is counteracted by forces of change. These forces are mutation on the gene level, and sexual reproduction and natural selection on the population level. Mutation is a change in a self-reproducing unit, such that the altered unit makes copies of its altered structure. Because of mutation, there exist in the world many different genes instead of a single kind. If the assumption is made that life arose from inanimate matter only once, then the entire diversity of genes must have resulted from sequences of mutational changes in the progeny of the same primordial gene or genes.[1]

The evolutionary role of sexual reproduction is quite different from that of mutation. In Darwin's day this issue was hopelessly confused. On the one hand, crossing and hybridization seemed to destroy the genetical variability present in a population, since the " blood " of the offspring was believed to be an alloy of the " bloods " of the parents. On the other hand, Darwin surmised that sex is a source of genetical variety. His surmise was correct, but the matter required a firmer basis than suspicion. Such a basis was provided by Mendel, who pointed the way out of the apparent dilemma.

Leaving technical details aside, the theory is really quite simple. Suppose that two parents differ in *n* genes ; their offspring are, then, hybrids (heterozygotes) for n pairs of variant gene structures (alleles). It can be shown that such hybrids are potentially capable of producing 2^n kinds of sex cells with different combinations of the parental genes. As the numbers of the genes (*n*) increase, the numbers of their possible combinations increase very much faster. For example, a hybrid for 5 pairs of genes will produce $2^5 = 32$ kinds of sex cells. With 20 pairs of genes the number of possible kinds of sex cells grows to 1,048,576; with 31 pairs of genes more than two thousand million kinds of sex cells may be produced. This number about equals the number of persons of the human species now living on earth.

What biological meaning can be attached to such calculations? Although there is no way at present to count the

1 [This topic is discussed from a different point of view in chapter 14.—Ed.]

numbers of genes for which a person or an animal or a plant are heterozygous, there are good reasons to think that these numbers run at least into dozens, probably into hundreds, in a vast majority of individuals of sexually reproducing and habitually outcrossing species. It follows that no two brothers or sisters (" identical " twins excepted) are at all likely to receive the same sets of genes from either of their parents. The probability of siblings receiving the same complement of genes from both parents becomes negligible; with each parent heterozygous for n genes, this probability is of the order of 2^{-2n}. This explains why siblings have almost invariably (again, except for " identical " twins) different genetical endowments. This is a circumstance which was by no means well understood a century ago.

Looking from another angle, we can see why sex is so prodigiously efficient a mechanism for generating organic diversity. Indeed, sexual reproduction would be biologically meaningless if the two parents that mate always carried the same genes. Although large colonies of individuals with identical genetical endowments (pure races or clones) are formed in some plants and among micro-organisms, these creatures reproduce chiefly by asexual means. Conversely, in sexually reproducing organisms there are no pure races (the doctrine of " racial purity " notwithstanding); no two individuals are likely to have the same genes. All individuals in such organisms are multiple heterozygotes. Man is one such organism.

GENETICAL BASIS OF DISCONTINUITY

A sexually produced individual receives one copy of each kind of gene (gene locus) from his mother and another from his father. Therefore, with exceptions which need not concern us here, an individual cannot carry more than two variants (alleles) of any one gene. But mutations can change a gene in a variety of different ways, and thus give rise to a multiplicity of gene alleles. Suppose, then, that in a world there exist only 1,000 kinds of genes, each represented by only ten different alleles (both figures are patent underestimates for the world in which we live). Sexual reproduction would, then, be potentially capable of engendering 10^{1000} gene combinations. This is an almost unimaginably large number. Physicists estimate the number of electrons and protons in the visible universe to be a mere 10^{73}.

Above we have argued that, since every sexually produced individual has a set of genes different from every other, the number of gene combinations which exist in the world is enormous, being about equal to the combined numbers of individuals of all sexual species. And yet, now we can see that the existing gene combinations constitute a very minute, really a negligibly small, fraction of the possible ones. The power of the sexual process to generate new genetical endowments is far greater than can be realized.

Inevitably a new question now presents itself: Is it simply a matter of chance which of the possible gene combination are embodied in living individuals and which are not? The answer is that it is not chance alone. The existing genetical endowments are, in the main, those which permit their possessors to survive and to reproduce successfully in some environments. In the language of evolutionary genetics, the genotypes which are perpetuated have adaptive values greater than zero, since they make their carriers fit to pursue certain ways of life. Here enters on the evolutionary scene the great force, the importance of which Darwin was the first adequately to recognize. This is natural selection. Mutation and sexual reproduction without selection are adaptively ambiguous: they engender new genetical equipments regardless of whether the latter may be useful in any environment. Natural selection is a deputy of the environment. One could imagine a world in which all possible combinations of the available genes would actually be formed and would survive. In such a world there would be no species, because there would be no clustering of individuals with similar genetical endowments and no gaps between the clusters. Systematists could then make no classification like our present one: about the only classification that would be possible would be a sort of multidimensional periodic system.

Classical evolutionists used to say that natural selection makes the fittest survive and the rest die off. We prefer less dramatic language: natural selection favours the perpetuation of adaptively more efficient, and decreases the chances of perpetuation of less efficient, genotypes. Natural selection was often compared to a sieve which retains the fit and lets the unfit be swept away. We prefer a less mechanical analogy: environment presents challenges to living populations, to which the latter may, through natural selection, respond by retention or by improvement of their adaptedness to the conditions of their existence.

Natural selection imposes a restraint on the adaptively promiscuous fabrication of gene combinations. Sewall Wright visualizes the situation with the aid of a helpful metaphor. The gene patterns which make their possessors fit to survive and reproduce in a certain environment occupy " adaptive peaks "; those which would make their carriers unfit to live correspond to " adaptive valleys ". Generally speaking, a cluster of related gene combinations, differing from one another in relatively few genes, occupies a single adaptive peak. For example, the human species is a cluster of gene patterns which make their possessors human and enable them to " inhabit " the adaptive peak which represents the human way of life. The dog is another cluster (or, rather, a cluster of clusters) of genetical endowments which fit their bearers to the canine way (or ways) of life. Still other adaptive peaks are tenanted by clusters of genotypes corresponding to the species wolf, coyote, jackal, red fox, arctic fox, and so on.

The seemingly endless diversity of living creatures is fascinating, and often staggering. A biologist sees in this diversity a resultant of the stupendous power of mutation and sexual recombination to generate ever new genetic endowments. More subtle, but in a way equally remarkable, is the fact that the diversity is everywhere combined with discontinuity. Living beings come in clusters of more or less similar individuals; the individuals who mate and produce offspring are, with rare exceptions, members of the same cluster. Dogs mate with dogs and rarely if ever with wolves, coyotes, or jackals; coyotes mate with coyotes, and probably never with wolves or jackals. These mating communities are biological species.

The existence of species means, then, that natural selection has confined the sexual union within the limits in which the gene recombination is likely to produce adaptively valuable genetical endowments. In other words, the gaps between species correspond to gene combinations most of which would yield low fitness, or which would make their possessors wholly unfit to survive. These gaps correspond, in Wright's metaphor, to the adaptive valleys.

SPECIES AS INCLUSIVE MENDELIAN POPULATIONS

We have, however, run ahead of our story and glossed over important complexities. Mating communities, or Mendelian populations, often show astonishingly intricate structuring. None can illustrate this better than the human species—mankind. Mankind is a single inclusive Mendelian population, since all races, classes, nations, and tribes can, and if given opportunity do, intermarry and produce viable and fertile progeny. But it certainly does not follow that the probabilities of marriage are uniform for any two persons of opposite sex and suitable age, regardless of where they live or who they are. In the first place, there are the geographical barriers; even in our era of automobiles and aircraft the mean distance between the homes of the bride and the groom is not very great. In addition, the likelihood of marriage is influenced by a multitude of linguistic, cultural, economic, religious, and caste and class barriers. A Mendelian population which is completely panmictic, that is, one within which the chances of marriage of any boy with any girl are equal, is sometimes referred to as an isolate. The existing estimates of sizes of human isolates are surprisingly low—only hundreds, rarely thousands, of individuals in modern industrial societies. The isolates were probably much smaller in primitive societies.

The situation is not too different from the above in sexually reproducing animal and plant species in the state of nature. Strains of the fly *Drosophila pseudoobscura* from California cross easily, in laboratory experiments, with strains from Texas or from Mexico. The hybrids, at least in the first generation, are fully vigorous and fertile. But outside biological laboratories the opportunities for such crosses of strains of different geographical origin arise seldom, if ever; the average distance between the birthplace of a fly and that of its offspring varies (depending on season) from about a hundred metres to slightly less than one kilometre. Every mountain and every valley in California has, therefore, its own Mendelian population of the fly, and the same is true in Mexico and in Texas. But these elementary Mendelian populations (sometimes called also demes) are components of larger Mendelian populations, or races, for examples those of the Pacific Coast, of the Colorado Plateau, the Rocky Mountains, of Texas, and of Mexico. These are, in turn, components of the most inclusive Mendelian population, that of the species *Drosophila pseudoobscura*.

RACES

The opinion is often expressed (of which more below) that species, as well as demes, races, genera, families, and all other categories are man-made groupings which have no existence apart from the minds of systematists engaged in making a classification of organisms. Indeed, in what sense can we consider the " species " *Homo sapiens* and *Drosophila pseudoobscura* biologically meaningful entities? At this point we shall try to give only a part of the answer. Mendelian populations which compose a species (isolates, demes, races, subspecies, or whatever names one may care to give them) are genetically open systems. A species is a genetically closed system. Although marriages may be concluded mostly within, for example, a caste or a religious community, there is always some gene diffusion also between communities. Although flies native to California do not usually mate with flies from Texas, the genes of the California populations may percolate, though slowly, through a chain of geographically intermediate populations, to Texas, or anywhere else in the species area. This is important, because an adaptively valuable constellation of genes arising in any part of the distribution area of the species may, propelled by natural selection, become the property of the species as a whole. A species is a unit of evolutionary change.

Anthropologists have studied the racial subdivisions of mankind for almost two centuries, only to discover that the number of races which can be recognized in our species is limited solely by expediency. This does not mean that race differences are subjective or fictitious; a glance at a group of persons native to, say, Central Africa and one native to Central Europe suffices to show that the differences are real and tangible. However, the number of races to which we may give names is arbitrary. The arbitrariness is a consequence of the fact that the Mendelian populations composing a species tend to form hierarchic systems. One may divide mankind into only two major races, one of which inhabits Eurasia plus the Americas, the other, Africa plus Melanesia and Australia. A little finer division gives the classical white, yellow, red, black, and brown races. A still finer division gives about thirty " races ". No argument except inconvenience can be adduced against a division into more than two hundred races. Finally, one may get down to perhaps millions of elementary panmictic populations or isolates. Sexually reproducing species other than man show basically similar conditions as far as race differentiation is concerned.

As a general rule we may say that the finer the racial subdivision chosen the vaguer become the confines of the Mendelian populations concerned. The isolates in a territory continuously inhabited by a species (or, in man, in a large urban community) have no boundaries at all; one may even say that the isolate of each person is different from that of every other person. The gene diffusion between such isolates is impeded by distance alone. Conversely, the most inclusive Mendelian population, the species is, with exceptions discussed below, a discrete entity. For example, mankind is, in a very real sense, one huge family in which everybody is related, however distantly, to everybody else. Every person is a potential mate of every other person of the opposite sex. Mankind has a common gene pool. It has evolved, and it will probably continue to evolve, as a single genetical system. In the long run, the fate of my genes depends upon the fate of everybody else's genes. Biology has given a new force to John Donne's famous dictum, " I am involved in Mankind ".

SPECIES FORMATION AND ISOLATING MECHANISMS

The definition of a sexually reproducing species as the most inclusive Mendelian population describes what species are, not how they arise. In other words, it describes only the statics of the situation. Dog is called a species because all varieties of dogs can interbreed, either directly or via intermediate varieties. Dog and coyote are assigned to different species, since they interbreed seldom or not at all. All horses belong to

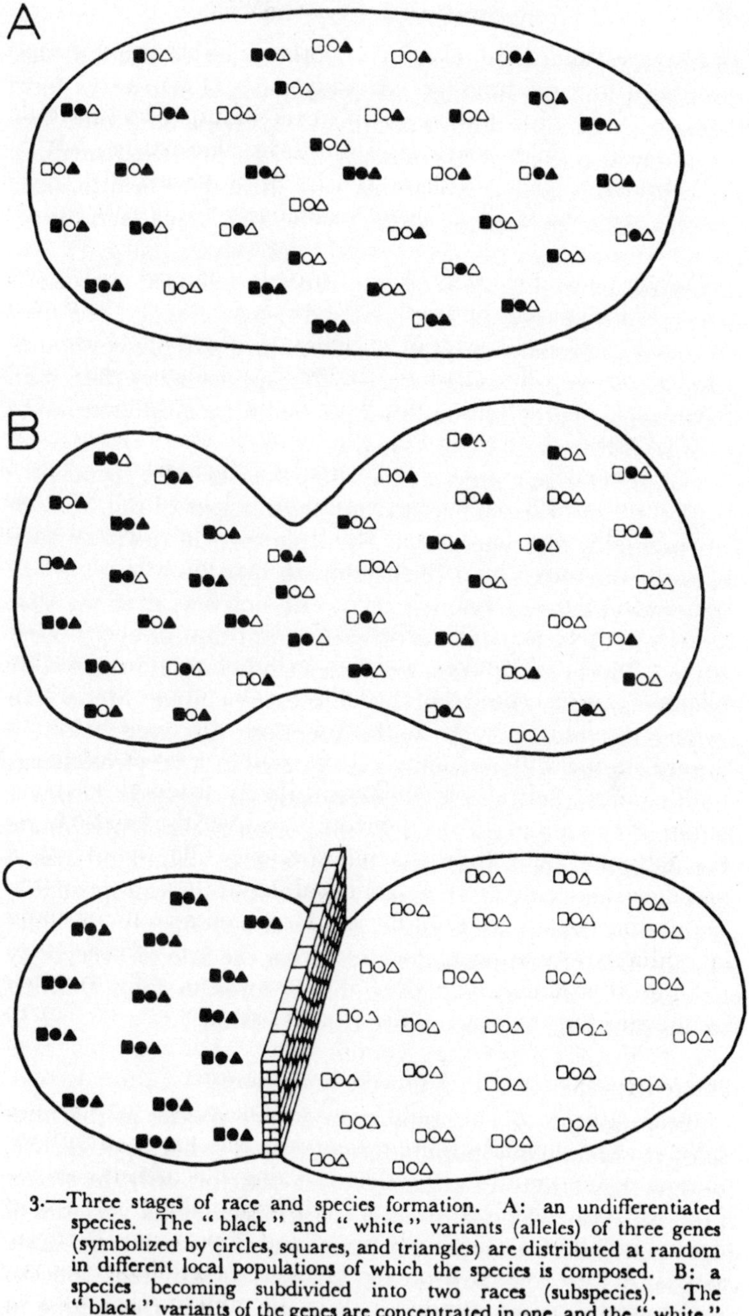

A

B

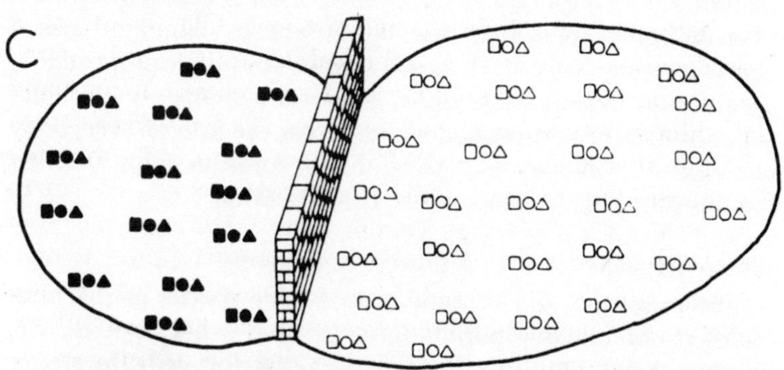

3.—Three stages of race and species formation. A: an undifferentiated species. The " black " and " white " variants (alleles) of three genes (symbolized by circles, squares, and triangles) are distributed at random in different local populations of which the species is composed. B: a species becoming subdivided into two races (subspecies). The " black " variants of the genes are concentrated in one, and the " white " variants in another part of the species. C: the two races are becoming separated by reproductive isolating mechanisms (symbolized by a wall), one species and all donkeys to another; although mules are produced in large numbers in many parts of the world, these hybrids are, with few doubtful exceptions, wholly sterile. Mankind is a single species, distinct from chimpanzee, gorilla, and orang. Each of these species is a genetically closed system, and there is no gene exchange at all between the systems.

Darwin has, however, demonstrated that species are not static but dynamic entities. There is no sharp distinction between species and races or varieties, for the good reason that species develop from races (figure 3). The hierarchy of Mendelian populations is not fixed in time: genetically open systems (races) diverge, usually gradually, to become genetically closed systems (species). This process of genetical closure of previously open Mendelian populations is known as speciation (a decidedly post-Darwinian term). We shall here attempt to indicate only the broad lines of the modern theory of speciation [1].

Why is there no gene exchange between the species dog and coyote, horse and donkey, man and chimpanzee? An uncomplicated answer could be that these pairs of species are too widely different. But this answer does not make clear the nature of the differences which exclude gene exchange. A bulldog and a setter look about as different from each other as either of them looks from a coyote; and yet the bulldogs and the setters belong to the same species while the coyote is specifically distinct. The genetical separation of species is maintained by a variety of means, for which a common name, reproductive isolating mechanisms, has been proposed.

Some reproductive isolating mechanisms interpose obstacles to the meeting of individuals of different species. When the species do not meet they cannot mate. For example, some related species of plants flower, and species of animals reach sexual maturity, at different seasons. Or species may grow in different soils, live at different depths in the sea, or in different biological communities. Other isolating mechanisms impede mating, cross-fertilization, or cross-pollination of species. Sexual isolation is quite widespread and important. It consists simply in that, when a choice of mates is available, matings occur mostly between individuals of the same species, with few or none between species. Females and males of a species "recognize" each other as suitable mates by the appearance, size, shape, and colouration of various body parts, by smell, or by behaviour during courtship and copulation.[1]

Biologists have often been baffled by the profusion of apparently quite useless but often exceedingly complex structures and colour patterns differentiating some closely related species. No less puzzling seemed the intricate courtship "rituals", which again may differ greatly in different species. Darwin ascribed the origin of these curious traits to a process which he called sexual selection.[2] It now appears probable that many such traits are simply "recognition marks", whereby the two sexes identify each other as members of the same or of different species. The more numerous are related species living together in the same territory, the more important become the recognition marks; this may account for the profusion of lavish, grotesque, and extravagant forms in the tropical lands with their rich and diversified animal and plant life.

Still other isolating mechanisms incapacitate the hybrids and thus block the gene exchange between species. Hybrids may be weak or inviable. Thus, the often claimed hybridization of sheep and goat has been shown to be impossible because the hybrid embryos die at an early stage of pregnancy. Hybrids of some species of marine and fresh-water animals die during the cleavage of the hybrid eggs. On the other hand, hybrids may be quite robust but sterile because they fail to produce sex cells. The hybrid sterility may affect hybrids of both sexes (as in mules), or of only one sex (as in the crosses domestic cattle by yak and cattle by bison, each of which produces fertile cows but sterile bulls). Microscopical examination of the sex cells reveals that hybrid sterility may have a variety of causes, but we cannot pursue this topic here despite its great interest. Finally, the inviability or sterility may afflict not the first generation of interspecific hybrids but rather their progeny. Such hybrid breakdown is observed, for example, in the second generation of hybrids between certain species of cottons, which give vigorous hybrids in the first generation.

[1] [See also chapter 9, pp. 213 *et seq.*—ED.]
[2] [Discussed in chapter 10.—ED.]

ORIGIN OF REPRODUCTIVE ISOLATION

It is clear, even from the above very brief account, that reproductive isolating mechanisms are really quite a heterogeneous assemblage. There is little in common between, for example, becoming sexually mature at a given season, adopting a certain posture during courtship, the ability of the sperm to remain alive in the sexual ducts of the female, bodily vigour or weakness of a hybrid, and its fertility or sterility. But this, at first sight chaotic, heterogeneity of the isolating mechanisms points towards an important conclusion. The biological function of all reproductive isolating mechanisms is essentially the same—inhibition and eventual stoppage of the gene exchange between populations. Speciation, as distinguished from simple divergence of races through accumulation of genetical differences, consists in the development of reproductive isolation.

Another significant fact which points towards the same conclusion is that the separation of related species in different groups of organisms is maintained by quite different isolating mechanisms, or groups of isolating mechanisms reinforcing each other's action. Failure to appreciate this fact has led to much misunderstanding. Over and over again attempts were made to define distinct species as forms which fail to yield viable and fertile hybrids when crossed. Some entomologists assumed, more often implicitly than explicitly, that species of insects necessarily differ in the structures of their genitalia, since these differences were held to guard against interspecific copulation. It has been claimed that pine trees which differ in the chemical composition of their resins belong to different species, since chemical differences are reputed to be somehow more important than differences in the appearance of the trees.

It would, indeed, be a great convenience to invent a simple criterion of species distinction but, unfortunately, facts prevent us. We now can see why this should be so: the gene exchange between species populations can be curbed by *any* isolating mechanism or mechanisms. And so it happens that hybrids between some species are fully viable but fully sterile, as with mules, zebroids and the male progenies of the crosses between the species of bovine animals mentioned above. Other species hybrids die if left to themselves, but grow vigorous and fertile if aided through a critical stage of their development. Good examples are the hybrids between species of jimson-weed (*Datura*) studied by Blakeslee and his collaborators; the hybrid seeds normally die, but the embryos may be saved by cultivation on artificial media, and may then grow into partly fertile plants. Closely related species of *Drosophila* flies usually exhibit strong sexual isolation (aversion for interspecific matings), but between some pairs of species such isolation is weak or absent. Patterson and his co-workers found the sperms of some species of *Drosophila* die in the sperm receptacles of females of related species, but this form of isolation is missing completely between other species of the same genus.

In the course of the foregoing discussion we have stated that the genetical equipment of the existing organisms cluster on the " adaptive peaks ", which symbolize the different ways of eking out a living from the environment. Gene combinations intermediate between the clusters do not occur in nature, and this is no accident. Some of the missing gene patterns might perhaps be fit to occupy as yet undiscovered adaptive peaks, but most of them would certainly be disharmonious, inefficient, or altogether inviable. In short, they would become extinct in the " adaptive valleys ". It is, then, no whimsy of Nature that it segregates its creations into species. Species are genetically closed Mendelian populations, attached each to its own adaptive peak. They are closed by reproductive isolating mechanisms; without reproductive isolation species would disappear, submerged in a mass of genetical debris.

The biological function of reproductive isolation is, then, to prevent wholesale proliferation of useless gene combinations that would perish in the adaptive valleys. This is a clue to an understanding of the mechanisms of speciation. Reproductive isolation is likely to be built by natural selection when the gene exchange between populations results in formation of many enfeebled gene combinations. Brilliant experiments of K. F. Koopmann and of J. R. Knight, A. Robertson, and C. H. Waddington have shown that under favourable conditions this process may go rather rapidly. Koopmann made, in special population cages, experimental populations consisting of equal numbers of females and males of two closely related species of flies, *Drosophila pseudoobscura* and *Drosophila persimilis*. There is an incomplete sexual isolation between these species, and in mixed population some hybrids are produced. These hybrids are effectively sterile, and in the experimental populations they were destroyed. In these experiments, then, selection promoted the genetical constitutions which led the flies to mate preferentially within their own species: a female which mated with a conspecific male produced fertile offspring, and thus transmitted its genes to the succeeding generations; a female which accepted a male of the foreign species yielded sterile progeny and its genes were lost. Within some ten generations, the experiment showed a spectacular intensification of the sexual isolation between the species used. The flies developed, so to speak, a " species feeling ", analogous to the " race feeling " which Hitler tried unsuccessfully to induce in his " Aryan Nordics ".

There are many unsettled problems about ways in which reproductive isolation may arise. Muller, and more recently Moore, have contended that isolating mechanisms are simply by-products of the genetic divergence of races, and do not arise because miscegenation of incipient species yields unfit genetical constitutions. Moore found that the race of the leopard frog (*Rana pipiens*) from Vermont gives inviable hybrids when crossed to frogs from Florida or from Texas. The hybrids are, however, viable if races from less distant parts are crossed (for example, Vermont with New Jersey). The species, therefore, preserves its unity: the genes may diffuse throughout the species, including the geographical extremes, via chains of geographically intermediate populations. Note how different is the genetical behaviour of these frog races from that of the races of man, in which hybrids between even the most distant races are fully viable.

Since Vermont frogs have no opportunity in nature to meet Florida or Texas frogs, the hybrid inviability must have arisen despite the absence of hybridization. The difficulty is due in part to a misunderstanding. Natural selection can initiate or intensify reproductive isolation between Mendelian populations only if the gene exchange between the latter produces hybrids of inferior fitness. Such a primary inferiority of hybrids will easily arise when races become adapted to distinctive environments of the countries which they inhabit. The race hybrids might, then, be deficient in fitness. The one species of the leopard frog might conceivably become split into two or several species if its Vermont and Florida races could meet in nature and hybridized: natural selection might then erect isolating mechanisms which would impede or contravene such hybridization. This has not happened in fact, probably because such hybridization did not take place. (Again contrast the races of the leopard frog with human races: hybridization of human races has been going on since time immemorial, but a reproductive isolation has failed to evolve, evidently because the hybrids are not lacking in fitness.)

DISCONTINUITY AND CLASSIFICATION

Aristotle, in the fourth century B.C., knew some 540 species of animals; with Linnaeus, in 1758, the number rose to 4,235; in 1946, E. Mayr estimated the described species to be close to 1,000,000; the actually existing number is anybody's guess. The number of plant species is between a third and a fourth of that of animals. The species of all organisms can hardly number fewer than two million. The magnitude of the organic diversity being so great the human mind can cope with it in only one way. This is classification. A large library is unusable if not properly arranged and catalogued. Historically, the science of life had to begin with making an inventory of living things, and the inventory must always be kept up to date. This is the function of systematic zoology and botany.[1]

A fortunate attribute of organic diversity facilitates the task of the biological systematist. Reference has already been made to the fact that living things come in clusters of individuals which act as mating groups, and which are called species. But the discontinuity of organic variation extends beyond species. Species come in clusters which are called genera; these cluster into families, orders, classes and phyla. This hierarchy of clusters lends itself admirably for the purpose of making a classification. If species be compared with unit volumes in a library, genera may perhaps be likened to series of volumes of a work or a periodical, families to groups of books dealing with

[1] [Further discussed in chapter 5.—ED.]

similar topics, orders to sections of a library devoted to different sciences, and so on.

Comparison of biological and library systematics should not be stretched too far. The number of subdivisons which one makes in a classification, both with organisms and with books, is largely a matter of convenience. However, the hierarchy of clusters of organisms is conspicuously irregular, so that nothing resembling a modern " decimal " classification used in libraries will probably be applicable to organisms. Not only do some genera have many species and some families many genera while others have only one or two, but some clusters contain more levels of subdivision than others. For example, some families are obviously composites of several subfamilies while others are not; some of these subfamilies are naturally divisible into tribes while others are not; some tribes contain several natural groups of genera while others show no such grouping. The number of categories of classification is, therefore, indefinite and arbitrary. Some systematists get along with categories of species, genus, family, and order, while others use also species groups, subgenera, tribes, subfamilies, superfamilies and yet more.

These are not petty technicalities. They teach an important lesson. Biological classification reflects objectively ascertainable facts, because the confines of the subgenera, genera, families, and other groups are made to coincide with the clusters of organic forms. The genus of cats, *Felis*, is a fact of nature because there are a number of cat-like animals which " naturally " belong to this genus. But it is a matter of convention and convenience whether we call a given cluster a subgenus, a genus, a tribe, or even a family.

As an illustration, consider the fly genus *Drosophila*, to which belong some of the insects much used in genetical and evolutionary studies. This is a large genus: by 1951 it contained, according to J. T. Patterson and W. S. Stone, about 613 described species (some dozens more have been described since). The genus breaks up naturally into eight subgenera, of which the largest has more than 200 species and the smallest a single one. The larger subgenera are composed each of several quite natural species groups, and within some of the species groups there are again two or several subgroups of closely allied species. Suppose however that some splitter will call the present species groups, genera; the present subgenera, tribes; and the present genus *Drosophila* a subfamily or a family. There is nothing to prove him wrong, except that to most people who work with these flies it seems more convenient to call them *Drosophila* rather than something else. But ornithologists have felt differently, and the genera of birds have been split repeatedly, until they contain now on the average only some 3·27 species (according to Mayr).

ARE SPECIES NATURAL ENTITIES?

This question has continuously been in the background of our discussion and we should now attempt to come to grips with it. Species were very "natural" to Linnaeus, who believed that each species represented a separate creative act of the deity. The opinion is current that Darwin showed species to be fictions contrived for the convenience of systematists. This is incorrect: Darwin demonstrated that species are not static but dynamic entities, but this detracts nothing from their authenticity. Species is one of the categories of systematics, together with race, genus, family, and the rest. But, as biologists have for a long time perceived intuitively, species are tangible units not only in the same sense as these other groupings; species possess special properties which make them biological phenomena as well as group concepts.

Biological classification, like classification of books in a library, is a contrivance made to facilitate the study of the real world. But it is more than a contrivance; as pointed out above, biological classification exploits the discontinuous character of organic diversity, makes its groupings coincide with the clusters of organic forms, and thereby becomes not only a catalogue but also a description of their evolutionary relationships.

Furthermore, the discontinuity and clustering of organic variation reflect the adaptation of living beings to different ways of life. Cat-like animals get their sustenance from the environment in a way different from that of dog-like animals; the adaptations of the carnivores are different from those of the rodents or bats or primates; the lives of mammals are designed differently from those of birds or of fishes ; and vertebrate animals live in an adaptive world quite different from those in which molluscs or infusoria live. To use Sewall Wright's metaphor again, the adaptive peaks of the races and species are grouped together into adaptive ranges of genera and families, these into adaptive systems of orders and classes, and so on. The peaks, ranges and systems are separated by ever deeper adaptive valleys.

The living world is, then, an immense series of discontinuous clusters of forms, hierarchically arranged. This is true of higher as well as of lower organisms, of animals as well as of plants, of sexually reproducing as well as of asexual forms. The clustering is not a fortunate accident which facilitates the work of systematic biologists: it is a consequence of natural selection which allows only adaptively coherent gene combinations to perpetuate themselves. But while the clustering is, thus, an objectively ascertainable fact, it is, as pointed out above, a matter of choice which clusters are called genera and which are tribes or families. Generic clusters have no inherent properties which tribes or families do not have, and conversely. On the other hand, species in sexually reproducing organisms do possess biologically distinctive attributes. Speciation is the stage of evolutionary divergence at which a Mendelian population becomes split into two or several Mendelian populations the gene exchange between which is impeded or prevented by one or by a combination of several reproductive isolating mechanisms. Species are, as stated above, the most inclusive Mendelian populations, or Mendelian populations reproductively isolated from one another.

BIOLOGICAL SPECIES AND SPECIES OF SYSTEMATICS

Modern biology has vindicated the insight of zoological and botanical systematists that species are important unit products of the evolutionary process. The evolutionary divergence of races is reversible, that of species is irreversible. Races may grow increasingly more distinct in response to the vicissitudes of the environments in the territories which they inhabit. But migration, mixing, and miscegenation may undo the divergence; in the melting pot of hybridization races merge into single variable populations. For at least several centuries this has been unquestionably the trend in the human species.

Human races are becoming less sharply distinct than they once were.

Races of sexually reproducing forms wax more distinct when they are allopatric, that is geographically segregated in different territories. There was little mixing between men of the Americas and of the Old World so long as intercontinental travel was hazardous and rarely attempted. Race differences wane when hybridization is facilitated. But reproductive isolating mechanisms reinforce and eventually supplant the geographical separation. When the reproductive isolation becomes complete, species may live together in the same territory, sympatrically, without gene exchange and without fusion (figure 3). (The fusion of human races, which are now in part sympatric, is slowed down by social regulation of marriage which is a novelty in the biological world.) Absolute reproductive isolation makes evolutionary divergence final and irreversible. The evolutionary courses of races are mutually contingent, those of species are independent. Every human being is involved in mankind; we may be interested in, but we are not in the same sense involved in, other species. Speciation is a critical phase of the evolutionary process. There is nothing of comparable importance taking place on the generic, or subfamilial, or familial levels of divergence [2].

One may marvel at the perspicuity of systematic zoologists and botanists who were usually able to discern correctly the biological species in their materials. A systematist working with mammal or bird skins, or with pinned insects, or with dried plants in a herbarium, has obviously no direct knowledge of whether the forms which he examines could or could not exchange genes. It would be preposterous to expect him to acquire such information before he classifies his specimens. Making a classification cannot be postponed; it is needed now, as an aid to all other biological studies. As Simpson put it; the species of systematists are inferences concerning the biological species which are the reality of nature. Yet these inferences are, by and large, surprisingly well grounded.

How, indeed, does a systematist, or for that matter a non-biological classifier of living things (like the Brazilian woodsman mentioned at the beginning of this article) establish their species? Let us consider again the examples of the wolves, coyotes, and red foxes. In some parts of North America these animals do (or did) live in close proximity, sympatrically. Supposing that we know nothing about their inability to interbreed, we still would find no specimens which are intermediate between, or which we could not place with assurance in one of the three groups—wolf, coyote, and red fox. These groups are separated by gaps, they are discrete. Now, their discreteness is prima facie evidence of their biological separateness; if they exchanged genes with any appreciable frequency intermediates probably would occur. (Here a caveat: some single gene differences, like those which determine the human blood groups O, A, B, AB, produce discrete clusters of forms and no intermediates, but of course, there is no reproductive isolation and no separation into different species. This is a pitfall into which systematists, beginning with Linnaeus, have occasionally fallen.)

Several dozen species of *Drosophila* may be collected on a few acres of a South American tropical rain forest: we recognize them as species because of the lack of intermediates between them. We can prove that they are indeed distinct biological species by testing them in laboratory experiments in which their behaviour will show them to be reproductively isolated. Sympatric species of birds, of butterflies, or almost any other sexually-reproducing creatures are sorted out because of the existence of gaps between them, whether or not we can prove experimentally that the gaps are due to reproductive isolation.

Geographically separate, or allopatric, forms are more difficult. How do we know that the populations of *Drosophila pseudoobscura* from California belong to the same species as those living in Texas? In this case we can prove the point experimentally, since these populations show not a trace of reproductive isolation. But where experiments are not practicable, we have to fall back to Darwin's dictum that the " sound judgment and wide experience (of the systematist) is the only guide to follow ". An experienced systematist knows that a certain kind and magnitude of difference between sympatric forms is usually specific, in the particular group of organisms which he studies. When he finds such or greater differences between allopatric forms he judges them to be probably specific there as well.

The biological species concept is, thus, not a novel doctrine which emerged from the genetical experiments of recent decades; it is rather a clarification of the classical concept inherited from Darwin and his predecessors and generally used by systematic biologists as a working tool. But this fact should not be used to conceal certain divergent attitudes with which systematists and experimental biologists approach the study of species. To a systematist the species is before all else a category of classification, like the genus, family, order, and so on, although the unique properties of the species category are readily recognized. A systematist attaches names to everything, from bacteria to mammals, and is perforce interested before all else in the nameability of his specimens. A name is necessarily static; but Darwin showed that species emerge, and an experimentalist is interested in the processes of their emergence; and these differ in different groups. We must now give consideration to these divergent attitudes, because they show better than anything else the direction of current research on the species problem.

ARE THERE SPECIES IN ASEXUAL ORGANISMS?

Reproductive isolation prevents, as we saw, the hybridization of populations between which gene exchange would yield inferior gene patterns. Speciation has, as its biological function, restriction of sexual unions within the bounds of the adaptive peak which a population occupies. But there is obviously no need for this where sexual reproduction has not evolved, or has been lost.

A single bacterium may produce, by repeated fission, a " clone " of countless individuals which have, barring mutation, the same genes. When reproduction occurs by fission (as in most microorganisms), or by buds or spores (yeasts, many fungi and some lower animals), or by roots, bulbs, stolons and similar means (some plants), gene exchange between the clones is excluded, and the integrity of the gene patterns which these clones carry is protected automatically. The same is true of most forms of parthenogenesis or apogamy; for example, many clones of the blue-grass (Poa) bring forth without fertilization seeds, all of which have the same genes as the mother plant. Lastly, the flowers of some plants are so constructed that the pollen grains fall on the stigma of the same flower, effecting self-fertilization (most wheats, oats, beans, some violets). Where the selfing is obligatory, no gene exchange can take place between the selfed lines.

Note should be taken of the recent work which has shown that organisms which reproduce only by asexual means are not nearly as numerous as was formerly believed. J. Lederberg, N. D. Zinder, G. Pontecorvo and others have discovered that even bacteria and lower (so-called "imperfect") fungi may achieve gene transfer between clones, with or without a full-fledged sexual union. Moreover, asexual reproduction, parthenogenesis and selfing as a rule alternate at shorter or longer intervals with sexual crossing. Where sexual (or "parasexual" as in the imperfect fungi) gene transfer occurs with any regularity, one is justified in speaking of Mendelian populations and of isolating mechanisms.

There is, nevertheless, no denying that the processes of speciation occur differently in organisms with different modes of reproduction. In extreme cases, where sexual crossing is absent or very rare, it is meaningless to speak of biological species. The full import of this situation, which has led to some misapprehensions in the biological literature, must be clearly understood. Systematists certainly have named species in asexual, parthenogenetic and selfing organisms; indeed, museum systematists often have no way of knowing by what means the creatures which they describe reproduce themselves. What, then, are these species? Reference has already been made to the fact that the clustering of gene combinations on adaptive peaks is the normal state of affairs in sexual as well as in asexual forms. Favourable gene combinations are preserved and unfavourable ones are cast out in all living things, and the clusters so produced are nameable.

The chief point is that in exclusively or predominantly asexual, parthenogenetic, or selfing forms it is a purely arbitrary decision which clusters are to be named as varieties and which as species or genera. Working with these forms, the systematist loses the guiding light which pilots him towards recognition of the biological species in sexual outbred organisms. In the latter, two populations living side by side without formation of intermediates are in all probability reproductively isolated, and should therefore be named as species. But two or several asexual clones may live together and preserve their distinctness. Should each clone be recognized as a species? Or should we regard as species only large clusters of clones well separated by gaps from other such clusters? If so, how large should the clusters and the gaps be?

We need not be surprised that different authorities have reached quite different decisions on the limits of species in plants such as hawthorn (*Crataegus*), raspberries and blackberries (*Rubus*) and some others, among which asexuality or apogamy are widespread. So extreme have been the differences of opinion that such plants have been dubbed the " crux et scandalum botanicorum ". Of course the systematics ot sexually reproducing forms has also produced some differences of opinion. The species of mammals and birds having been for the most part described, systematists who study these groups now devote their attention to investigation of the racial (subspecific) divisions of these species. There was, chiefly during the first quarter of the twentieth century, a tendency to treat some of these races as species. C. H. Merriam divided the American grizzly bear (*Ursus horribilis*) into seventy-nine species and some additional named varieties, a procedure matched for absurdity only by the more recent attempt ot R. R. Gates to divide living mankind into four separate species. The difference is, however, this: it can be demonstrated that all men belong to a single inclusive Mendelian population, and presumably that all grizzly bears belong to another (these animals have become extinct in many regions), but there is no way to fix, except by convention, the number of hawthorn species.

SIBLING SPECIES

A systematist likes to have his species easily distinguishable by the inspection of specimens preserved by customary techniques—bird skins, insects dried on pins, plants on herbarium sheets. But nature has not been altogether obliging; some insect species are distinguished by the structure of their genitalia, some worms are recognized by microscopic examination of sections of their bodies, and bacteria and yeasts generally require biochemical tests for identification. Note, however, that (except for a brief resistance of conservatives) the necessity to examine the genitalia did not cast doubt on the specific status of those insects which are identified only in this way. Biological species being facts of nature, it is up to the biologist to devise techniques whereby they can be perceived.

In recent years more and more instances are being brought to light of sibling species. These are pairs, or groups of several related species, which are recognizable with difficulty (if at all) by inspection of their easily visible structures. Two examples of this fascinating phenomenon will have to suffice here. In tropical America there occur five species of *Drosophila* flies (*willistoni*, *tropicalis*, *equinoctialis*, *paulistorum* and *insularis*) which, at least until the recent discovery by B. Spassky of minute differences in the external genitalia of the males, could not be safely identified in single living specimens. Nevertheless, these species show so strong a sexual isolation that cross-copulation rarely occurs in laboratory experiments, even when the flies have no choice of mates. In the rare instances of interspecific insemination no hybrids are produced (excepting that *Drosophila insularis* does give, in laboratory experiments, wholly sterile hybrids with the other species). Over much of their geographical distribution areas, four of these species occur in the same localities; and although they can be found feeding on the same decaying fruit, the females and males of each species " recognize each other " as the proper mates. No hybrids at all have been found in nature, and the species can be identified unmistakably by inspection of their chromosomes, by genetical experiments, and now, thanks to Spassky's work, by careful examination of the male genitalia.

The work of T. M. Sonneborn, T. T. Chen, R. F. Kimball, D. L. Nanney and others has disclosed the existence of groups of sibling species among infusoria. The classical " species ", *Paramecium aurelia*, *P. bursaria*, *P. caudatum*, *Tetrahymena pyriformis* and probably some others, are actually groups of several (up to fifteen) sibling species. The infusoria reproduce normally by division of their bodies into two, that is, asexually. But from time to time and in certain conditions there occurs a sexual process, called conjugation, during which individuals come together in pairs and exchange division products of their nuclei, which then undergo sexual fusion. Different sibling species usually fail to conjugate but, in exceptional instances, when they have been observed to do so, the hybrids proved to be inviable. Each species seems, on the basis of the evidence published, a reproductively isolated Mendelian population.

Some differences of opinion have arisen concerning the evaluation of sibling species. Should they be treated as species, or should one recognize only species which are easily distinguishable? Two issues are involved, and it is well to keep them separate.

First, one may ask whether sibling species are comparable as biological phenomena to the ordinary "good" species of sexual organisms. This must be answered in the affirmative. The sibling species of *Drosophila* referred to above are each genetically closed systems, independent in evolutionary course from the other siblings. Sibling species often live together, sympatrically, without losing their identities. Calling them races of one species results in nothing but confusion. They do not behave in the least like races, subspecies, varieties or breeds of either wild or of domesticated species.

Second, and quite separate, is the question whether sibling species should be given species names in Latin. This seems advisable in some but not in other instances. The existence of five *willistoni*-like sibling species of *Drosophila* is a nuisance to a systematist working with pinned specimens in a museum, since he cannot write simple and unambiguous determination labels for some of his specimens. However, these species are used as materials for experimental work, and the experimentalists will have more occasions to talk and to write about them than will museum systematists. It is, hence, convenient to have these species separately named. The same is true for the sibling species of anopheline mosquitoes, since some of them do and others do not transmit malaria. Malariologists and epidemiologists will have frequent occasions to talk and to write about these insects; naming them is necessary. But the students of infusoria have not so far found it advisable to name their sibling species. The nameability is a matter of convenience, the biological status of species is a matter of ascertainable fact.

UNCOMPLETED SPECIATION

Discrete, clear-cut, easily distinguishable species are unquestionably most " convenient " to a classifier. Linnaeus thought he was able to divide the whole living world into such uncomplicated species. Yet it appears that even Linnaeus in his old age came to doubt the universal discreteness of species, for in the late editions of his *Systema Naturæ* he quietly deleted the famous dictum that there are as many species as were produced on the day of Creation. In any case, the two centuries of systematics since Linnaeus have clearly established the fact that, although it can usually be ascertained beyond reasonable doubt whether two groups of forms are or are not distinct species, organisms are not quite always so accommodating. More and more difficult cases have been discovered, in which the decision whether two populations are distinct species, or races of the same species, is hard or impossible.

One of the sources of the difficulties is simply that some species do not differ in external appearance enough to make them easily distinguishable to a human observer. This difficulty is greatest among sibling species. The limiting situation is a pair or a group of sibling species which are identical in appearance. Some sibling species of *Drosophila* approach this limit rather closely. But, as pointed out above, the difficulty with sibling species is essentially a technical one. Many siblings are, without doubt, biologically distinct species, no less so than, for example, the horse and the ass. It is simply that, to distinguish them, special methods of investigation have to be used. Thus, siblings of the *willistoni* group of species of *Drosophila* are easily distinguishable by the appearance of the chromosomes in the salivary glands of their larvæ.

Another, much more fundamental, source of difficulties in drawing the lines of separation between some species was discovered by Darwin. In fact, it is the *leitmotiv* and the core of *On the Origin of Species*. Species arise through divergence of races, and the divergence is usually a gradual process. It follows that situations must be expected, and they do occur, in which two groups of populations are too distinct to be regarded as races, but not distinct enough to be considered species. Whether they are named as races or as species, is,

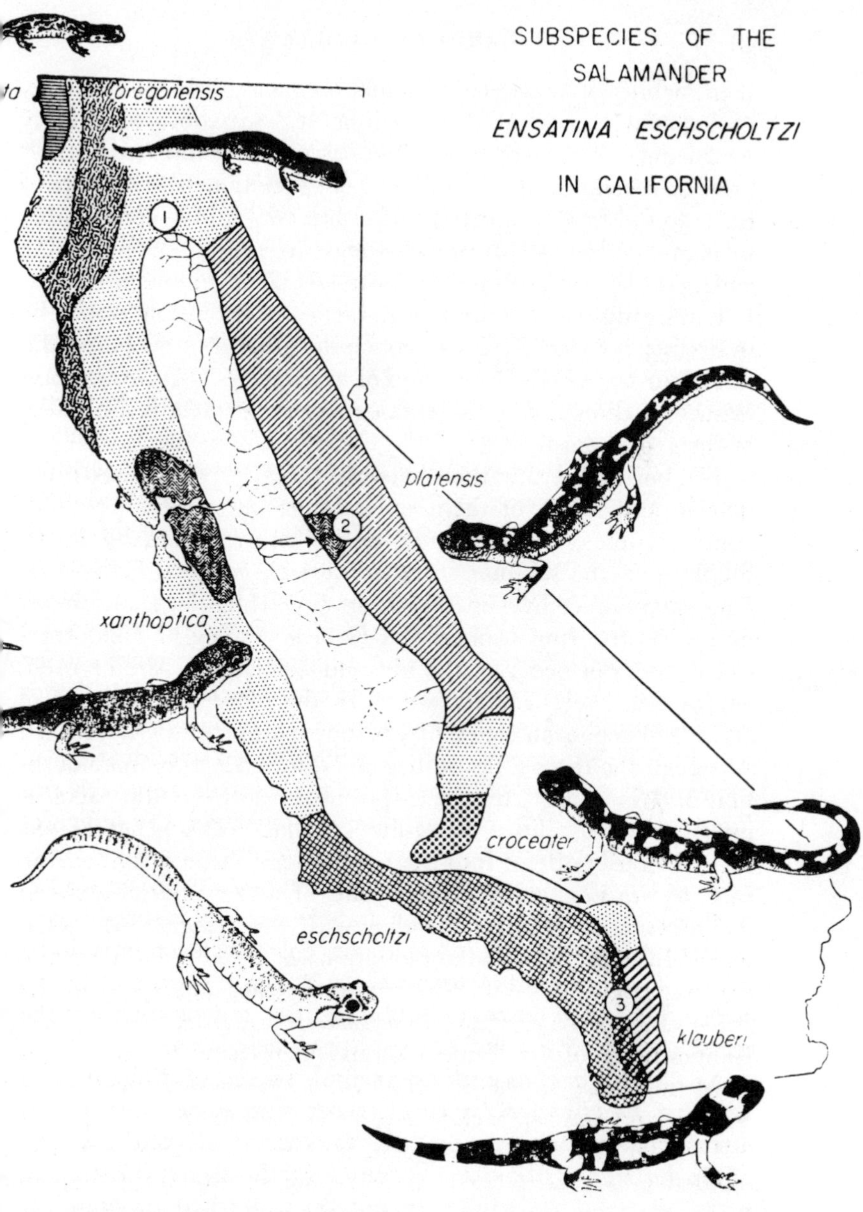

—The races of the salamander *Ensatina eschscholtzi* in California. The geographical areas occupied by the seven races are shown by different types of shading, and the areas of transition (intergradation) between these races by stippling. (1) an area of smooth intergradation between races; (2) an area in which two races, living in closely adjacent places, hybridize infrequently; (3) an area where two "races" or incipient species are sympatzic (occupy the same territory) but do not interbreed. (Drawn by Gene Christman, courtesy of Professor R. C. Stebbins.)

then, arbitrary and decided solely on grounds of convenience. Biologically, however, these " difficult " situations are highly significant. They are borderline cases of uncompleted speciation, of speciation in the process, of speciation which happens to be unfinished on our time level but which may conceivably be completed in the future (see the figure on page 47). It is no exaggeration to say that if no instances of uncompleted speciation were discovered the whole theory of evolution would be in doubt; we would have to conclude either that evolution did not occur or that the formation of new species is instantaneous. What is a difficulty to the cataloguing systematist is a blessing to the evolutionist.

The borderline situations between races and species are quite diverse and many of them quite interesting. Space permits consideration of only a single example, discovered by R. C. Stebbins in the salamander *Ensatina eschscholtzi* in California. This salamander lives in the mountains which form an almost uninterrupted ring around the Central Valley of California, but it does not occur in the hot and dry Central Valley itself, nor in the deserts farther east. In the mountains a series of races replace each other from north to south (figure 4). Although the races differ sometimes quite strikingly in colouration and in other traits, the transitions between the races in intermediate localities are quite gradual. The genes of one race obviously diffuse into the populations of the neighbouring races by crossing, and extensive zones of intergradation between the races are present. Starting with the typical *eschscholtzi*, which lives in the coastal mountains of Southern California, one passes gradually northwards to the race *xanthoptica* in the San Francisco Bay region, to *oregonensis* and *picta* in the Coast Ranges north of San Francisco, and then eastward and southward to *platensis* in the Sierra Nevada Mountains, and then to *croceator* and *klauberi* back in Southern California.

Thus far the situation suggests nothing more than a species divided into a series of races (like, for example, the human species). But it happens that in Southern California the three most distinct of all " races " (*eschscholtzi*, *croceator* and *klauberi*) meet without intergradation. In at least one locality they have been found living together, and there is no evidence of gene exchange between them. In other words, in Southern

California occur two reproductively isolated populations, which might, accordingly, be considered species. And yet, these species are connected by an unbroken series of intermediate population living to the north of them. They *can* exchange genes, not directly but by a long circuitous route, through the other races. In this respect they resemble races of a single species (compare this case with that of the leopard frog, *Rana pipiens*, discussed above). Moreover, although the salamander does not at present live in the Central Valley, it evidently succeeded in crossing it at some time in the past, since the race *xanthoptica* occurs not only near the San Francisco Bay but also in the Sierra Nevada, across the Central Valley directly to the east of San Francisco. But, in contrast to the situation in Southern California, the race *xanthoptica* meets in the Sierra Nevada the race *platensis*, and there some gene exchange does occur. These races have not yet reached the status of reproductively isolated species. They are races of a single species.

The diagram in the figure on p. 50 may help the reader to visualize how the process of speciation occurs in time. If we happened to live at time A, we would find a single species made up of populations or races which differ from each other in the frequencies of some genes, but between which the gene exchange is impeded only by geographical separation. (The populations are symbolized in figure 5 by the strands which compose the bundle.) At time B, the bundle tends to become divided into two bundles, but the separation of the strands which compose them is not yet complete. This symbolizes the splitting of the species into two groups of races, adapted to different environments, but only just beginning to evolve a reproductive isolation which prevents gene exchange between them. At time C, there are two separate bundles. Here reproductive isolation has become complete, and the two species are closed systems, genetically independent of each other during their further evolution. At time A we have, then, a single species, at time C two species, but at time B we observe a borderline case.

Not enough is known at present to tell how many examples of uncompleted speciation there are in different groups of organisms. It obviously depends on how rapid is the passage

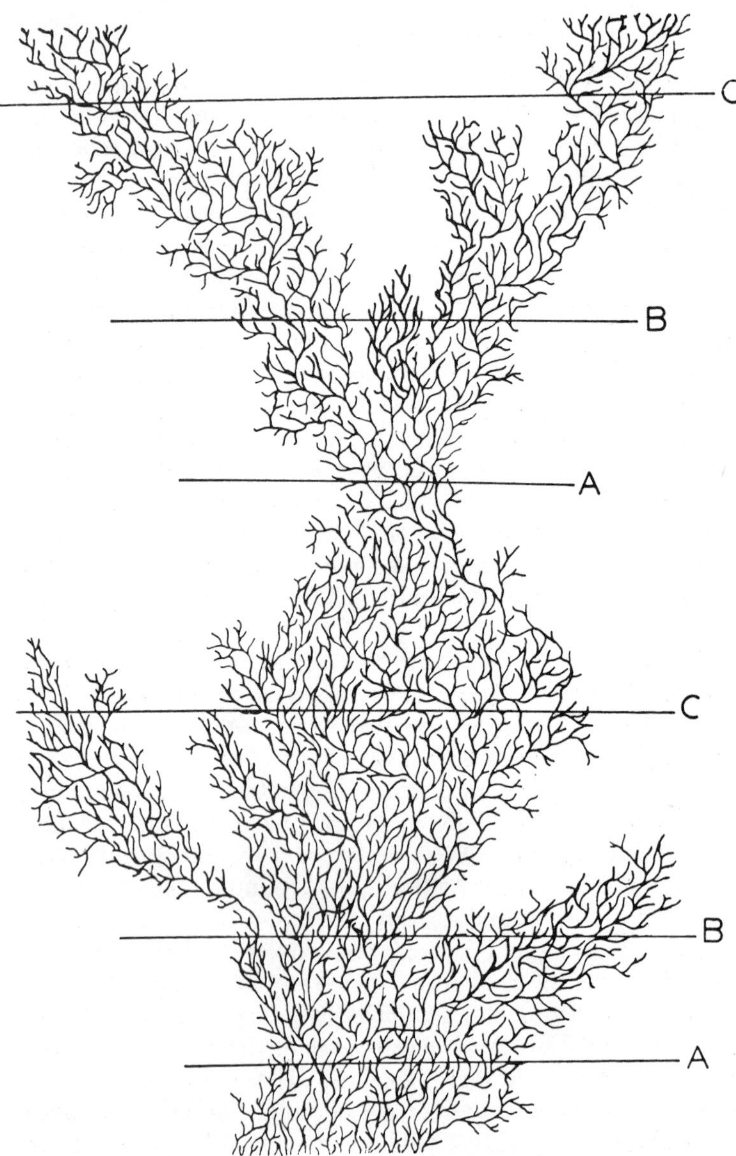

5.—The process of species formation in time. An originally single species (below) becomes divided into two derived species (above). The Mendelian populations which compose the species are symbolized by black strands: some of them end blindly (become extinct), others unite with other strands (populations fusing owing to intermarriage), while still others split up (populations becoming subdivided). At the time levels A we have a single species; at the time C there are two derived species; at the level B the speciation is not complete, and it is arbitrary whether a single species or two species are recognized. (Drawing by L. Van Valen.)

from stage A to stage C represented in the figure. If this passage were instantaneous, no borderline cases between race and species would exist. Such instantaneous origin of reproductive isolation does occur when new species arise through allopolyploidy, that is, by doubling the chromosome complement in a hybrid between two previously existing species. This kind of speciation is important in some groups of plants but, considering the living world as a whole, it represents a special case. More frequently, reproductive isolation takes a long time to develop. If, in some group of organisms, the splitting of species is actively going on, then borderline cases may be found in relative abundance.

By and large, borderline cases seem to be rather rare, although more is written about them than about "good" species. This is as it should be: when speciation is in progress its study is most profitable. A mistaken impression has, however, gained some currency among biologists not possessing a first-hand knowledge of systematics, to the effect that considering two populations as distinct species or races is in general an arbitrary matter. This is not true at all: with most species in most groups of organisms there is no dispute at all concerning what is a species and what is not one. According to Mayr, among the 1,367 species and subspecies of North American birds, only in ninety-four, or in 12·5 per cent of the species, can there be any doubt whether one is dealing with a separate species or with a race of another species [2].

INTROGRESSION

Races are sometimes referred to as incipient species (Darwin thought them so), meaning that races may grow more differentiated, develop reproductive isolation, and then develop into species. It does not, however, follow that every race will at some future time be a species. Speciation is not inevitable; it is a rather special form of adaptation. In the course of the foregoing discussion reference has been made to natural selection promoting reproductive isolation when the gene exchange between populations yields gene combinations of low fitness. When the interbreeding of populations does not lower the fitness, the stimulus for the development of reproductive isolation is lacking. We have seen that this is probably the reason why not a trace of reproductive isolation between human races has evolved: the interracial hybrids do not lack fitness. The rigidity of the reproductive isolation between species may, then, be greater when they are exposed to the risk of hybridization and this hybridization diminishes the fitness, than when this risk is small or when the hybrids are relatively well off. Occasional hybridization of species is, indeed, recorded in fairly numerous instances among animals, and still more among plants. When the hybrids are in addition fertile, the hybridization results in diffusion, or introgression, of genes of one species into another.

The two species of California oaks studied by J. M. Tucker may serve as examples. *Quercus douglasii* is a common deciduous tree of the foothills of the mountains; its leaves are large, lobed or entire, bluish-green in colour, with more or less sparse stellate hairs having usually four rays on the lower surface. *Quercus turbinella* is an evergreen shrub, rarely a small tree, growing mostly on semi-arid slopes bordering desert areas; its leaves are small, spinose dentate on the margin, greyish-green, with dense stellate hairs on the lower surface, having usually eight rays. Where these two species grow separately they appear to be very clearly distinct. But their distributions overlap in some localities in the Coast Ranges south of San Francisco, and there a hybrid is found intermediate in most characteristics between the parental species. This hybrid has even been described as a separate species and given a name, *Quercus alvordiana.* The populations of either species which grow near the localities in which the other species, or the *alvordiana* hybrids, occur are modified. They contain various proportions of aberrant individuals which are most probably derivatives from the hybrids crossed back to one of the parental species. Although the hybridization of these oaks has not been tested experimentally, the weight of the evidence given by Tucker is overwhelmingly in favour of the view that a channel for gene exchange between *douglasii* and *turbinella* continues to exist, despite the very striking differences in their appearance.

The occurrence in nature of introgressive hybridization between some species is evidence that the process of speciation has in them not reached completion. Such species are not irrevocably committed to follow separate evolutionary paths.

Two species of toads, *Bufo americanus* and *Bufo fowleri*, frequently breed in the same localities in the eastern United States. Hybrids between these species have been found by A. P. Blair, E. P. Volpe and others in a number of places, and in some localities hybrid populations have become established. The hybrids are fertile and in no obvious ways inferior to the parental species. There is some reason to think that the hybridization of these species is of rather recent origin. The two toads have rather different preferences as regards the kind of environment in which they like to breed. This may have kept them separate and isolated, so long as the clearing of forests and modification of the old and construction of new bodies of water by man did not create habitats which were equally acceptable for breeding both to *B. americanus* and to *B. fowleri*.

There is no way to predict what will be the eventual result of the introgressive hybridization between these species of toads. It is possible that they will end by fusing into a single species, and their past evolutionary divergence will be thus reversed. Man's changing the face of the earth has become a powerful factor in the evolution of not a few biological species. Some evidence that introgression has become more frequent, or has even begun, because of human interference with the habitats of wild species, is available in many, if not in most, carefully studied cases of introgression. H. P. Riley found introgression between two species of irises, *Iris fulva* and *Iris hexagona*, chiefly in man-made habitats in the delta of Mississippi in Louisiana. Indications of man's activities causing, or at least stimulating, introgression between two species of birds in Mexico (the towhees, *Pipilo erythrophthalmus* and *P. ocai*) have been detected by C. J. Sibley.

Just how frequent is introgressive hybridization, with or without human interference, is a controversial matter. Some authors have made rather wild claims that it is ubiquitous, both in the plant and in the animal kingdoms. These authors have confused instances of intergradation between subspecies or races (which is certainly a common situation) with gene exchange between partly isolated species. More critical studies are obviously necessary.

The evidence now available suggests that introgression is

generally more frequent among plants than among animals, and more frequent among plants which are long-lived or able to reproduce both by asexual and by sexual means than among short-lived and purely sexual annual plants. Ledyard Stebbins has given a brilliant analysis of why this should be so. Consider a tree which exists for decades, or even for centuries, and brings forth during its lifetime very large numbers of seeds. To maintain a constant population, on the average only one of these seeds must give rise to another adult tree. A species of such a tree may, then, afford to sacrifice a part of its seeds to introgressive hybridization with other species. There exists a finite possibility that some, perhaps a small minority, of the products of such gene exchange with related species may possess a high fitness in some environments. There is, thus, little advantage in making reproductive isolation between the species complete, and consequently little pressure of natural selection to that effect. The same is true of species which reproduce chiefly asexually: in such species the few sexually produced seeds are, so to speak, genetical experiments, the biological meaning of which lies in the occasional emergence of new and adaptively valuable constellations of genes. Annual plants, and animals which produce relatively few young, can scarcely afford to sacrifice their progeny to genetical " experiment ". Reproductive isolation confines, then, the breeding within the species limits, where the chance of production of fit offspring is greatest.

CONCLUDING REMARKS

A century ago Darwin wrote:

> on the view that species are only strongly marked and permanent varieties, and that each species first existed as a variety, we can see why it is that no line of demarcation can be drawn between species, commonly supposed to have been produced by special acts of creation, and varieties which are acknowledged to have been produced by secondary laws.

The development of biology has borne out Darwin's argument. No static definition of species, which might be used as a yardstick to distinguish all species from all races, has emerged. Borderline cases, which are instances of uncompleted speciation, bridge the gaps between species and races. The processes of speciation are not uniform in different kinds of organisms. Where reproduction is chiefly or exclusively by asexual means, or by parthenogenesis, or by selfing, species do not exist in the same sense as they do in sexual cross-fertilizing organisms. This does not prevent systematists from naming species according to the same rules throughout the world of life. But the nameability of species is, as J. B. S. Haldane has so correctly pointed out, " a concession to our linguistic habits and neurological mechanisms ".

Species as biological phenomena are quite another story. Their existence would have to be recognized even if biologists were uninterested in classifying living things. Speciation is a corollary, a supplement and a corrective to sex. Sexual reproduction is a master evolutionary adaptation, since it makes other adaptations more easily accessible than they would have been without sex. This is because sexual reproduction permits favourable gene changes, arising in different lines of descent, to be combined in a single genotype. Crossing individuals or populations A and B, having adaptations X and Y respectively, leads, by way of Mendelian segregation, to the emergence of C endowed with the adaptation XY. But sex is just as efficient in combining adaptations as it is in breaking them up. Unlimited interbreeding of populations adapted to different ways of life would result in disintegration of adaptive gene complexes set up by natural selection. Speciation is, then, an adaptive instrument which avoids this hazard. Species consolidate the evolutionary gains of the past and thus facilitate further evolutionary progress.

3

The Third Stage in Genetics

By

DONALD MICHIE

A HUNDRED years after the publication of *On the Origin of Species*, Charles Darwin's ideas on genetics are still not known, with the exception of one or two morsels (such as his account of " telegony ") which are offered to the student as objects of ridicule. My contention in this essay is that their neglect is simply explained: one hundred years ago these ideas were one hundred and ten years ahead of their time. By 1859 the evidence that evolution occurred, and that it occurred mainly through the conservation by natural selection of favourable heritable variants, was strong enough to justify joining battle. The evidence for Darwin's conceptions of the mode in which heritable variation might arise was not. These conceptions, with their strong environmentalist emphasis, passed into the shadows, eclipsed in turn by Weismann's " immortal germ-plasm ", by Mendel's " uncontaminated gametes ", by Johannsen's " pure lines ", and by Morgan's " chromosome theory ".

TWO PHILOSOPHIES

To confront the two genetical philosophies one with the other, Darwin's and that of his successors, I have redrawn in the figure opposite part of an admirably clear diagram by Darlington [6]. I should hasten to excuse Darlington from any complicity in the ideas which I shall later develop, for they are diametrically opposed to his own. In the upper picture the germ of the new generation is a direct product of the body and hence carries the imprint of bodily influences. Such a picture allows for the " inheritance of acquired characters " usually associated with the name of the French naturalist Lamarck, and promises the experimental biologist the opportunity of modifying the nature of plants and animals through the deliberate control of their nurture. The connection between germ and body is reciprocal, and for this reason I have termed it the "two-way theory" of heredity. This was Darwin's picture, although many will be surprised to learn that Darwin was to this extent a Lamarckist. Statements linking Darwin's name with the *second* theory, which is illustrated by the lower of the two pictures in the diagram and which is in fact that of August Weismann, abound in biological literature. They can

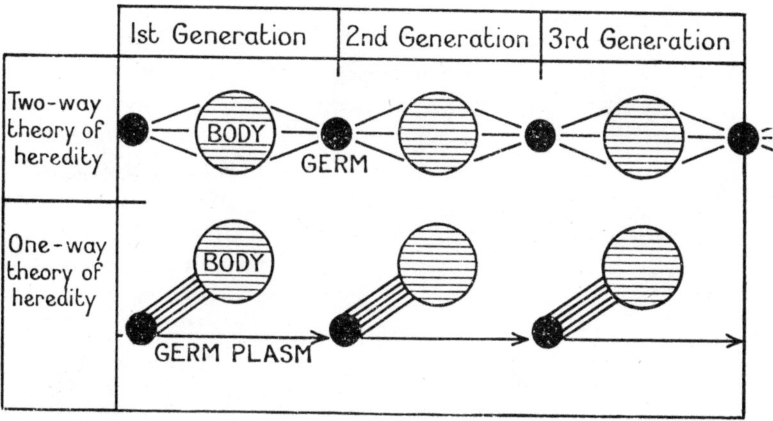

6.—Two theories of heredity. *Above*: a system by which influences acting on the body can influence the genetical material (as in the hypotheses of both Lamarck and Darwin). *Below*: the type of system postulated by Weismann and by most present-day biologists. (Redrawn by kind permission of Messrs George Allen and Unwin from *The Facts of Life*, by C. D. Darlington.)

be disregarded by those who have even a slight acquaintance with Darwin's writings.

"Under certain conditions organic beings even during their individual lives become slightly altered from their usual form, size, or other characters," Darwin writes in his essay of 1844 (9, p. 57) "and many of the peculiarities thus acquired are transmitted to their offspring." We find him reiterating this view throughout the *Origin* and with the publication in 1868 of *The Variation of Animals and Plants under Domestication* he had built it up into the elaborate theory of pangenesis, of which I shall have more to say later. Even when, in 1876, he reported a long and extensive series of experiments on fifty-four different plant species demonstrating the beneficial effects of cross- as opposed to self-fertilization, he attributed the phenomenon to the difference in the environmental rather than the genetical histories of the parents entering the cross [8].

Now turn to the lower picture of figure 6. Here the germ creates the germ, budding off a mortal body in each generation to act as its temporary vehicle. The traffic of causation is thus strictly one-way, and I have termed the theory the " one-way theory of heredity ". Heredity determines bodily form and

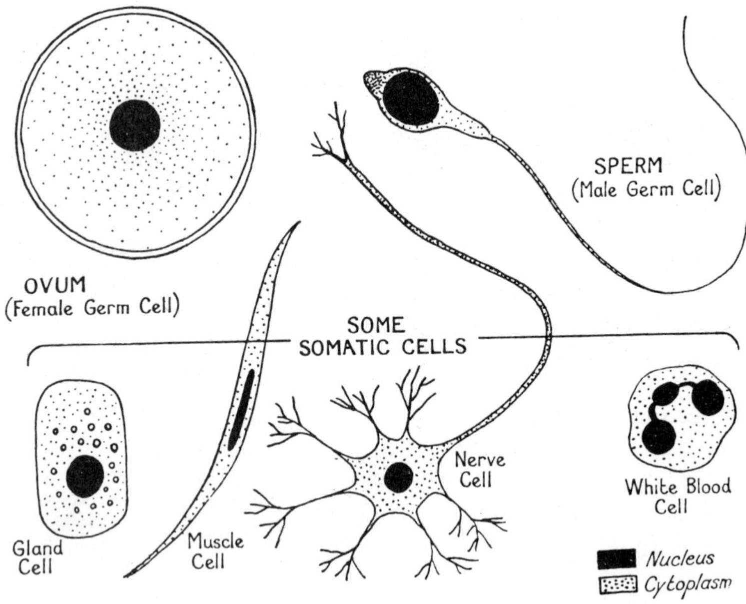

7.—Some mammalian cells (schematic, not drawn to scale).

function, but modifications of the latter cannot leave any trace in heredity. New variation can arise only by intrinsic errors in the process by which germ copies germ down the generations. In this process, apart perhaps from accelerating or diminishing the rate of origin of random errors of copying (mutations), the environment is powerless to intervene.

I cannot overstress the importance of the simple diagrammatic contrast presented here between the opposing genetical philosophies. Without it one cannot begin to understand the history of genetics in the present century, some leading features of which I now propose to summarize.

In order to define terms which will crop up in the course of what follows, I have drawn the pictures opposite. The tissues of the body are composed of myriads of microscopic *cells*. Each cell is a partially self-contained unit, usually specialized to do some particular job. The *cytoplasm* of each cell is bounded by a cell-membrane. The *nucleus* which each cell contains is separated from the cytoplasm by a nuclear membrane. In the nucleus are thread-like bodies called *chromosomes*. The nuclei of the *germ cells* contain only half the number of chromosomes found in the cells of the rest of the body (*somatic cells*) and are said to be *haploid*. The full *diploid* set is restored in the fertilized egg, from which all the somatic cells are derived by successive cell-division in which the process of nuclear division is called *mitosis*. The germ-cell nuclei are produced by *meiosis*, a kind of nuclear division in which the number of chromosomes is halved. The term *clone* will also be used in later sections. It describes the aggregate of cells derived by successive mitotic (i.e. non-sexual) divisions from a single ancestral cell. It is commonly used in contexts where each cell is a separate organism, as with bacteria, or one-celled animals like *Amœba*.

THE CLASSICAL STAGE

In a wide range of biological situations the one-way theory of Weismann is so close to the truth that it was possible for a generation of experimental geneticists to build upon this foundation. Weismann himself amputated the tails of twenty generations of mice, yet found the tails of the twenty-first undiminished (22, p. 397), and by this feat made a considerable impression on his contemporaries. By the beginning of this century, when the rediscovery of Mendel's experiments heralded the birth of a new science soon to be christened " genetics ", most biologists were aware that the bodily effects of mutilations, disease, and the use and disuse of organs are in general *not* transmitted through sexual reproduction.

This was the root assumption of T. H. Morgan and the school which he founded. Working on the *Drosophila* fruit fly, he was able to particularize Weismann's vaguely conceived " germ-plasm " into visible structures, the chromosomes, subsisting in the nucleus of the living cell. By a brilliant combination of breeding experiments with microscopical observation of cell nuclei, Morgan and his colleagues were able to decipher the elaborate manœuvres and exchanges by which the chromosomes of the reproductive cells ensure the orderly distribution and recombination of inherited differences among the progeny. Thus arose what has been termed the " chromosome theory of heredity ". It was the dominant theory of what I have called the " classical stage " of genetics, which I conceive as extending roughly from the beginning of the century up to the Second World War.

For illustrative purposes I have reproduced in the adjoining

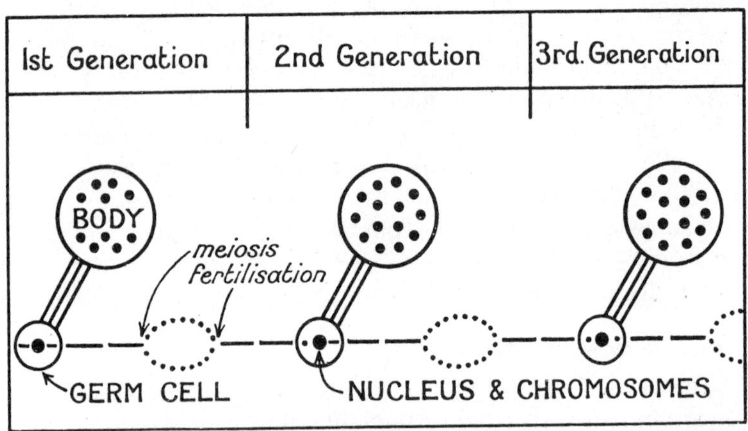

8.—The " chromosome theory " of heredity—a one-way theory. (Compare figure 7, p. 58.) (Redrawn by kind permission of Messrs George Allen and Unwin from *The Facts of Life*, by C. D. Darlington).

figure another selection from Darlington's diagram, in which the " chromosome theory " is shown to be a sophisticated brand of the one-way, or Weismannian, theory. In this theory the single nucleus of the fertilized egg gives rise by repeated and faithful replication during the course of development to the millions of nuclei which reside in and which control the skin cells, the bone cells, the nerve cells, the blood cells and the rest; but, however infinitely variegated in function and delicately responsive to their biochemical milieu the several million cytoplasms may be, the nuclei which they ultimately obey are pictured as identical replicates of the nucleus of the ancestral fertilized egg, at least as concerns their biologically potent and durable contents, the chromosomes. Darlington

A "normal" moth (*Biston betularia*), with a black or melanic form of the same species (variety "*carbonaria*") on a typically lichenous tree trunk in Dorset—an area of the South of England free from atmospheric pollution. Compare the next photograph. On this background the light form is the less visible.

Here the moths are of the same types as in the other photograph, but are on an oak trunk in Birmingham, where filth from chimneys pollutes the atmosphere and kills the lichens. The "new" or "mutant" black form is the less visible. (Photographs by Michael Lister, Department of Zoology, Oxford.)

PLATE I. "INDUSTRIAL MELANISM" IN MOTHS. (SEE PAGE xiv.)

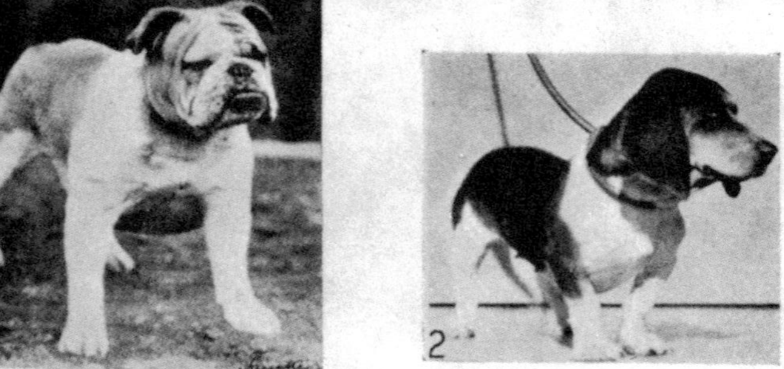

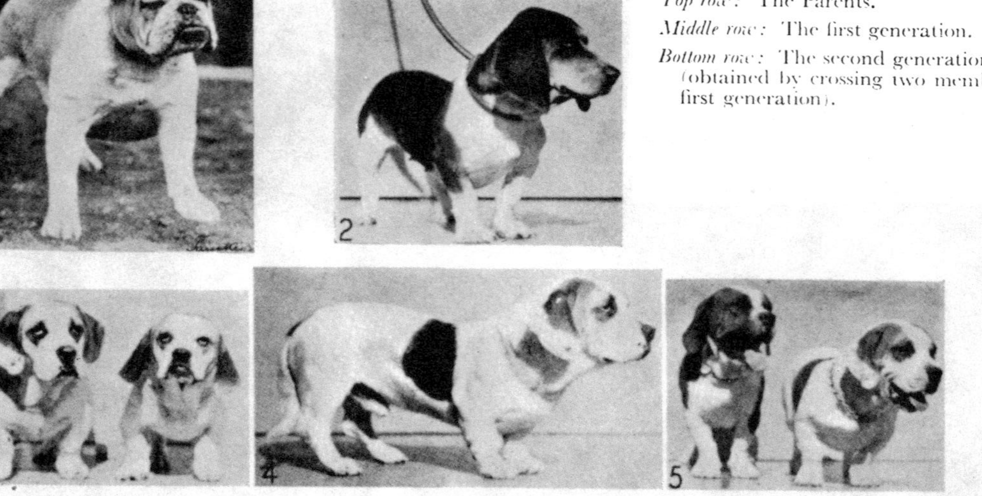

Top row: The Parents.

Middle row: The first generation.

Bottom row: The second generation (obtained by crossing two members of the first generation).

By courtesy of the Wistar Institute of Anatomy and Biology

PLATE II. STOCKARD'S BULLDOG-BASSETHOUND CROSS. (SEE PAGE 62.)

(from original of plate 19, Charles R. Stockard, *The Genetic and Endocrinic Basis for Differences in Form and Behavior*, 97.)

and Mather (7, p. 189) have termed this " the principle of the genetic uniformity of the parts of an individual ".

The number of chromosomes in the set is characteristic for each species, and in many organisms the different pairs of chromosomes can be separately identified under the microscope. For example, figure 9 shows the relative sizes and shapes of the members of the diploid chromosome complement in *Drosophila melanogaster*, and these are reproduced more or less recognizably in the cell nuclei of most of the tissues of the fly's body. Of each pair, one member is the lineal descendant (and in the " chromosome theory of heredity " is presumed to be an exact copy) of a chromosome contributed by the sperm at fertilization; the other is the lineal descendant of a chromosome contributed by the egg. The aggregate of paternal chromosomes carry the father's quota of hereditary determinants, or genes, and the maternal chromosomes carry the mother's quota.

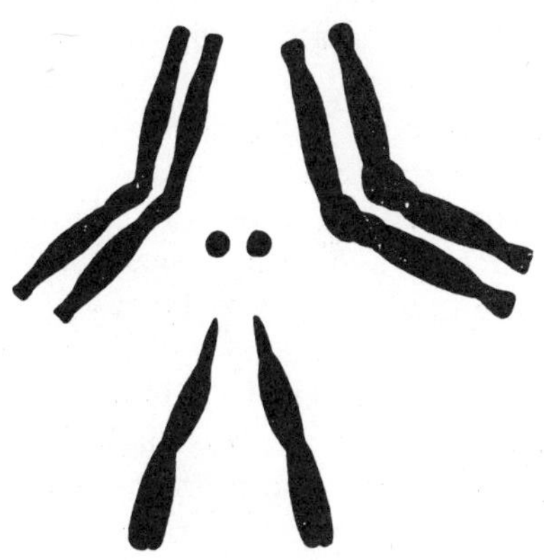

9.—The chromosomes of the fruit-fly, *Drosophila melanogaster* (female).

So it is, according to this theory, that when the sex cells (each containing half the somatic chromosome number) are formed, novelty arises solely in the manner in which genic material is selected from the full diploid set, apart from rare and " random " mutation. The nuclear apparatus is not permitted in this scheme to bear any marks of the vicissitudes of development and differentiation, so that, as indicated in the diagram of figure 8, development in each generation is an offshoot of heredity, and is without reciprocal effect on the hereditary process.

These one-way assumptions were of immense value for interpreting the results of the traditional kind of genetical experiment in the classical stage. The technique employed was typically the mating of related individuals with the minimum of environmental interference. The phenomena to be interpreted were restricted to the reappearance and recombination of innate traits in succeeding generations. As an entertaining illustration of these phenomena, I reproduce a photograph of dogs obtained by Stockard [21] after crossing a bulldog with a basset hound. Note the reappearance in the second generation of the bulldog's length of leg in some, but not all, members of a litter (plate facing page 61).

Yet there are all manner of ways in which animals and plants reproduce themselves *without* mating, that is vegetatively. The simple amœba splits in two, the hydra can generate a new hydra by budding, the potato is commonly propagated by tubers. Even the process by which a newt regenerates an amputated leg is a form of reproduction, although partial. All these are channels of hereditary transmission, yet variation between cells or individuals supposed to have identical chromosome sets cannot be brought within the confines of the chromosome theory, the more so since such variation is found to be peculiarly subject to environmental influences. The geneticists of the classical stage prudently regarded these phenomena as beyond the pale of their theory, and as being, all in all, much better left to puzzle embryologists.

In the end the artificially restricted territory of experimental genetics was worked to near exhaustion, and investigators began to break out from its cramping bounds.

SECOND STAGE

During the next stage, which I identify rather arbitrarily with the post-war period, the focus of attention swings to all those hereditary phenomena which will *not* fit into the classical picture—to the transmission of non-chromosomal effects in sexual reproduction and, in asexual reproduction, to the study of the role of the cytoplasm in cell-heredity as manifested in embryonic differentiation, or the propagation of clones of single-celled organisms, whether protozoa, bacteria or fungi.

This stage, which is only now ending, has seen the reign of a dualist philosophy in biology. " There are two systems of heredity," says the dualist. " One is the chromosomal system, more or less fully elucidated during the classical stage. In this system, all is according to the one-way theory of Weismann;

so much so, that whenever we come upon the inheritance of an acquired character, the transmission of an heritable trait by grafting, or an effect on the offspring of the parents' intrinsic processes of bodily development or decay, we can assign it without further evidence to the operations of that other system of heredity, the cytoplasmic system. The cytoplasmic system, by contrast, obeys the two-way theories of Lamarck and Darwin. On the rare occasions when their paths cross it is the chromosomal system which gives the orders, never the other way round."

PARAMECIUM: NUCLEUS AND CYTOPLASM

To show how neatly the dualist view can fit the facts unearthed during the second phase, I shall say something about the single-celled " slipper animalcule " *Paramecium aurelia.* If a few thousand paramecia from a pure strain are injected into the blood stream of a rabbit every few days for two or three weeks the rabbit manufactures specific antibodies against paramecia of that particular strain: actually, against the proteins of their swimming organs or cilia. We call such proteins " antigens " because of their power to provoke a specific antibody response. An antiserum prepared in this way immobilizes and kills all paramecia which have the same antigen as the strain with which the rabbit was immunized, but is powerless against paramecia lacking the antigen.

In one variety investigated by Beale [1], the type of antigen formed by any given stock is dependent on the temperature of the environment. At moderate temperatures (say, 23° C) the antigen is of one type, let us call it the " cool " type, but at high temperatures (31° C) another antigen—call it the " warm " type—takes its place. If paramecia are transferred to an intermediate temperature (27° C) they and their descendants continue to manifest the antigen appropriate to the temperature from which they have been transferred. For example, animals brought from 23° C to 27° C continue to have " cool " antigens, and so do their vegetative descendants. But members of the same stock which have acquired the " warm " antigenic type through sojourn in a 31° C climate will breed true to it indefinitely when transferred to 27° C. This frankly " Lamarckian " behaviour can be shown to be due to the modification by temperature of the cytoplasm, which is able to propagate the modified state under certain "neutral" conditions of temperature.

On the other hand the *kinds* of "warm" and "cool" antigens which the animals make show subtle differences from one stock to another. For instance, the "cool" antigens of stock 60 differ from those of stock 90, and the "warm" stock 60 antigens differ from the "warm" stock 90 antigens. These differences are determined by chromosomal genes of the

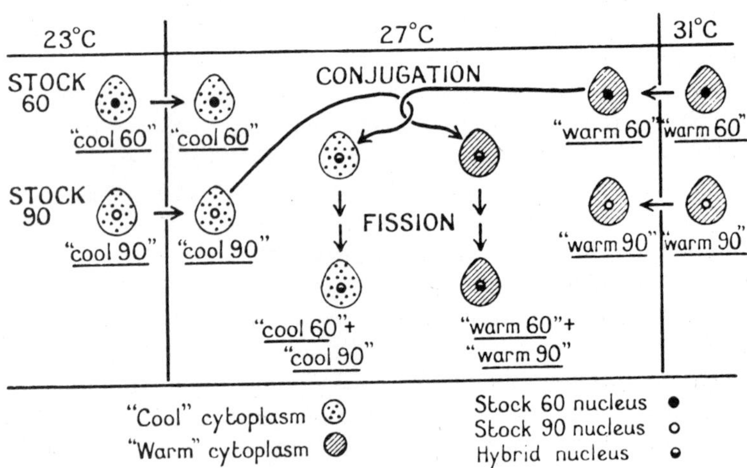

10.—Types of antigen produced by *Paramecium* in a typical experiment (see text).

conventional type. The dual determination of antigens by nucleus and cytoplasm is beautifully illustrated in the experiment shown in the accompanying diagram.

If we transfer to the neutral temperature of 27° C a cool-kept paramecium from stock 90 and a warm-kept paramecium from stock 60, the vegetative descendants of each will continue to make "cool 90" and "warm 60" antigens respectively, for at this temperature the cytoplasmic system of each is stable. Now cross them. The two lie with oral surfaces apposed, and each receives from the other a migratory nucleus carrying the haploid chromosome set. Migratory and stationary nuclei fuse to form the new zygotic nucleus and reconstitute a diploid set of chromosomes. The two "ex-conjugants" (as they are called at this stage) separate, and each gives rise to a clone of genetically identical daughters by successive binary fission. What is the antigenic type of each clone? The answer, which is illustrated in figure 10, can be given only through a knowledge of *both* systems of heredity, the chromosomal and the cytoplasmic, and of the environmental as well as genetical histories of the conjugants' ancestors. The ex-conjugant which retains cytoplasm of cool-kept stock 90 origin will follow its cytoplasmic parent in continuing to make " cool " antigen. But its genes are no longer pure 90, for half of them have been derived from the migratory nucleus of the stock 60 parent. Accordingly, its descendants will be found to manufacture a mixture of two antigens, " cool 60 " and " cool 90 ". In the same way, the clone descended from the other ex-conjugant, although chromosomally identical, will make a mixture of " warm 60 " and " warm 90 ".

In this instance it is probably unnecessary to look in the cytoplasm for self-replicating gene-like bodies or particles to explain the cytoplasmic determination, transmission and stability of the temperature-induced differences. The stability of environmentally induced traits through many cell-generations could theoretically be assured by the existence of stable equilibria in a complex system of linked chemical reactions. A view of this type has been energetically advanced by C. Hinshelwood in connection with the " training " of bacteria. He has shown that various kinds of bacteria will " learn " to cope with a new food or poison, and will transmit their acquired biochemical wisdom more or less durably (according to the number of cell-generations for which the treatment has been applied) to their remote descendants.

THE CHILDREN OF ELDERLY ROTIFERS

In the same way the hereditary transmission of somatic effects in a many-celled organism, a rotifer, or " wheel animalcule ", *Philodina citrina*, does not require us to postulate self-replicating cytoplasmic particles, although they may be involved. This animal reproduces parthenogenetically by laying eggs and normally has a life-span of about twenty-five days. Lansing (15, pp. 14–19) found that when he propagated rotifers entirely from eggs laid by elderly females they developed more quickly and lived shorter lives in each generation until the lineage became extinct. Propagation from adolescent mothers (for example, five days of age) had the opposite effect, and resulted in an unusually long-lived stock. Clearly the potentialities with which the egg is endowed depend here upon cumulative changes occurring in the parent's body. If the reader will turn back to the diagram of the two genetical philosophies in figure 6, he will see that Lansing's rotifers rudely violate the second (one-way) theory, for the germ of the new rotifer is in part determined by what has happened to the body of the old. In such a case, the disciple of dualism would tend to assign the phenomenon to the cytoplasmic category, where such irregularities are permitted. Whether he favoured the accumulation in the cytoplasm of a toxic substance, or the intemperate multiplication of a self-replicating particle, it is to the cytoplasm that he would direct his search, rather than tear aside the nuclear membrane and expose the *sanctum sanctorum* of chromosomal heredity to the forces of development and decay.

FRUIT FLIES AND GUINEA PIGS

In some examples of germinal transmission in many-celled organisms, the cytoplasm *has* been directly implicated, although never yet in so exact and satisfying a fashion as in Paramecium. L'Heritier [17] and Teissier observed that some stocks of Drosophila flies reacted with a characteristic and fatal paralysis to small concentrations of carbon dioxide. The susceptibility was inherited strongly through the female line, but weakly and irregularly through the male line. This is a reasonable indication of cytoplasmic transmission, since the sperm, equal in nuclear equipment to the egg, has relatively little cytoplasm. Injection of a resistant fly with body fluids from a susceptible fly caused some of the offspring to be susceptible. In addition, something essentially like Darwin's " telegony " occurred when a resistant female was mated in quick succession to a susceptible and to a resistant male. The hypothesis of telegony states that a previous mating may have an influence on a female's offspring by a subsequent mating. Female flies which had been double-mated, as described above, produced some offspring which on genetical evidence had certainly been sired by the resistant male and yet were susceptible. Supernumerary sperm from the susceptible male must have penetrated the cytoplasm of eggs fertilized by the resistant male. A self-replicating particle with infective properties seemed to be involved. In many ways it resembled a virus, but the symptoms caused were disagreeable to the patient only in the presence of unusual quantities of carbon dioxide.

" PLASMAGENES "

Wide currency during the second stage of genetical history was gained by the idea that cytoplasmic genes (or " plasmagenes " as they have been called) of this kind play an occasional and secondary part in hereditary transmission in many-celled organisms, but a constant and necessary part in the differentiation of cells during embryonic development. Several cases of variation within clones of one-celled creatures were tracked down to the action, multiplication and loss of cytoplasmic particles, thus offering an attractive model for the appearance of differences within the clone of cells which makes up a multicellular individual. Billingham and Medawar demonstrated that the gradual spread of dark pigmentation into the white patches of piebald guinea pigs, which is in essence a case of differentation, though a simple one, was due to cell-to-cell infection, in all probability by a virus-like particle with the power to catalyse pigment-formation [3].

The temptation seemed very strong to attribute the appearance of differences between the different tissues of a developing embryo to the fact that cytoplasmic particles, in contrast to the chromosomal genes, are not obliged to multiply in step with cell division. In response to subtle differences of the physical and chemical environment from one part of the embryo to another, different types of " plasmagenes " may multiply at different rates; so that cells in different sites may come by a " sorting out " process to lack characteristic types of cytoplasmic determinant, and hence to breed true for the properties of the given tissue.

In order to illustrate this terminal state of the dualist conception, I reproduce in the figure on page 68 yet one more version of the Weismannian diagram of heredity. It is intended to show how the essentially Lamarckian phenomena of cell-heredity, operating via the cytoplasm, can be fitted into a scheme which still denies the bodily environment a say in what is transmitted via the nucleus to the next generation.

DIGRESSION ON PANGENESIS

In the background of all this speculation the chromosomes have remained constant from cell to cell and conservative from generation to generation. This constancy and conservatism, relative to cytoplasmic processes, is real and important. If Darwin had possessed a more lively awareness of the conser-

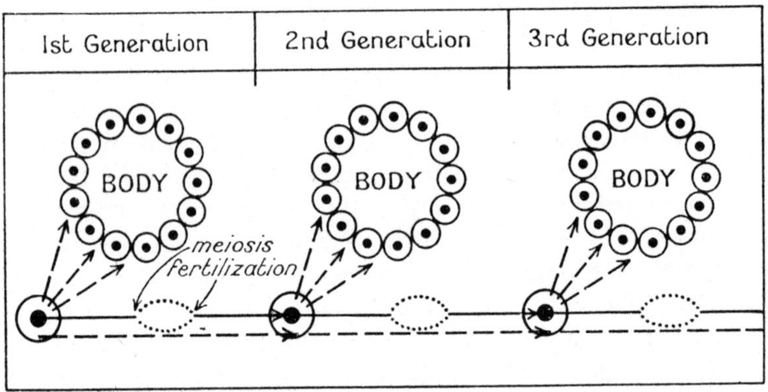

11.—The final state of the "one-way" theory of heredity: "dualism" (see text).

vatism of heredity he might have moderated the exuberance of his theory of pangenesis. Under this term he sought to unify the conflicting phenomena of heredity and development, and explain ". . . how it is possible for a character possessed by some remote ancestor suddenly to reappear in the offspring; how the effects of the increased or decreased use of a limb can be transmitted to the child; how the male sexual element can act not solely on the ovule, but occasionally on the mother-form; how a limb can be reproduced on the exact line of amputation, with neither too much nor too little added; how it comes that organic beings identical in every respect are habitually produced by such widely different processes, as budding and true seminal generation."

Pangenesis, although quaint and naïve in many respects, is in others an arresting and prophetic hypothesis. Since it may serve as a goad to new attempts to find a unitary theory, I shall pause here in order very briefly to consider the main outlines of Darwin's conception.

During the proliferation of the various tissues of the body, their cells " throw off " in Darwin's words " minute granules or atoms, which circulate freely throughout the system, and when supplied with proper nutriment multiply by self-division." These granules, or " gemmules " as he called them, are transmitted from parent to offspring, and " in their dormant state ", says Darwin, they " have a mutual affinity for each other, leading to their aggregation either into buds or into the sexual elements."

This model of differentiation and heredity allows for cases of the inheritance of acquired characters, since environmental modification of a bodily tissue during its formative phase will result in the production of altered gemmules which, passing through the germ-cells into the next generation, " would naturally produce the same modification ". This we know to be far-fetched. Yet there are two things about the theory which make Darwin's pangenesis theory seem remarkable for its time. First, his postulation of self-replicating particles: fifty years were to pass before the discovery of the chromosomal genes proved that such entities existed and served as agents of hereditary transmission; another twenty-five years before the possibility was seriously mooted that similar " gemmules " (plasmagenes) might play a part in embryonic development. Second, pangenesis represents a heroic attempt to unite the heredity of the germ-line and the development of the body in a single scheme, in which the germ-line is not set apart, remote and inviolable, but is subject to the forces of development and decay which govern the rest of the body. For a hundred years such theories have been resolutely excluded from the pale of biology. In the remaining pages I shall suggest that the day of their return has dawned.

THIRD STAGE

The present dualist fashion cannot last. In the end genetical workers are bound to seek a unitary theory of the two-way type.

It must embrace both development and heredity and find room for a reciprocal chain of command between them. Hence the germ-plasm must be conceded some vulnerability to influences arising from the mortal body, or " soma " which houses it. With the reinterpretation of germ-plasm and soma in concrete terms as nucleus and cytoplasm, the new genetics will consist in the rejection of the former dualism and the search for pathways by which the chromosomal genes may be reached and modified by cytoplasmic action. Evidence that these pathways exist comes from some of the most exciting programmes of research in contemporary biology. Most of this work is still in progress at the time of writing. I shall summarize below some of the results which have been so far obtained, without attempting a " rounded-off " interpretation. In making an arbitrary choice from among a number of researches which I might have described, I have been guided solely by personal familiarity and predilection.

IS THE NUCLEUS AFFECTED BY DIFFERENTIATION?

I spoke above of the doctrine that the somatic nuclei are interchangeable, being descended through a succession of exact replications from the diploid nucleus of the fertilized egg. Evidence that *some* somatic nuclei preserve the full potentialities of the fertilized egg is provided by the regeneration of a whole plant from a mere fragment of leaf, as is observed in *Begonia* for instance. Evidence that the nuclei of different tissues are identical with respect to chromosomal genes of at least one important category is provided by the work of Medawar's school on skin grafting in the mouse. To bring home the force of this evidence will require a short excursion into " immunogenetics ".

If tissue, for instance skin, is grafted from one warm-blooded animal to another of the same species, the recipient, after an initial quiescent phase, exhibits a characteristic inflammatory reaction and eventually sloughs the graft off. In the case of skin grafted from mouse to mouse, the survival time is about ten days. If a second graft is made *from the same donor* it receives much rougher treatment than the first, and is sloughed off inside a week. The recipient has become immunized against specific antigens characteristic of the donor's genetical make-up.

The fact that it is the donor's *genotype* that is responsible for the antigenic constitution is shown by two facts: first, a mouse immunized as described above is equally immune to skin from a second genetically identical donor, for example, a member of the same highly inbred strain; second, grafts between genetically identical mice, for instance, two sisters or two brothers from an inbred strain, always " take ". They are neither themselves attacked nor do they provoke a state of immunity. Many additional facts show that the antigens of tissue transplantation are determined by the chromosomal genes.

The first observation bearing on the doctrine of nuclear equivalence, was that it was not necessary to use the donor's *skin* in order to immunize a mouse against a skin graft. Any tissue consisting of nucleated cells, if injected into the intended recipient, would provoke immunity against a subsequent skin graft, or, for that matter, against the graft of any other nucleated tissue from the same donor. The inference from this is that there are *some* transplantation antigens common to all cells, and hence presumably that the corresponding genes are common to all cell nuclei. This, of course, does not exclude the possibility that in addition to individual-specific antigens which are common to all tissues there are organ-specific transplantation antigens which are not, rather as the naval ratings of a given country all wear the same uniform, but differ in the name of the ship embroidered on their caps. The disproof of this supposition was obtained as follows:

We have seen that, according to the rules of transplantation immunity, the recipient reacts against any antigen in the graft which it does not itself possess, tolerating only those antigens which it possesses in common with the donor. Billingham, Brent and Medawar [2] found a way of tricking new-born mice into breaking the rules, basing themselves in part upon the " self-recognition " hypothesis of Burnet and Fenner. This hypothesis supposes that at some stage of development, before the body's defence system against foreign bodies has matured, the organism must take stock of all the antigens present in its body in order, so to speak, to teach its cellular Home Guard to recognize native antigens so that they may reserve their fire for intruders.

Medawar and his co-workers accordingly introduced tissue from members of one inbred strain into members of another at a period of development during which they hoped that the recipients would be engaged in this stock-taking process. The critical period was found to be around the period of birth. If for example new-born mice of strain A received an injection of strain B cells into their blood stream, they became tolerant for the rest of their lives to skin grafted from B strain mice. Confronted with the B antigens while engaged in stock-taking, the new-born mice had mistakenly and irretrievably classified the foreign proteins as their own.

It was shown that spleen cells, kidney cells, white blood cells —indeed every cell type of which trial was made—all had the property of conferring on the new-born mouse a lasting tolerance to skin, cornea, bone or any other tissue tested by grafting. This can only mean that there is no transplantation antigen possessed either by skin, or by cornea, or by bone, and so forth which is not also possessed by spleen, kidney, white blood cells and the rest. In terms of our naval analogy, not only do all the ratings from a given country wear the same uniform: they also carry the same legend on their caps. As far as concerns the antigens of tissue transplantation, it seems that the principle of the genetical uniformity of differentiated tissues is upheld. Can the principle safely be generalized? Are all the differentiated cells of an organism really identical in chromosome make-up? To obtain an answer proof against quibbling would require the transfer of chromosomes from one cell to another. Briggs and King [4], working in Philadelphia, have done the next best thing by transferring cell nuclei from differentiating tissues of frog embryos into fertilized frog eggs from which the nuclei have been removed. Their results, as we shall see, indicate that in important respects differentiated tissues are *not* genetically uniform.

NUCLEAR TRANSPLANTATION

If nuclei are unaffected by the differentiation of the cells which they inhabit and control, then the result of Briggs and King's experiment should be the development of perfectly normal embryos, and ultimately tadpoles and frogs. For the nucleus which is put into the fertilized egg is *ex hypothesi* identical in biological potentialities with the one which is taken out. But if the nuclei of somatic cells suffer irreversible changes during differentiation, then we would expect eggs which have received transplanted nuclei to fail or become abnormal at some stage in their subsequent development.

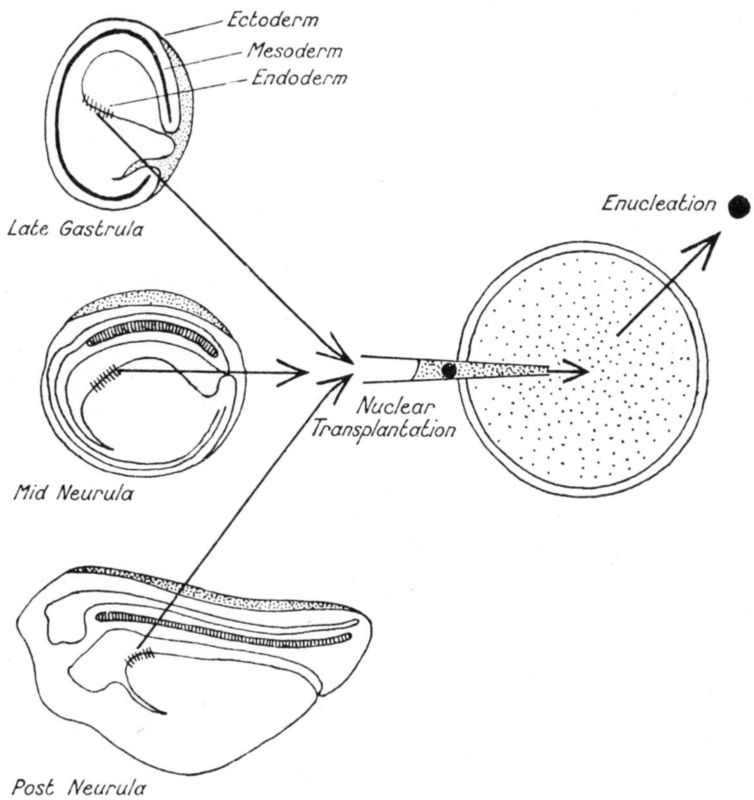

12.—Transplantation of cell-nuclei from embryos into eggs in the frog.

The early stages of the development of the frog's egg consists of the multiplication of cell-number by repeated cleavage, so as to form a hollow sphere (*blastula*) consisting of round cells without visible signs of differentiation apart from the division into an upper *animal hemisphere* with more darkly pigmented cells and a lower *vegetative hemisphere* with rather larger, more yolky cells. This stage is terminated by a radical rearrangement of material known as *gastrulation* during which the three main categories of embryonic tissue, namely, ectoderm, mesoderm and endoderm, are laid down and irreversibly determined.

Briggs and King transplanted nuclei from endoderm cells o[f] three post-gastrulation stages, as shown in the figure on page 73. In addition they performed " control " transplantations from

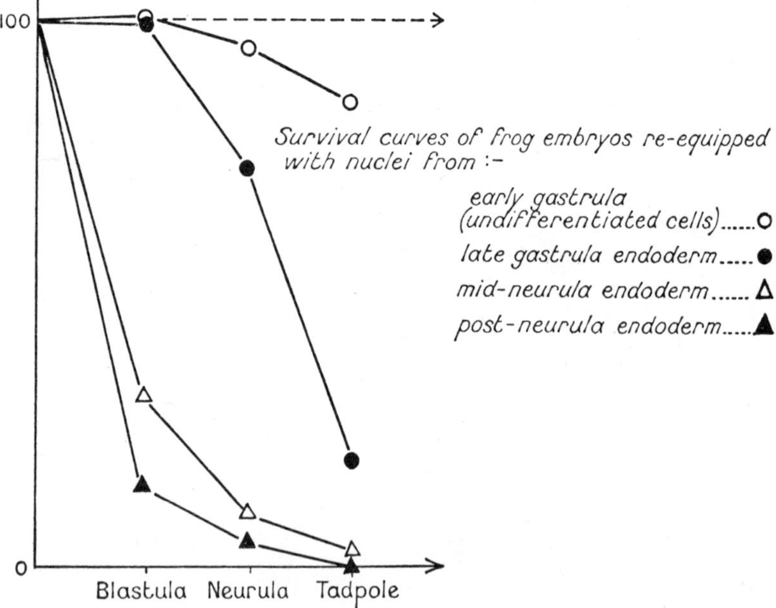

13.—Nuclear transplantation: survival rates. The broken line at the top indicates 100 per cent survival. (The data have been corrected to allow for losses before the blastula stage attributable to technical hazards.)

undifferentiated cells from the animal hemisphere of early gastrulæ.

Those transplantations in this last category which surmounted technical hazards and developed as far as the blastula stage went on, with few exceptions, to form normal embryos which grew into tadpoles. This can be seen by inspecting the open circles in the figure above. But the fate of recipients of " endoderm nuclei " was very different. First we will consider the proportion which managed to reach the blastula stage.

It is evident from the graph that endoderm nuclei from late gastrulæ have not lost the power to bring the fertilized egg through early cleavage as far as the blastula stage. But there is a marked and progressive loss of this power in nuclei transplanted from the later stages.

All the blastulæ looked the same on external inspection whatever the origin of their nuclei. But their subsequent fates showed striking and dramatic differences, as shown in the survival curves of figure 13. The later the stage of origin of the nuclei, the earlier the embryonic stage at which breakdown occurs and the fewer are the survivors which complete the course. *Post-mortem* examination of arrested embryos showed that this progressive limitation of the potentialities of the somatic nucleus is *specific*. Those " endoderm embryos " which completed gastrulation and failed at some later stage were found to suffer from defects of the organ systems derived from ectoderm (for instance, nervous system, skin) and mesoderm (for instance, embryonic kidney, heart), but to be perfectly normal as regards endodermal derivatives. The final proof of the specificity of the nuclear changes awaits the transplantation of nuclei from differentiated ectoderm or mesoderm. But the fact that the differentiation of cells brings about a progressive and irreversible modification of their nuclei has been demonstrated, it seems, beyond possible doubt.

" PSEUDO-HYBRIDS " IN AMŒBA

During the same period the method of nuclear transplantation has been used by Danielli and his colleagues [18, 5] on very different material to attack fundamentally the same problem. Instead of transplanting nuclei from one cell to another in a many-celled organism, they have transplanted nuclei between two species of the single-celled Amœba, *Amœba proteus* and *Amœba discoides*. The resulting compound animals usually survived the operation and resumed locomotion and feeding. In some cases they underwent a few divisions, but only very rarely was a clone obtained which could be propagated further. The " pseudo-hybrids " as they might perhaps be termed, were intermediate between the nuclear and the cytoplasmic " parent " species in shape. In the other character studied, the diameter of the nucleus, the cytoplasmic species predominated. More detailed observations were made on a " pseudo-hybrid " clone which had been propagated for six years. Shape was still intermediate and nuclear diameter unchanged, showing that the cytoplasmic systems involved are self-propagating. A number of immunological characters were examined and these were found for the most part to follow the nuclear species. But most interesting of all were the results of back-transfers of nuclei from the " pseudo-hybrids " into the cytoplasm of members of the " parent " species, and of replacement of the nuclei of " pseudo-hybrids " with nuclei derived from the " parent " species. This method of analysis revealed that both nucleus and cytoplasm had been altered by their six years' enforced coexistence, and that the alterations were specific; that is to say, an *approximation* between the species in both nuclear and cytoplasmic properties had occurred.

One way in which the process of approximation manifests itself is in the frequency with which inter-species clones are obtained. The clone which I have just been describing was the only one obtained out of sixty-one inter-species transfers, although 66 per cent of transfers between individuals of the same species gave clones. There seems to be some kind of incompatibility between nucleus and cytoplasm reminiscent of the barriers which prevent or make difficult the sexual hybridization of different species. It is intriguing therefore that Danielli and his co-workers found that " pseudo-hybrid " clones could be obtained with greater ease if " approximated " rather than ordinary nuclei were used. The same held good for " approximated " cytoplasm.

This finding is vividly reminiscent of the use of grafting by the Russian plant hybridizer, Michurin, in order to facilitate difficult sexual crosses. He termed the method " vegetative approximation ". I shall illustrate it with an experiment first done by Pissarev and Vinogradova and later confirmed by Hall [13] in Sweden. Wheat and rye are very difficult to hybridize. Hall found that only 2 to 3 per cent of wheat flowers pollinated by rye gave seeds. But of 2,897 wheat flowers borne by plants which, as embryos, had been grafted on to rye endosperm, 400 (or 14 per cent) gave seeds when pollinated by rye. Haldane [12] has pointed out that whatever is acquired from the rye endosperm must be self-propagating, since the mature wheat plant has many thousand times the bulk of its embryo. If the phenomenon is general, as it probably is (at least in plants) its analysis in biochemical terms should offer one of the most promising approaches to the problem of the interaction between nucleus and cytoplasm.

At this point my biological dualist whose philosophy is illustrated in figure 11 becomes restive. "Admittedly," he interjects, "there is a somatic effect on the *germ cells* in the wheat–rye experiment, but it could be entirely an effect on the cytoplasm: there is no evidence that the nucleus is affected. And admittedly Briggs and King have shown a somatic effect on the *nucleus*, but this involves the nuclei of somatic cells only. To justify high-flown talk about the rejection of dualism, the new synthesis, and so forth, it is necessary to show that there is some prospect of obtaining somatic modification not just of germ cells, but of the *nuclei of germ cells*."

This sounds like a tall order. If such a thing could be demonstrated an avenue would have been opened towards the fulfilment of the fantastic dream that man might remould at his pleasure the hereditary nature of plants and animals, and even of his own species. I shall conclude this essay by describing one or two lines of recent work which in my estimation place the modification of the chromosomes of the germ line within the bounds of possibility.

"AGEING" IN A ONE-CELLED ANIMAL

First we return to the slipper animalcule, *Paramecium aurelia*. This creature, like other ciliates, has two kinds of nucleus, a micronucleus and a macronucleus. In vegetative reproduction by binary fission both nuclei divide. In sexual reproduction, however, in which (as described earlier) micronuclei are exchanged, the macronucleus disintegrates and a new one is formed by the action of the micronucleus. Deprived of a micronucleus, a paramecium will survive apparently unaffected over long periods of vegetative reproduction, but without a macronucleus the animal soon dies. Beale has summed the matter up as follows:

> Thus the macronuclei may be described as somatic nuclei, which are physiologically active during periods of asexual multiplication, while the micronuclei serve as germinal nuclei which play

an essential part in the fertilization process and the formation of new macronuclei, but have little if any direct effect on the phenotype of the organisms. [1]

Asexual clones of paramecia cannot be propagated indefinitely. They require from time to time to be " rejuvenated " by passing through a generation of sexual reproduction. Sonneborn has demonstrated that crossing as such is not necessary for the rejuvenation of senescent clones. " Autogamy ", in which the animal mates with itself by an internal reorganization of micronuclei, is sufficient. In autogamy, as in sexual crossing, the old macronucleus is scrapped and a new one refashioned under the influence of the rearranged micronucleus. Sonneborn saw in this the key to the phenomenon of clonal ageing [20]. He showed that the deterioration is due to the accumulation of defects in the macronucleus, in part at least associated with the random loss of individual members of the macronuclear chromosome complement during their distribution to daughter cells at successive fissions. But if the process of somatic deterioration is allowed to go far enough, the rejuvenating power of autogamy begins to fade, and no longer suffices to restore a senescent clone to full vigour. Evidently the micronucleus itself has been corrupted by prolonged cohabitation with the products of a defective macronucleus.

This particular alteration of nuclear heredity may involve a gain, a rearrangement, or a loss of material. The last two are the more probable, though less exciting, alternatives. But there is a well-analysed instance in bacteria which demonstrates a clear-cut and durable *acquisition* of new material by the cell's genetical apparatus. To my mind the importance of the example which I shall give, however remote the application to other and higher organisms may seem, lies in its proof that not only impoverishment but also deliberate enrichment of an organism's genetical endowment is in principle possible.

" TRANSDUCTION "

Bacteria do not have nuclei, and probably do not have chromosomes, in the usually accepted sense. They do, however, contain the characteristic chromosomal material, desoxyribose nucleic acid, generally referred to as DNA. After the discovery of sexual reproduction in a strain of *Escherichia coli* by Lederberg and Tatum it was possible to show that the sorting out of hereditary characters in crosses behaves as if the genetical factors are situated in a definite place in the cell and arranged in a linear order. *E. coli* is one of the bacteria in which bacterial viruses, known as bacteriophages ("phages" for short), have been studied. The *normal* mode of existence of these viruses is in the form of "prophage", a harmless constituent of the bacterial cell which multiplies in step with its host and hence is invariably inherited. But after certain environmental treatments the prophage "runs wild" and multiplies at the expense of its host. The host bacterium ultimately bursts, liberating phage particles into the medium. These may be "virulent" and after infection of a fresh host repeat the cycle of multiplication and burst. Or they may be "temperate", in which case a proportion will settle down in new hosts to lead a quiet life as symbiotic prophage once more.

The key facts which I wish to stress are the following. First, phage is made of DNA plus a protein skin which it sheds on entering a host bacterium; hence prophage is mainly DNA. Second, infection of some bacterial strains with temperate phage results in the acquisition and orderly hereditary transmission by the host of some of the genetical traits of the previous host; the phage is said to "transduce" the trait. Third, Lederberg and Lederberg [16], and also Wollman [23], took advantage of the fact that *E. coli* is not only susceptible to infection by phage, but also, in one of its strains, given to sexual crossing, thus permitting orthodox "chromosome mapping" techniques. They found that prophage "lambda" was inherited in sexual crosses as a gene, being closely linked to a marker-gene, *Gal*, controlling the fermentation of a sugar, galactose. Fourth, Morse and Lederberg [19] showed that lambda would transduce the gene *Gal*.

If these facts are taken together the conclusion emerges that prophage is part of the host's genetical apparatus, situated at a definite site in whatever passes for a chromosome in bacteria, and that this chromosomal locus can be transferred, together with some of the nearby genes, into a new host, where it is assimilated into the genotype.

Before becoming overexcited about the possibilities of general applications of what one might call "gene transfusion", we must remind ourselves that we are dealing with bacteria, that realm of creation in which the genetical material and genetical mechanisms bear least resemblance to the higher organisms. On the other hand, recent experimental findings in higher plants encourage me to draw the long bow of speculation on the possibility that "transduction" of hereditary traits from one plant variety to another may occur in much the same way as in bacteria, thus opening a path to the direct and durable implantation of genetical material in the chromosomes themselves.

INFECTIVE INHERITANCE IN TOMATOES

The formal genetics of tomatoes has been fairly thoroughly studied, and the main varietal characters (form of leaf, colour and shape of fruits and so on) have been shown to follow the classical Mendelian rules. Hence they can be presumed to be controlled in the usual way by chromosomal genes. However, Russian investigators have repeatedly claimed that when seedlings of one variety are grafted on to a root-stock of another variety some of the varietal characters of the stock reappear among the progeny of the graft. (See Glushchenko [11], and references cited by Hudson and Richens [14].) Except that transfer has not yet been achieved by injecting cell-free extracts in place of grafting, the process seems formally identical with the infective acquisition and hereditary transmission in *Drosophila* of sensitivity to carbon dioxide. But there is an additional difference on which I propose to lay great emphasis: the tomato characters concerned are, as stated above, good "Mendelizing" chromosomal characters as regards their behaviour in sexual crosses.

As illustration I choose an extensive series of experiments done in Belgrade by R. Glavinic. My choice is in part guided by a desire to steer clear of the fog which has been engendered in the field of tomato grafting by the intrusion of questions unrelated to scientific enquiry. I refer to the outburst of political and personal calumny which became known during the post-war decade as "the Lysenko controversy". In common with some other British geneticists I am personally acquainted with Glavinic, who reported her work to the

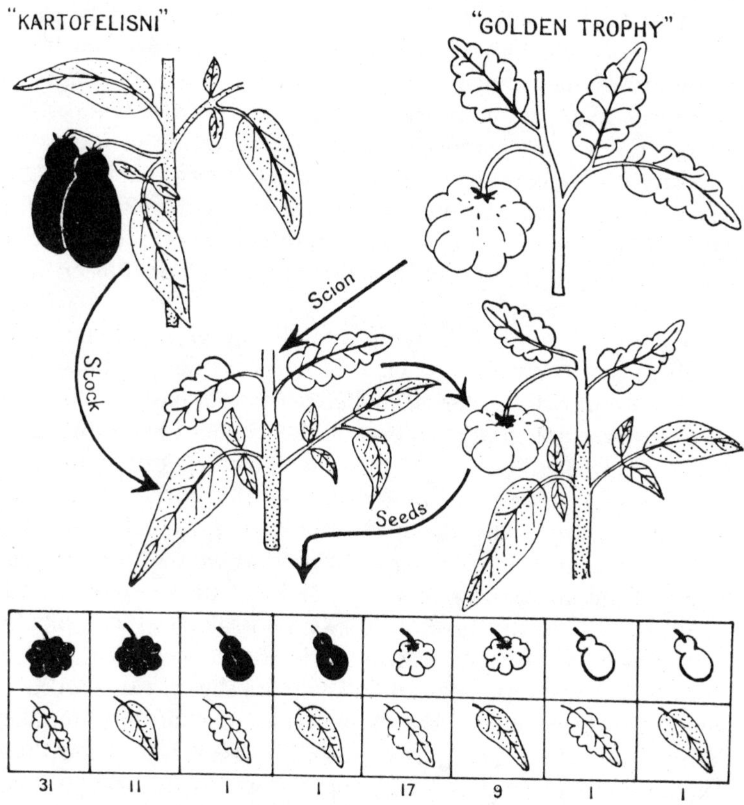

14.—Glavinic's experiment with two varieties of tomato. The numbers denote the frequencies of the various types, obtained by sowing seed taken from grafted Golden Trophy plants.

14th International Horticultural Congress in the Netherlands in 1955 [10].

Seedlings of the tomato variety "Golden Trophy" (cut leaf rather than potato leaf, yellow rather than red fruit and short rather than long fruit) were grafted on to stock of the variety "Kartofelisni" which contrasts with "Golden Trophy" in all three characters, as shown in the figure above. I have omitted a fourth character-difference, number of locules in cross-section, because its classification is known to be uncertain and strongly affected by the " long fruit " character. The buds on the scion were bagged to prevent cross-pollination and, when these eventually formed fruits after obligatory self-fertilization during the flowering stage, seeds were taken from them and planted out.

Instead of breeding true to the characters of the ancestral " Golden Trophy " genotype, the seed generation contained a majority of plants with one or more characters of the " Kartofelisni " variety in various combinations, as depicted in the figure. When one plant of each combination was selfed, there was a tendency for like to beget like, but a good deal of further segregation occurred.

The interpretation of Glavinic's result is at present extremely perplexing, the more so in view of irregularities in the segregations obtained in her sexually crossed tomatoes. But it seems impossible to doubt that in her material an infective serially propagated system in close association with the chromosomal genes was transferred from the stock to the germ cells of the scion. The analogy with phage transduction is very provoking, and it is tempting to think that we may be dealing with chromosomal fragments which have broken away and " run wild ". But if this is so, they do not, as temperate phage does, revert to their proper stations in the host's chromosomal system. For in Glushchenko's work the " transduced " characters were more strongly inherited through the female than through the male parent, usually a good indication that the determinant involved is in the cytoplasm rather than in the nucleus. As my colleague Anne McLaren has pointed out to me, it would be interesting to know whether such characters can ultimately be persuaded to " settle down " and resume a respectable existence in the nucleus after selection of the " transduced " forms through a sufficient number of generations of self-fertilized plants. As far as I know no attempt has yet been made to answer this crucial question. If the answer is found to be in the affirmative then the door to active control of chromosomal heredity, left ajar by the work on Paramecium and bacteriophage, will have been well and truly opened.

DARWIN'S PROPHECY

I have dwelt upon Glavinic's work because the Mendelian nature of the characters which she used and the extensive segregation data which she has published constitute a direct challenge to the alert botanical geneticist to apply to plants the principles derived from bacterial studies. But it must not be thought that the primary phenomenon of heritable influences of stock on graft is new. In the years between the wars the French botanist Daniel published a series of papers claiming to have obtained "graft hybrids" in this sense. Darwin himself took an active interest in the possible "formation of hybrids between distinct species or varieties without the intervention of the sexual organs", and his belief that the phenomenon occurred contributed to his formulation of the theory of pangenesis.

In the second edition of his *Variation of Animals and Plants under Domestication* he writes:

> For if, as I am now convinced, this is possible, it is a most important fact, which will sooner or later change the views of physiologists with respect to sexual reproduction.

"Sooner or later" has proved to be a long time, and the fulfilment of Darwin's conditional prophecy is not yet complete. I hope that this essay has made plain some, at least, of the reasons why I estimate that genetics has about ten years to go before it can claim fully to have caught up with Darwin.

DUCKS AND DNA

Since this essay was written, an experiment has been published which, on the face of it, is so sensational as to demand the addition of a postscript. J. Benoit, P. Leroy, C. Vendrely and R. Vendrely extracted DNA from Khaki Campbell ducks and injected it into ducklings of the Pekin breed. The majority of the treated birds, *and of their offspring*, developed a constellation of characters (pigmentation of beak, morphology of feathers, shape of head, size and conformation of body) apparently derived from the donor breed. If the authors have correctly interpreted their observations, there had occurred, to use Darwin's phrase, a "formation of hybrids between distinct . . . varieties without the intervention of the sexual organs ". Genetical transformation of *bacteria* by experimental treatment with DNA has been known for many years. The similar claim now made for ducks rests at present on this single " pilot experiment " and requires confirmation under more rigorous conditions. But whether it is confirmed or not, I must now withdraw the 10-year estimate of my last paragraph. The claim made by Benoit and his colleagues has fired experimental biologists throughout the world. The third stage in genetics has begun.

4

Darwin and Animal Breeding

By

JOHN HAMMOND

DARWIN'S works have had a profound influence on the breeding of domestic animals and his theories on selection still form the main basis for the improvement of livestock. When Mendel's research was first rediscovered in 1900 there were great hopes that it would revolutionize the procedures used in animal breeding. It has certainly helped greatly where the character has been due to a mutation or "sport", such as the "fancy" points of colour and horns, and in cases of abnormalities, such as bulldog calves, hernia, cryptorchids and so forth; but it has done little to enable the breeder to produce animals which give more meat, milk or wool, and these are the qualities of an animal which are his main concern. Mendelism and the gene concept have certainly led to the idea of "progeny testing", that is, judging the breeding qualities of the animal by the value of its offspring rather than by its own appearance or performance. In trying to do this in practice, however, environmental influences are found, as stressed by Darwin, to have profound effects; and it is not until these have been eliminated that progeny testing can become effective.

Darwin's dictum, that like begets like, is in general true for the commercial qualities of livestock, although there are many exceptions; this is shown by the records of performance tests which have been made (under uniform feed conditions) on beef cattle, and in which the progeny of bulls that grew fastest had the progeny with the highest growth rates.

While Mendelism explains the mechanisms of inheritance and the occurrence of mutations in haphazard directions it does not explain the course of evolution and enable the breeder to steer the direction of the changes in his animals towards a specified goal, such as can be done by selection under defined environmental conditions.

ISOLATION OF POPULATIONS

Especially as a result of his observations on finches in the Galapagos Islands, Darwin commented on the differences which develop within a species when populations are isolated.[1] Perhaps one of the most potent instruments in livestock breeding is the isolation of populations that has been brought about by the institution of the pedigree herd or flock, with studbooks for different breeds each developed for a different purpose. While this is not geographical isolation it brings about genetical isolation, since no animals are accepted into the breed unless both parents are already in the herd book. This enables a group of animals, a breed, to be selected and bred true for a specific type and purpose. It has been possible by this means to develop different types of one species within a single geographical area. However, it has been found that, even within a breed which has been genetically isolated for a long time, differences in size and conformation take place gradually through successive generations when they are removed from their original habitat to a new environment. An example is the increase in size of the Ayrshire cow when it is removed from its high-rainfall, calcium-deficient native area to the dryer chalk downs of England or to the maritime provinces of Canada. The same thing occurs when Jersey cattle are removed from their native island to New Zealand or the United States: the type changes, just as Darwin found in the finches in the different Galapagos islands, but much more quickly.

MATERNAL INFLUENCES

Since the changes in size and type, which occur when the individuals of a breed are moved from one environment to another, are gradual over several generations and not sudden as they would be if caused by mutations or by a change in diet, some explanation is required to account for them. Recent work on maternal influences supplies a possible explanation. When reciprocal crosses are made between the large Shire horse and the small Shetland pony, the crossbred foal from the large dam is at birth three times the size of that from the small

[1] [See also chapters 1 and 2.—Ed.]

dam; and, since the leg from the knee and hock downwards is fully grown at birth the adult sizes of the reciprocal crosses are very different. This is seen also in reciprocal crosses between the small ass and the large horse, the mule being larger than the hinny. Reciprocal crosses between large and small breeds of cattle and sheep [6] are also different in size at birth but, because they are born at an earlier stage of development than the horse, the differences in adult size are less. That such differences are not due to differences between the eggs is shown by the fact that, when fertilized ova are reciprocally transplanted between large and small breeds of rabbits, or sheep [6], the differences in birth weight due to maternal influences are similar. In sheep the effect of maternal influence gradually becomes less as the animal grows older [6]. Not only is the maternal influence shown in the size of the young at birth, but this is followed by the influence of the dam's milk supply. Since it has been found that within a breed the larger the animal the larger on the average is the milk supply, there is some reason to think that maternal influences can bring about a gradual change in size and type through successive generations. Similarly, influences on birth weight and rate of growth are caused by reduction in the number of young at birth, so that reduction in the number in a litter tends to increase size in the breed or species.

CONVERGENT AND DIVERGENT EVOLUTION

Within a closed population, such as the individuals registered in the herd or flock book of a breed society, selection has a defined aim. This is usually described in words in a " standard of excellence " for the breed in question. Shows are held at which judges pick out and give prizes to those animals which most nearly reach the standard. Animals that win prizes are in great demand for breeding purposes and are used as widely as possible throughout the whole breed. By this artificial selection there is evolution, by the accumulation of small variations, towards a type which is suited to a particular use [12]. Occasionally a breed society may decide to split into two sections, each with a different objective. This has happened in the Shorthorn Cattle Society : one section has made milk, and the other section beef, their main objective. Here we have an example of divergent evolution, and two different types have arisen accordingly. The beef Shorthorn is short-legged and blocky, to give a high proportion of valuable joints such as loin and thighs compared to head, neck and leg, which are of lower edible value. The dairy Shorthorn has longer legs, so that there is room for a large udder to hang clear of the ground. The normal age changes in body form are pushed forward in the beef type so that it attains the adult body proportions of the dairy type by fourteen months. On the other hand there is converging evolution in breeds with different origins which have been selected for the same purpose: the Aberdeen Angus breed, of different origin but selected for beef, is closer to the conformation of the beef Shorthorn than the latter is to the dairy Shorthorn of the same origin. Similarly the Friesian selected for milk is more similar to the dairy Shorthorn than the latter is to the beef Shorthorn.

Similar examples of divergent evolution due to artificial selection are to be found in the horse [5]. The natural evolution of the horse has been for speed, with increases in the relative proportions of the length of leg [7]. This is repeated in the fœtal development of the individual whereas, in the unimproved Welsh pony, there is a post-natal development of the muscles of the body for power. The natural evolution of the horse for speed has been followed by the thoroughbred, whereas the diverging evolution of the heavy draught horse for power, as shown by the Suffolk, has involved a further development of powerful body muscles.

NATURAL SELECTION AND ENVIRONMENT

Darwin's theory was based on the action of natural selection in bringing about adaptation of an organism to its environment. Recent work on the physiology of farm animals has provided examples of the mechanism of this process. The improved breeds of beef and dairy cattle have been almost exclusively of European origin. Wherever men have migrated they have brought their native breeds of livestock with them. While the improved breeds of European cattle have been successfully imported to the temperate zones of North America and Australasia, they have been a dismal failure in the tropics. The basic cause has been found to be their lack of heat tolerance which leads to lower production, slower growth, degeneration and lack of disease resistance [1]. First crosses between tropical Zebu cattle of low production but with good heat tolerance, and European cattle, with high production but with poor heat tolerance, give high production in the tropics [1]. These have formed the basis for the development of new high-producing breeds for tropical areas such as the Santa Gertrudis for beef and the Jamaican Hope for dairying. Experiments have also been made to develop heat-tolerant strains of British breeds of beef cattle by selection for this character in the tropics (in Queensland), for it is not possible to do this in temperate climates. Hybrid vigour obtained by the crossing of two European breeds in the tropics has been found ineffective against the degeneration which occurs there.

In addition to temperature, many species and breeds of animals which have originated in the more northern regions of the north temperate zone have seasonal responses to daylight hours, which, being more constant from year to year than temperature, regulate functions which lead to temperature control or to birth of the young in the most favourable period of the year for their survival. Thus moulting and the growing of a winter coat in mink and in cattle have been shown to be controlled by daylight hours rather than by temperature, this applies also to the rate of growth of wool in sheep.

The stimuli through which daylight acts may be compared to morse code signals in slow motion. With ferrets which breed in the spring and summer, long light alternating with short dark periods acts, via the eyes and the receiving centres of the brain, on the hypothalamus and then on the anterior pituitary gland to cause it to increase the output of follicle stimulating hormone into the blood stream ; this hormone acts on the ovaries and causes the development of follicles containing egg-cells and so brings the female into the breeding state. On the other hand long dark periods alternating with light short ones cause a decline in the activity of the anterior pituitary gland and so put an end to the breeding season. The sheep responds to changes in the light ratio in the opposite way : long dark alternating with short light periods stimulate the onset of the breeding season, and long light, short dark ratios end it [13]. The end result is the same however : the young are born in the spring. Thus the ferret with a six

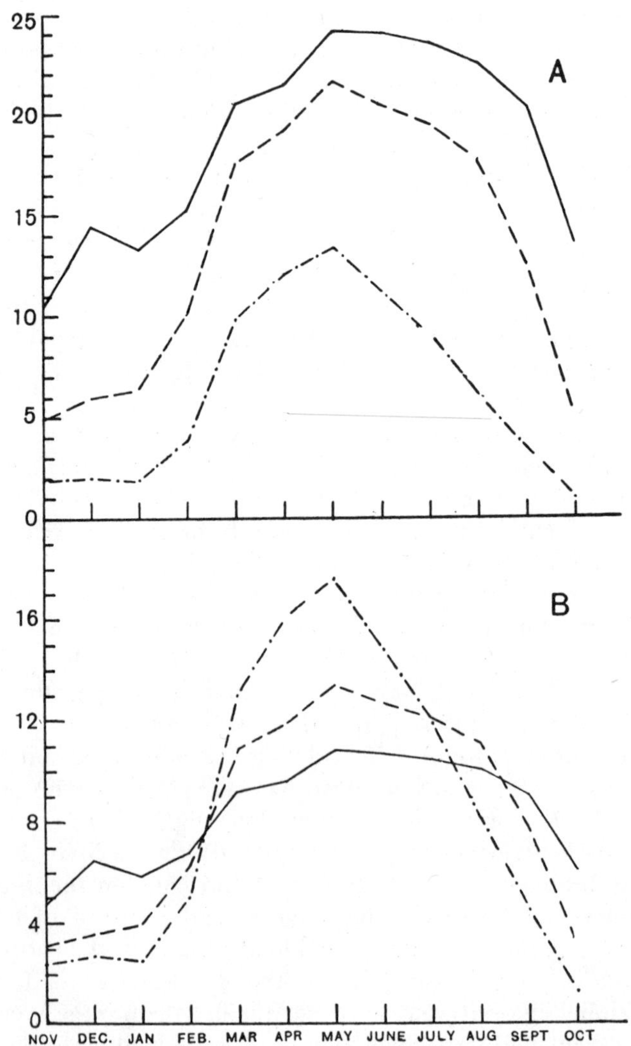

15.—*A*. Seasonal egg production (eggs per month) of high, moderate and low producing hens. *B*. Percentage of the yearly total laid each month for the three groups. Selection for high production has decreased the dependence of the fowl on increased daylight hours. [12]

weeks' pregnancy and the horse with an eleven months' pregnancy both respond to long light, whereas the sheep and goat, both with a five months' pregnancy, respond to long dark periods.

Within a single species, the sheep, the breeds which have originated in the higher latitudes and altitudes have a longer latent period and so a shorter breeding season than those originating at lower latitudes and altitudes; while in the tropics breeding occurs throughout the year. Crosses between breeds with long and short breeding seasons have breeding seasons intermediate in length.

The fowl has increased its rate of egg production considerably under domestication and part of this increase has been caused by better nutrition. The wild jungle fowl lays only a few eggs per year, and these mainly in the spring when there is an increasing light/dark ratio. Better feeding alone will not cause, in unimproved breeds, a high level of egg production at all times of the year. With all our improved breeds of fowls there is a seasonal variation in the rate of egg production, and this variation is closely associated with seasonal changes in the light/dark ratio. It has been shown by direct experiments that it is the light/dark ratio which is the cause of these seasonal egg-production curves, and electric lights in fowl-houses are now used to increase the rate of egg production during the winter months. Given good nutritional conditions, selection for high egg production is in fact selection to free the fowl from its dependence on particular light/dark ratios; as will be seen from figure 15, birds which produce a high annual output of eggs have a less marked seasonal production than low producers.

INFLUENCE OF NUTRITION ON THE EXPRESSION OF CHARACTERS

With animals under domestication it is possible to control the environment so that it becomes a suitable one in which to make selection for a given character. This perhaps has been the main means by which improvements for production in farm animals have been brought about. A very striking example of this is seen in selection for the colour of butter fat which is due to carotene produced from the green or yellow colouring matter of plants. If cows, such as Guernseys or Jerseys, which normally produce butter fat of a deep yellow colour, are fed on substances free from plant pigments their butter fat becomes dead white in colour. If to such cows increasing quantities of green-stuff are fed the colour of the butter fat gradually becomes more yellow. With some individuals, however, there comes a time when further increases in green-stuff feed fail to increase the shade of yellow in the butter fat, this ceiling value being a limit imposed genetically [5]. It is therefore only under maximum green-feed conditions that slight genetical variation in butter-fat colour can be selected for with any effectiveness, and the breed improved for this character by the accumulation of small variations in the desired direction.

Another example of the importance of nutritional environment for the expression of desired characteristics is that of the meat-producing animal, in which the development of a certain body conformation with a high proportion of the valuable parts of the body is required. As Darwin pointed out, many changes in conformation, such as the short-legged Ancon sheep and the many body forms found in pigeons, appear as " sports " or, as we now say, are due to mutations; these large variations seldom occur; when they do, they are usually transmitted as unit Mendelian characters. The gradually accumulating small changes in conformation which have been selected for in breeding for meat proportions are superficially blending in inheritance and all intermediate forms occur. How nutritional conditions work to enable these small variations to be selected for a definite purpose has been demonstrated by McMeekan for the pig [8] and Pálsson and Verges for the sheep [10]. In the case of meat-producing animals, body forms and proportions have been produced differing considerably from those found in nature and which would be quite unsuited for life in natural conditions.

The body proportions and composition of the animal change as it grows up. At birth the young animal is all head and legs while the body, including the valuable loin, is short. After birth the body begins to lengthen ; later it deepens. These changes are brought about by waves of maximum rate of growth spreading from the extremities early, to the loin in later life, and giving an orderly form of growth. These waves of growth apply not only to the different parts of the body but also to the different tissues, so that the composition of the animal changes as it grows. Nerve tissue has its maximum growth-rate early in life; this is followed by bone and then muscle, while fat is the last tissue to develop greatly; it produces in many animals, as in man, the middle-age spread. Experiments have shown [8, 10] that the plane of nutrition at which the animal is reared can materially affect the proportions not only of the different parts of the body but also of the various tissues and even the shape of the bones. This is due to the fact that the incoming nutrients in the blood

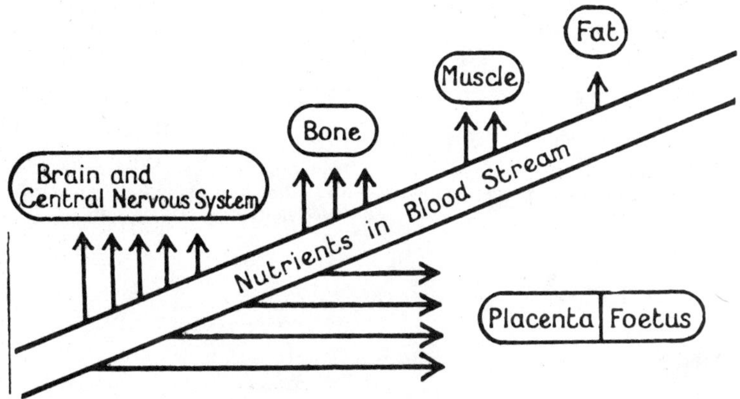

16.—Diagram showing how the different tissues of the body compete for food from the blood. Priority varies according to their order of development of tissues and their metabolic rate (see text).

stream are partitioned among the various parts and tissues of the body in accordance with their metabolic rate, which in general follows the order of their development. This is illustrated diagrammatically in the figure above. The number of arrows denotes the priority which each tissue has for nutrients from the blood stream. If the animal is on a high plane of nutrition, and the nutrients in the blood stream are high, each tissue will grow as fast as its inherent capacity allows. But if the plane of nutrition is reduced (take one arrow from each) and there is not sufficient for all, then while the other tissues grow at a reduced rate, fat ceases to be put on. If the animal is put on a lower plane still (take two arrows

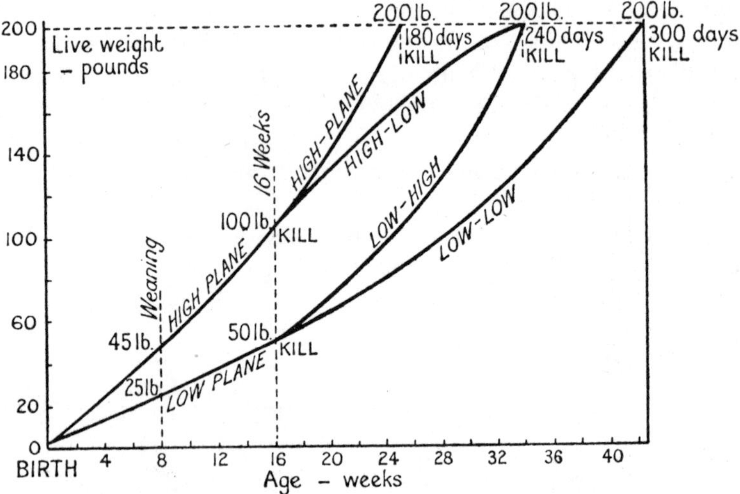

17.—Plan of McMeekan's experiment. Pigs from the same litter were made to grow along predetermined growth curves by quantative regulation of their food intake. For the results see figures 18 and 20. [8]

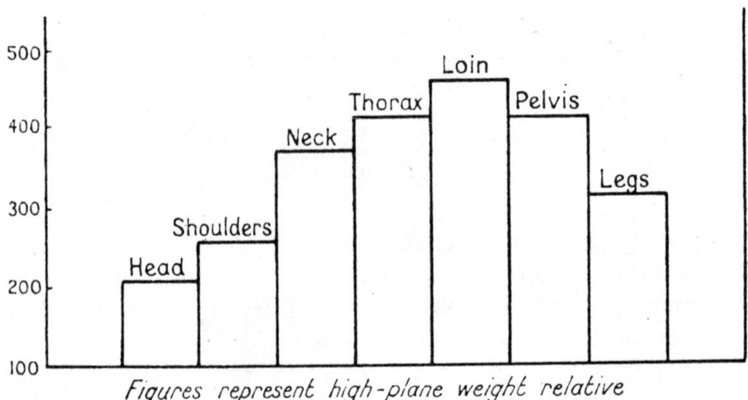

18.—Differences in the body proportions of pigs at sixteen weeks, after being fed well and poorly. The weight of the part of the body in the high-plane pigs is shown as a percentage of the weight in the low-plane pigs. In the low-plane pigs the later developing parts of the body, such as the loin, are much more affected than the early-maturing parts, such as the head. The body proportions of the low-plane pigs are more like those of the wild boar than that of an improved domestic pig (see figure 19). [8]

from each) then nerve and bone tissue continue to grow at a still slower rate, muscle ceases to grow and the arrow for fat is reversed so that fat is removed from the body to help to grow nerve and bone. This has shown to be true for pigs, sheep and cattle.

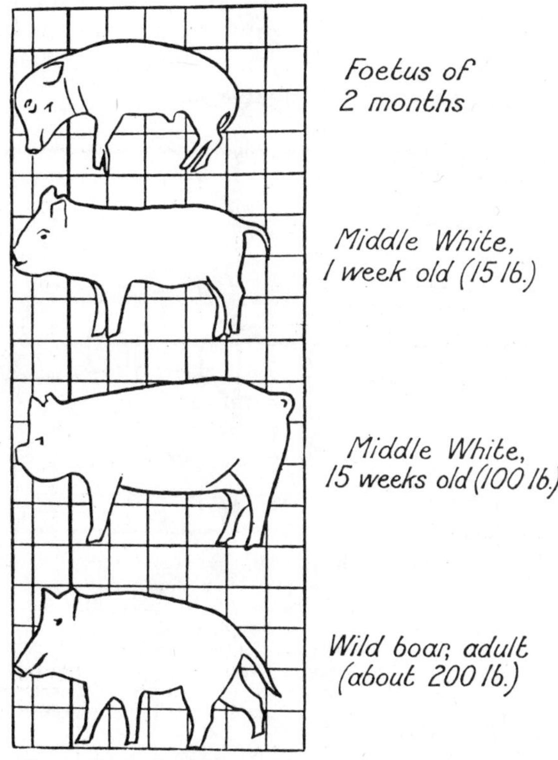

19.—Changes in the proportions of the pig brought about by selection under high-plane nutrition. Each animal is shown reduced to the same head size. As an improved breed grows up, the proportion of loin to head and neck increases greatly, but an unimproved type such as the wild boar grows up without much change in body proportions. Compare with the effects of low-plane nutrition in figure 19. [5]

How it is possible to control the conformation and composition of the body in pigs by changes in the plane of nutrition as they grow up has been demonstrated by McMeekan's experiments [8]. He rationed pigs from birth so that they grew along predetermined growth curves to 200 lb. live weight, making changes in the plane of nutrition in some of the individuals of a group when they were sixteen weeks old (see the figure below). Some low-plane and some high-plane-nutrition pigs were killed at sixteen weeks, and figure 18 shows the weights of the different joints in the high-plane pigs as a percentage of the weight of each joint in the low-plane pigs. The growth of the latest growing and most valuable part of the body, the loin, has clearly been reduced much more than the early-growing and less valuable head.

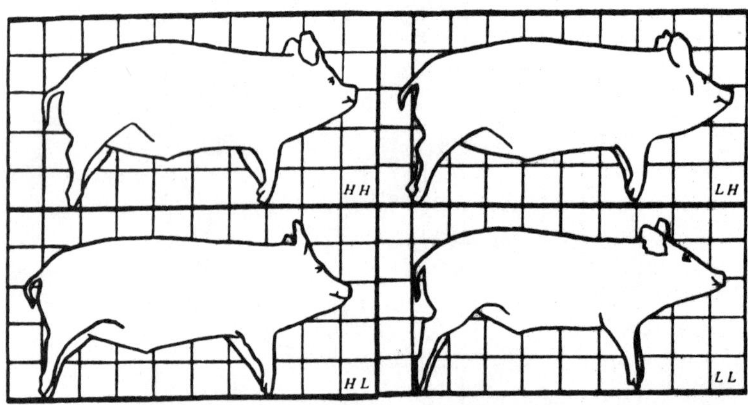

20.—Pigs of the same live weight (200 lb.) but reared along differently shaped growth curves brought about by quantative changes in the plane of nutrition (see also figure 18). The different body proportions of the pigs so produced resemble those of other breeds of pigs given similar food. The high-high (H-H) pig is similar to the pork type of pig in Britain. The high-low (H-L) pig, is similar to the bacon pig of Denmark, while the low-high (L-H) pig is like that of the lard hog found in the corn belt of U.S.A. or Hungary. The low-low (L-L) pig is like the unimproved pigs found in forest areas of Germany or Spain. [8]

In the low-plane pigs the expression of genetical qualities for a large development of the loin in early life, one of the main objectives for which the pork type of pig is bred, would render accurate selection for this character almost impossible. The way improvements in this respect have been brought about is by rearing the animals on a high plane of nutrition and selecting for breeding those which show the best development of the later-developing and more valuable parts of the body. What has been done in this respect is shown in figure 19, in which the body conformation of an improved pork pig is compared with that of the wild boar. Comparison with figure 20 shows that the changes in body proportions brought about by selection under domestication are similar to those which occur under high-plane nutrition, while the low-nutrition pigs are much nearer to the wild boar in conformation.

McMeekan's [8] four planes of nutrition (figure 17, p. 94) produced pigs of 200 lb. live weight, which, as shown in figure 20 opposite, paralleled in their body proportions and composition those breeds which had been produced in different parts of the world under similar nutritive conditions. The high-high are like the early maturing pork types of pig (Middle White and Berkshire) which have been produced in England by good feeding from birth onwards. The low-low are leggy and lean, like the unimproved breeds (Spanish, Hanoverian) which still largely graze in woodlands like the wild boar. The low-high are like the lard hogs (Mangalitza in Hungary and Poland-China in the corn-belt of U.S.A.) in which the young are poorly nourished because of the shortage of animal protein supplementary feeds and then later shut up and fed heavily on maize. In contrast the high-low are like the bacon pigs (Danish Landrace), which are well fed when young on large quantities of skim milk, supplemented with restricted amounts of cereals later.

The changes and improvements made in the pig by selection in a high-plane nutritional environment under domestication are reversible over a number of generations when the improved pig is allowed to return to natural conditions. In New Zealand there are no wild pigs. When Captain Cook went there in 1772 he brought domestic pigs from England and let some go wild. These pigs today are very similar to the wild boar in conformation and composition, and quite unlike the present-day domestic pig there. During the war some of these wild pigs were shot and sent over together with domestic pig carcasses to England for food. From the carcasses it was easy to see how short is the body and how long are the legs of the domestic pig which has gone wild as compared with the pig which has been kept and selected under improved feeding conditions.

These different nutritive environments not only affect the conformation of the animal as a whole, but also the shape of the bones. Just as low-plane nutrition inhibits the growth of the later developing parts (loin) and tissues (fat) of the body much more than the earlier growing parts (head) and tissues (bone), so it inhibits the later developing thickness of the bones more than it does the earlier developing length of the bones. Consequently the wild types have long thin bones while the improved meat types have bones relatively thick for their length. That this also is a reversible process is seen by comparing the shape of the tibia of the wild pig of New Zealand with that of the domesticated one. Comparison of the relative thickness of the cannon bones (metatarsals) of the improved meat breeds of sheep (Hampshire, Suffolk) developed under improved conditions of feeding with those of unimproved breeds (Soay, Shetland) and with breeds (Merino) improved for wool, both kept under poor feed conditions, shows how the thickness and shape of the bones have been developed in response to the environment [3].

Within an improved breed Pálsson and Verges have shown how it is possible to change the shape of the bones at any given live weight by changes in the plane of nutrition as the animal grows up [10]. They fed lambs on different planes of nutrition as McMeekan did for pigs (figure 17) but changed the planes of nutrition at six weeks and killed the animals at 30 lb. carcass weight. The metacarpal bones of the lambs on the high-high plane of nutrition, which reached this weight in 56 days, were short and thick in comparison with those on a low-low plane of nutrition, which took 294 days to reach the same weight. Under a low-plane nutrition, which cut down the rate of increase in the tissues as a whole, the skeleton had priority of supply and so the metacarpal was larger than the high-high bone at the same carcass weight. In spite of this larger size, however, the low-low bone is not as thick in proportion to its length as the high-high bone because, since the thickness of bone is later developing than length, the growth in length has had priority for nutrients from the blood stream (see figure 16). The same thing is illustrated by comparison of the bones of the high-low and low-high lambs which were both killed at the same age and body weight but had differently shaped growth curves. In early life, when the growth of bone in length is more rapid, the high-low grew rapidly in length; but later in life, when growth in thickness should have developed, the animal was put on a low plane of nutrition, and so the bone continued to grow in length rather than thickness. On the other hand in early life, when growth in length is at its maximum, the growth in length in the low-high bone was stunted; but when this animal was put on high-plane nutrition in later life growth in thickness was encouraged and a relatively short, thick bone was produced.

PHYSIOLOGICAL CONDITIONS LIMITING EVOLUTION

Although cattle have been bred and kept for milk production for thousands of years in India, in general little improvement in milk yield has been effected as compared with that which has occurred in European cattle under selection for this purpose. When deliberate selection has been applied to Zebu cattle however, it has been shown that gradual improvements in milk yield can be brought about, as shown in the figure below. These improvements are gradual : no sudden rises in yield occur when a higher nutritive level is first introduced. The way in which this evolution works is now becoming clearer and can be illustrated by milk yield in the cow. Milk is " let down " or ejected from the fine ducts and alveoli of the udder by a nervous reflex combined with hormonal action [2]. The sucking of the nipple by the calf sends nervous impulses to the brain, and the posterior pituitary gland lying at the base of the brain is stimulated by the hypothalamus to secrete a substance, oxytocin, into the blood stream; this causes muscle cells surrounding the alveoli to contract and to force the milk down. In primitive conditions of domestication the calf is allowed to suck one side of the udder to cause the milk to be " let down " while the other side is milked for man. In unimproved and primitive cattle, or where improved cattle are subjected to fright, a nerve block occurs which prevents the response, and so little milk can be drawn off. The first step for improvement in milk production is therefore to select for a placid temperament in the cow, so that she will let milk down to man without the intervention of the calf.

The amount of glandular milk-producing tissue which will develop depends on the amount of a hormone, œstrogen, circulating in the blood towards the end of pregnancy, and the amount of milk given after parturition depends in turn on the glandular tissue. If the animal is fed badly before parturition, the development of this tissue is reduced [11] and so the genetical possibilities of the animal are not realized and proper selection cannot be made. After parturition the high-yielding cow so developed must have food which does not contain too much fibre or bulk, since otherwise she will be unable to eat

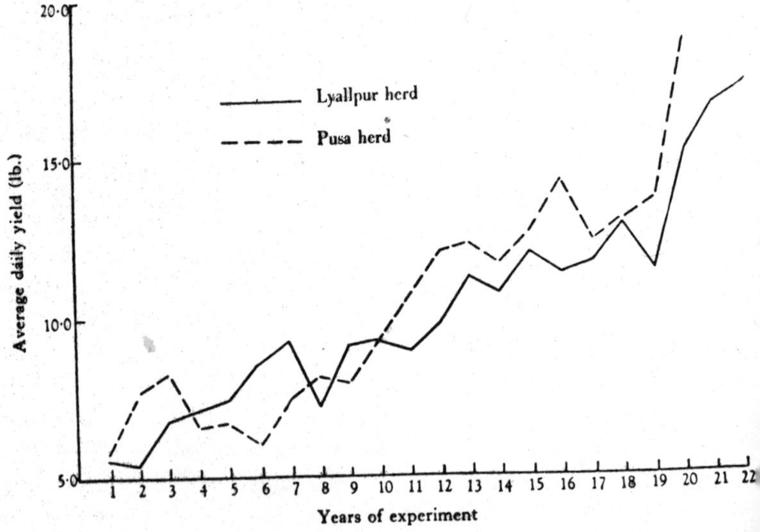

21.—Changes in the average milk yields of cows in two government farms in India, after modern methods of feeding, management and selection had been adopted. [4]

enough to support a high yield. When all this has been done it is found that the rate of milk secretion may be slowed down by milk pressure within the udder if the cow is milked only twice a day and so milking three and even four times a day may be necessary. In this way the milk yield of the cow has been raised to levels far beyond those which could be obtained merely by shuffling genes in the unimproved stock.

The extent to which the production characters of an animal have been improved in our domestic animals since Darwin's day, through the application of scientific knowledge concerning the nutrition and physiology of the animal, helps to confirm his evolutionary theories; for it has not been brought about merely by the fortuitous occurrence of mutations but by selection, under suitable environmental conditions, based on the variability available in the various domestic species.

5

Darwin and Classification

By

R. A. Crowson

Our classifications will come to be, as far as they can be so made, genealogies; and will then truly give what may be called the plan of creation. The rules for classifying will no doubt become simpler when we have a definite object in view. We possess no pedigrees or armorial bearings; and we have to discover and trace the many diverging lines of descent in our natural genealogies, by characters of any kind which have long been inherited. Rudimentary organs will speak infallibly with respect to the nature of long-lost structures. Species and groups of species which are called aberrant, and which may fancifully be called living fossils, will aid us in forming a picture of the ancient forms of life. Embryology will often reveal to us the structure, in some degree obscured, of the prototypes of each great class. [1, p. 401]

From these remarks it will be seen that I look at the term species as one arbitrarily given, for the sake of convenience, to a set of individuals closely resembling each other, and that it does not differ essentially from the term variety, which is given to less distinct and more fluctuating forms. The term variety, again, in comparison with mere individual differences, is also applied arbitrarily, for convenience's sake. [1, p. 39]

When the views advanced by me in this volume, and by Mr. Wallace, or when analogous views on the origin of species are generally admitted, we can dimly foresee that there will be a considerable revolution in natural history. Systematists will be able to pursue their labours as at present, but they will not be incessantly haunted by the shadowy doubt whether this or that form be a true species. This, I feel sure, and I speak after experience, will be no slight relief. The endless disputes whether or not some fifty species of British brambles are good species will cease. Systematists will only have to decide (not that this will be easy) whether any form be sufficiently constant and distinct from other forms, to be capable of definition; and if definable, whether the differences be sufficiently important to deserve a specific name. This latter point will become a far more essential consideration than it is at present; for differences, however slight, between any two forms, if not blended by intermediate gradations, are looked at by most naturalists as sufficient to raise both forms to the rank of species. [1, pp. 399–400]

THE above quotations from *On the Origin of Species* epitomize Darwin's mature views on classification, in what must have appeared to his contemporaries as a daring and challenging manner. He also considered the subject at greater length, though less forcefully, in the first part of chapter 14, and in relation to the famous diagram of chapter 4 (reproduced on page 104). In order to study his practice in this field, we shall need to turn to his earlier work on barnacles, the *Monograph of the Cirripedia* [2], published well before the *Origin*, but after the idea of evolution had taken firm shape in Darwin's mind. With this book, he established himself as one of the best systematic zoologists of his time; nobody seems to have regarded the work as controversial, and it was highly praised by all competent reviewers. Indeed, Darwin seems to have gone out of his way to avoid open or covert reference to evolutionary theory at any point in the book, at a time when tentative speculations of this kind were becoming common in the works of other systematists. Possibly he was anxious to avoid any anticipation which might weaken the impact of the great work he was even then preparing.

Our first quotation embodies perhaps the earliest exposition of the principles of what is now called phylogenetic (genealogical) classification. The word taxonomy is often used instead of classification in this and similar connections; it properly denotes the general theory and rules underlying classification, rather than classification itself, just as morphology should mean the general abstract theory of structure, but is commonly used as a synonym for anatomy or structure itself. This particular perversion of language is one which modern biologists have probably imitated from the technical jargon of the older medical profession. Darwin's formulation would be accepted today without serious qualification by many, perhaps a majority, of systematists, at least in zoology. The botanists of today tend to adopt what seems to the zoologists

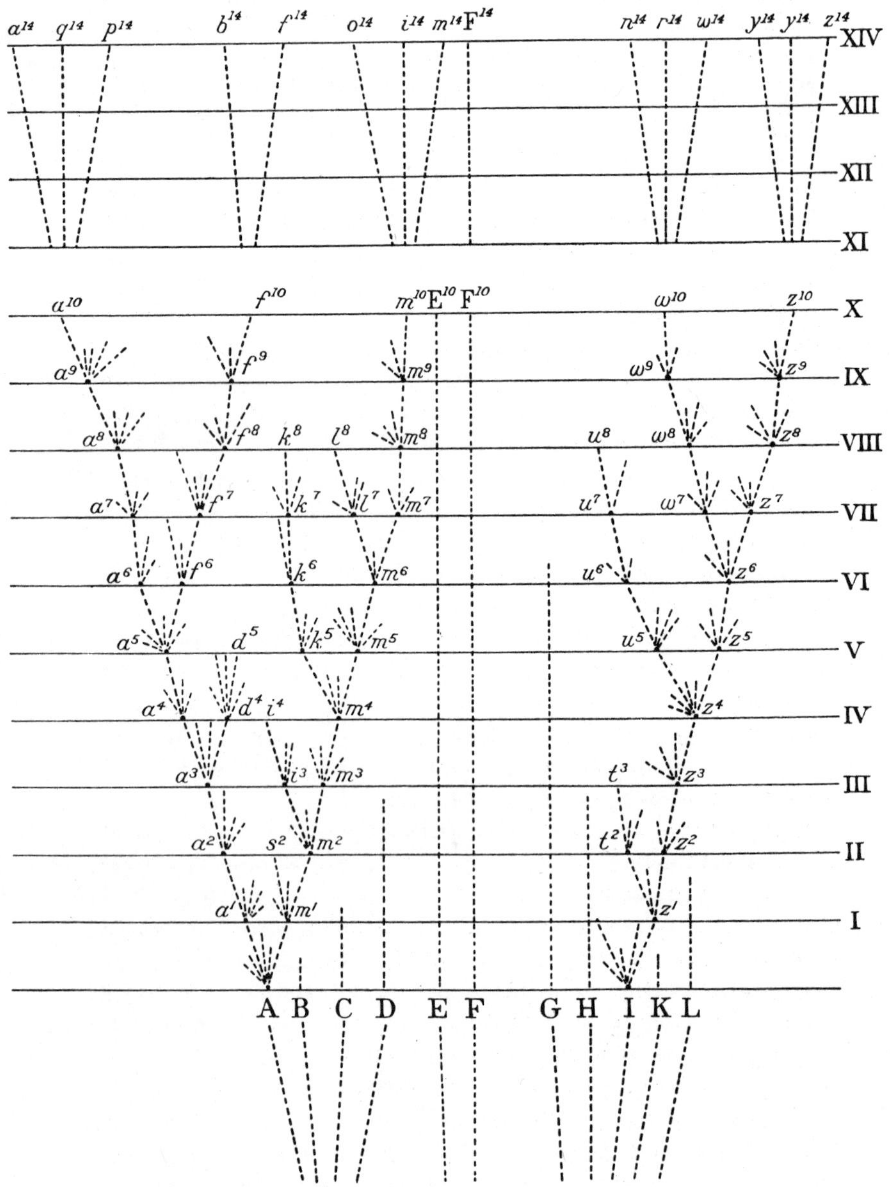

22.—The phylogenetic diagram from *On the Origin of Species* (redrawn).

a defeatist attitude; plant genealogies, many of them will say, are not known nor even in the foreseeable future knowable, therefore we should not pretend to base our classifications on them, but rather stick to what is practically convenient.

The opposing school to the Darwinian one is that of the "formal" or "artificial" systematists. Its exponents maintain that the problem of classifying animals or plants should be considered in the same light as that of arranging the books in a large library, where the system is aimed at maximum speed and facility in locating any particular book. Such an attitude is likely to commend itself to those responsible for the building up and use of large reference collections, and may facilitate the construction of analytical keys for easy determination of animals or plants. The classification of organisms has, however, other functions than to serve the convenience of curators. For one thing, the non-museum zoologist or botanist looks to it as a guide to similarities or differences to be expected in characters quite other than those normally used in museums. In particular, the modern disciplines of experimental physiology and biochemistry involve difficult and costly techniques, and it is out of the question to carry out each of their types of investigation on all the species of animals and plants. What they need is a system which will provide the best basis for generalization from a limited number of observations. The pursuit of this ideal leads us inevitably towards a natural or phylogenetic classification.

The last sentence of our first quotation, "Embryology will often reveal to us the structure, in some degree obscured, of the prototypes of each great class", would no doubt be criticized by an influential school of modern embryologists.[1] A leading figure among them has written: "It is a confusion of categories to assume that a character which is embryonic in ontogeny must also be primitive in phylogeny, or that a character which develops early in ontogeny must also have evolved early in phylogeny. The possession of an embryonic character may be, and very often is, the result of a secondary retention or prolongation of embryonic features, and a character which appears early in ontogeny may have been evolved recently in phylogeny" [3, p. 140]. By inserting the word "must" at two points in

[1] [Their view is expounded in chapter 7.—ED.]

this statement its author has put himself in the position of denying a statement which has never to my knowledge been made by a responsible zoologist; this would be rather pointless, and I think most readers would take de Beer's formula as meaning rather more than it says, that is, as denying that the early appearance of a character in ontogeny establishes any great likelihood of its early origin in phylogeny.

The question of the relative amounts of truth embodied in Darwin's formulation and de Beer's is of great importance in relation to phylogenetic classification. If, in the development of the individual, characters tend to appear in the same order as that in which they were evolved in its ancestry, then the study of development (embryology in its wide sense) will provide a valuable guide to the relative importance to be attached to different characters in phylogenetic classification. It follows too that the development of the individual will to the same extent tend to approximate to " climbing up its family tree "; that is, its successive immature stages will tend to resemble the adults of successive ancestral forms. The general acceptance of the reality of this tendency was responsible for the tremendous development of descriptive and comparative embryology in the later nineteenth century, and for the important place then assigned to it in general systematic zoology. The role of the twentieth century in relation to this, as in relation to so many other of the great constructional achievements of the Victorians, has been essentially negative and destructive. The modern embryologists who deny this " principle of recapitulation " have attempted to replace it by the appeal to physico-chemical causation, but have signally failed to produce an intelligible general theory of the subject on this basis; coincidentally with their growing influence, the position of embryology in university zoology courses and in the general body of zoological knowledge has steadily declined.

The second sentence of the quotation from de Beer refers to what is called pædomorphosis; evolution by the carrying over into the adult stage of characters previously found only in the young. Such a phenomenon hardly seems to have been envisaged by Darwin, and its real occurrence in large-scale evolution, in which category we can hardly include the Axolotl and perennibranchiate newts (referred to further below), has yet to be proved. The lack of direct fossil evidence that any important evolutionary step has taken place in this way is rather ingeniously dealt with by the same author on the same page: " Pædomorphosis explains gaps in the fossil record ". If pædomorphosis has played any important part in large-scale evolution, then our ideas on the genealogies of the major groups of animals may have to undergo drastic modification, and this in turn must affect our phylogenetic (or would-be phylogenetic) classifications. The influential propagation of the pædomorphic concept of evolution has already had a serious effect in undermining faith in the older " classical " views on phylogeny, without leading to the formulation of a firmly-based and generally accepted alternative to them. The effect of this has been to create an atmosphere of uncertainty and irresponsible speculation, which has played into the hands of those who oppose phylogenetic classification, and is certainly more harmful to the teaching of systematic embryology and general zoology than ever the alleged tyranny of the law of recapitulation could have been.

The classificatory position of the previously mentioned perennibranchiate newts is worthy of a little study in view of the far-reaching conclusions which some have attempted to draw from them. Newts belong in the Amphibia and, like the frogs and toads, pass through a fish-like tadpole stage in their development; during this larval stage they breathe dissolved oxygen like a fish by means of gill-pouches opening from the pharynx to the outside, often bearing feathery external gills (shown on page 108). The adult newt normally loses these gills altogether and relies on breathing atmospheric air into its lungs, helped out by some direct oxygen-absorption (either from the air or water) through its soft, permeable skin. Adult newts spend at least a part, often a major part, of their life on land, while the tadpoles are purely aquatic. The adults of the so-called Perennibranchiata retain the aquatic habits of the larvæ, and also some of the larval structural features, notably the fish-like external or internal gills, when they become sexually mature. The group includes a number of diverse genera, and has both Old and New World representatives.

Here, it is suggested, we have quite an extensive group which has admittedly evolved pædomorphically. But as is noted of these animals by J. Z. Young [5, p. 345]: "This process of pædomorphosis has developed to various extents, and independently in a number of different groups". It is not a case of pædomorphosis having produced a new "rejuvenated" stock which has then undergone a burst of "adaptive radiation", but of an apparently regressive or degenerative

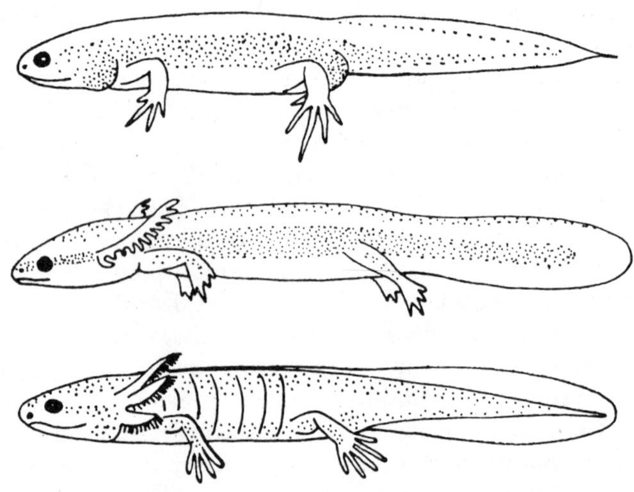

23.—Amphibians, illustrating pædomorphosis, or the retention of larval characters in the sexually mature adult. *Top*: a normal newt (*Triturus pyrrhogaster*) with no trace of gills; *middle*: an adult perennibranchiate newt (*Necturus maculosus*) showing external feathery gills; *bottom*: late larva of a newt, *Amblystoma paroticum*, showing gills resembling those of the adult *Necturus*. (The generic name of the third-mentioned animal first appeared in print as *Ambystoma*, an evident misprint which was soon corrected by Agassiz with no objection from the original describer; the edicts of the International Commission on Zoological Nomenclature have recently restored the misprint as the official spelling.)

change cropping up as a dead end in a number of evolutionary lines. It may well be regarded as analogous to certain phenomena commonly regarded as signs of "racial senility" in the latter days of some fossil groups in the past. De Beer, on the contrary, claims that pædomorphic evolution leads to racial rejuvenation and evolutionary vigour. If these perennibranchiate newts were to be the harbingers of a new evolutionary outburst, it is difficult to see what they could be giving rise to except fishes—and a derivation of fishes from

Amphibia is one speculation which has been conspicuously absent from the writings of exponents of pædomorphic evolution.

THE PROBLEMS OF PHYLOGENETIC CLASSIFICATION

In spite of his excellent enunciation of the principles of genealogical classification, nowhere in the *Origin*, or for that matter in his other works, does Darwin really get to grips with the difficulties that may be met in the attempt to transform a family tree into a classification. In the consideration of his phylogenetic diagram (page 104) in chapters 4 and 14, the question is treated in a very oblique way. Incidentally, in relation to this diagram he makes a curious error on p. 347, where he says, " Now all these modified descendants from a single species, are related by blood or descent in the same degree; they may metaphorically be called cousins to the same millionth degree." Reference to the diagram will soon show that forms a^{14}, q^{14} and p^{14} are much more closely related to each other than they are to F^{14} or y^{14}, that there are in fact many degrees of relationship among the fifteen present-day forms along the top line—as Darwin himself implies in suggesting that they could be grouped in two orders and several families.

The main difficulties recognized by phylogenetic classifiers at present are three: first, the imperfection of the fossil record and consequent uncertainty of our knowledge of evolutionary relationships; second, uneven rates of evolution; and third, the problem of fitting fossils into the same system as modern forms. For many groups of organisms a fossil record is nearly or quite lacking, and phylogeny can then be deduced only from the comparative study of modern forms, that is, from the same characters as those which form the basis of ordinary (not specifically genealogical) classification. It is these instances which are seized on by the opponents of phylogenetic classification (particularly among the botanists) as making nonsense of the claim to base classification on phylogeny: " manifestly ", they will say, " your phylogeny is a theory based on the facts of classification, and not *vice versa* ". Yet even in the absence of a fossil record we still have at least one guide to evolutionary relationships over and above the characters on which classification is based. This guide is Dollo's Law (with which we may associate Darwin's "rudimentary organs"). The opponents of phylogenetic classification usually ignore or deny Dollo's Law, which can be put simply as follows: "Complex organs, once lost in the course of evolution, are rarely if ever regained in the same form." Darwin's rudimentary organs are those which, while retaining some recognizable trace of their original structure, have lost their original function and so much of their original structure that it is impossible, or at least very unlikely, for them ever to regain that original function.

An example of the application of Dollo's Law may be quoted from the reptiles. It is well known to those who have studied Reptilia at all that snakes (Ophidia) and lizards (Lacertilia) are closely and directly related to one another; Dollo's Law implies that it must be the snakes which are derived from lizard ancestors and not vice versa. The fore and hind limbs of lizards have the typical structure of limbs of terrestrial vertebrates, and are certainly complex enough organs to come within the scope of the law. Darwin's rudimentary organs can also be brought into the consideration of this case, since a number of the more primitive types of snakes have concealed rudiments of the hind limbs. Dollo's Law may also be extended to apply in the newer fields of comparative biochemistry and physiology, from which much new and valuable evidence on the relationships of organisms is now coming—and to which Darwin's "characters of any kind which have long been inherited" might be prophetically related. For example, it seems likely that the ability to carry out certain complex chemical processes in the body, such as those involved in the synthesis of particular enzymes, once completely lost, is rarely if ever regained in subsequent evolution.

The second difficulty, the unevenness of the rates of evolution, is perhaps a more serious one. We may define a strict phylogenetic system as one in which all the members of a given group (or "taxon") are more closely related genetically than they are to any forms not included in the group; they should all be descended from a common ancestral species which they do not share with any form outside the group. That our present-day accepted classification of the Vertebrata is not strictly phylogenetic may be seen from the case of the Dipnoan and Crossopterygian lung-fishes: almost all authorities place these in the same class, Pisces, as the higher bony fish (Teleostei), while the evolutionary diagrams in the same books generally show them as sharing a common ancestor with the Amphibians more recently than with the Teleosts, [for example, 5, figure 212]. But undoubtedly the Australian lung-fish *Epiceratodus* (a member of the Dipnoi, shown at the top of figure 24) and the recently discovered Cœlacanth *Latimeria* (deriving from the Crossopterygii, figure 24) are much more obviously fish-like than amphibian-like (see figure 23). This can be seen as the result of different rates of evolution: the amphibian line diverged in the late Palæozoic era much more rapidly from the ancestral form than did the lines leading to *Epiceratodus* and *Latimeria*.

A comparable case could be quoted from the reptiles. The class Reptilia of conventional classification includes *inter alia* the tortoises and turtles (Chelonia), the snakes and lizards (Squamata), the crocodiles (Crocodilia) and the extinct dinosaurs (Dinosauria). Almost all modern authorities are agreed that the crocodiles and dinosaurs had a common ancestor with birds more recently than they had with the lizards or the tortoises; if this is so, then in a phylogenetic classification the Crocodilia and Dinosauria should be grouped together with birds rather than with the other reptiles. Both the Pisces and the Reptilia, it will be noticed, violate the canons of genealogical classification by not possessing a common ancestor which they do not share with other groups. Birds (Aves) and mammals (Mammalia) on the other hand are examples of truly natural groups; the ancestral bird and mammal have given rise to nothing but birds and mammals respectively. A strict phylogenetic system will be bound at times to group together forms apparently differing more from each other than one or other of them does from some form outside the group; this will not I fear be popular with common-sense naturalists.

A further corollary of strict phylogenetic classification is that the giving of a particular status to a group, for example genus or order, will imply a fairly definite age for its ancestral form. This ancestor must have been far enough back in time to have co-existed with the ancestral forms of the other groups of the

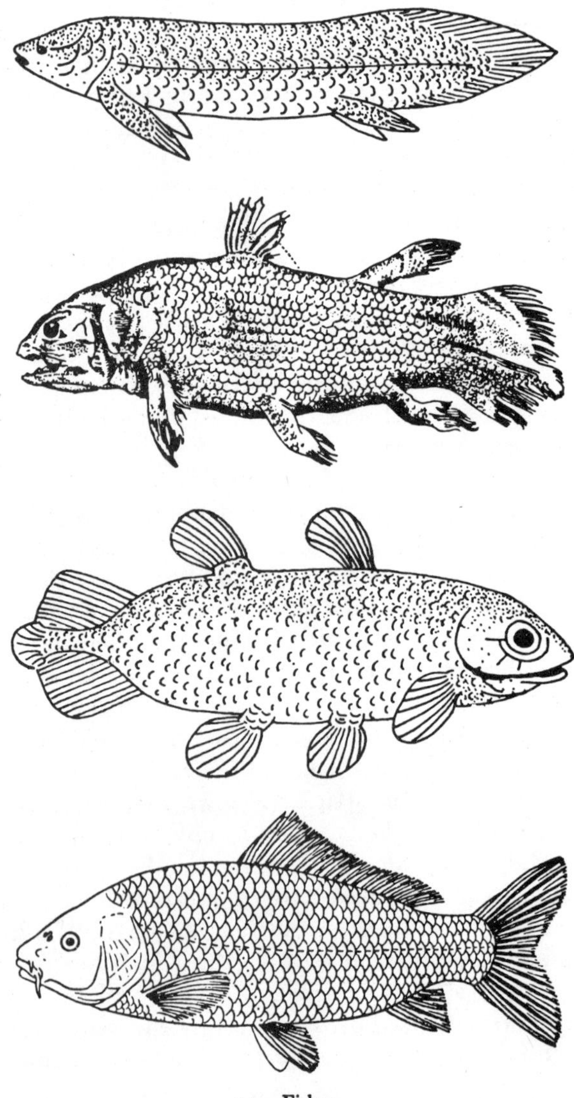

24.—Fishes.

Top: the Australian lungfish (*Epiceratodus forsteri*) which closely resembles fossils that appear to have become extinct in the northern hemisphere some one hundred million years ago. *Second*: the recent cœlacanth (*Latimeria chalumnæ*). *Third*: a fossil cœlacanth (*Undina* sp.) of the Jurassic period, some 130 million years ago. *Bottom*: the carp (*Cyprinus carpio*) a typical modern bony (*Teleostean*) fish.

same status within the same next higher grouping (for example, the other genera within the tribe). By our criteria it could not have been derived from any of them, yet it must have shared a common ancestor with all of them at one stage further back.

If we try to follow out the consequences of this, we might on the analogy of human genealogies suggest that all the forms within a genus, for example, must have descended from a common ancestral species not more than, say, a million generations back. Yet if this seemingly reasonable principle were adopted, the results would be in the highest degree disturbing to our present system. Suppose we take as one example a typical insect line, with an average generation length of one year, this would then give us a common ancestor for a genus not more than a million years ago, while for slow-breeding mammals like the elephant line it might be more reasonable to assume twenty years for an average generation and thus to postulate a generic ancestor something like twenty million years back. Now for the elephant line we have an unusually good fossil record, from which we can estimate the age of generic ancestors with some precision, and there is just enough fossil and distributional evidence to enable us to make estimates, though less precise ones, of the ages of many insect genera. It would appear that it is in elephants that the groups called genera are apt to have had a common ancestor about a million years before, whereas ancestors of present-day insect genera would generally go back more than twenty million years. If the category " genus " is taken as the measure of a definite amount of evolution, it would appear that the amount of evolution per generation in elephants is something like four hundred times greater than it is in insects.

The disproportion is excessive, and I suggest that the true meaning of these figures is that the generic category in elephants is not really comparable with the similarly-named one in insects. Hybrids between species attributed to different genera have been produced in several instances in mammals (and much more frequently in birds), whereas such a phenomenon is unknown in insects.

The difficulty about generation lengths remains, however, even if we restrict our comparisons to the mammals; for the rodent line, whose average generation length is probably no greater than that of insects, does not seem to have evolved faster than the elephants during the Tertiary era, and its genera would seem to be at least as old as those of elephants. Consideration of this case might prompt an alternative suggestion—to base the genus on a specified geological age for the common ancestor, irrespective of generation lengths. Though seemingly less reasonable than the first suggestion, this one would be more convenient from other points of view (see later discussion on classifying fossils) and less subversive of present practice; however, from the quoted examples of insects and mammals it will be seen that it would still involve serious alterations in our present system.

ON CLASSIFYING FOSSILS

The third difficulty of genealogical classification, that of fitting fossils into the same system as modern forms, represents in a manner the obverse of the first: if there are no fossils, how do you discover the evolutionary history of the group, and, if there are many fossils, how do you fit them into your classification? Darwin, in his later discussion of the phylogenetic diagram (page 104), evades this problem. He postulates eleven ancestral genera, all closely related, in the Silurian period (lettered A to L along the bottom line of the diagram), and fifteen present-day genera deriving from three of the Silurian ones, eight genera from A, six genera from I, and one genus from F. He groups the modern genera into two orders (one for the descendants of A, one for the descendants of I) with two or three families in each, but does not say what he would do with the remaining one, the " living fossil ", which he supposes to have persisted with very little change since the Silurian period. He does not say how he would place the Silurian forms, let alone the fossils from intermediate periods, in relation to his two orders. The most logical way of dealing with this difficulty might be to have a separate classification for each geological period, with persisting lines being promoted one degree in classificatory rank for each additional period of their existence. Then we should be obliged to define each classificatory category by a specified degree of antiquity for its common ancestor, with the attendant difficulties which we have seen this would involve; further trouble is likely to be encountered from the unequal lengths of the geological periods at present accepted (see the table), not to mention the uncertainty of their correlation and demarcation in many places.

A table is appended showing how the categories *as applied in Insecta* might be systematically related to geological ages. To make the scheme a little clearer, let us suppose that we were dealing with fossil insects from the lower Eocene deposits, dating from some sixty million years ago. The Insecta would then have the status of a subclass only, the ancestors of our present subclasses would be treated as supra-orders, those of our present orders as superfamilies, those of our present families as genera, and so on; there would probably also be some representatives of groups at various levels which have left

Category (Taxon)	Age of Common Ancestor (*millions of years*)	Geological Age of Common Ancestor *of present-day groupings*
*Species group	10	Lower Pliocene
*Section	20	Miocene
Subgenus	30	Upper Oligocene
Genus	40	Early Oligocene
*Subtribe	50	Upper Eocene
Tribe	60	Lower Eocene
Subfamily	80	Upper Cretaceous
Family	100	Lower Cretaceous
Superfamily	120	Upper Jurassic
*Series	140	Lower Jurassic
Suborder	160	Upper Triassic
Order	180	Lower Triassic
*Supra-order	220	Lower Permian
Subclass	260	Lower Carboniferous
Class	300	Lower Devonian
*Supra-class	340	Lower Silurian
Subphylum	420	Late Cambrian
Phylum	500	Lower Cambrian

* Taxonomic groups marked with an asterisk are not in general use.

no descendants at the present day. It will be noted that no time-value is attached to the lowest group, the species; this category alone can be defined more or less objectively without reference to any specific time-value.

LIVING FOSSILS

The existence and importance of the "living fossils" to which Darwin made reference has been abundantly demonstrated since the *Origin* first appeared; many of the most

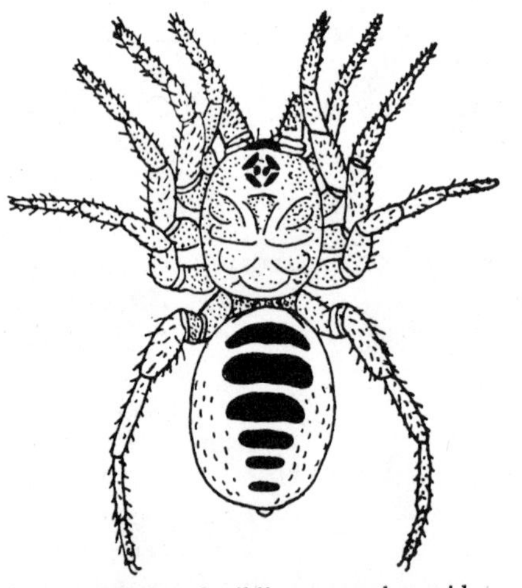

25.—A "living fossil" among the spiders, *Liphistius malayanus*, from tropical Asia. The segmented abdomen, found only in this small group among living spiders, is characteristic of the earliest fossil (*Carboniferous*) spiders.

26.—A fossil lamp shell (*Lingula quadrata*) from the Upper Ordovician deposits, some 350 million years old; the shell is remarkably similar to those of modern species attributed to the same genus.

notable of them, for example the Australian lung-fish *Epiceratodus* (figure 24, top), the Cœlacanth *Latimeria* (figure 24, middle), the Liphistiomorph spiders of the Far East (figure 25, above), the deep-sea Pterobranchia (living relatives of the Palæozoic Graptolites, ancient members of the Chordate stock from which the Vertebrata sprang), and the primitive leafless Psilotaceæ among vascular plants, have been brought to light since Darwin first wrote. It is now generally believed that no animal or plant can persist for very long periods without undergoing *some* evolutionary change; for example many of the most structurally conservative of present-day plants (such as the above-mentioned Psilotaceæ) have remarkably high chromosome numbers and presumably have evolved some very unusual genetical properties. The assignment of a modern animal to the same genus as its Silurian ancestor, as is suggested by Darwin for the " living fossil " of his genealogical diagram, is now regarded as rarely if ever likely to be justified. The best-known case where this has been done is that of the Brachiopod *Lingula*, one of the lamp-shells, a little-known group of marine invertebrates. In this case the fossils consist of nothing more than a shell of very simple form (figure 26, page 116), and it is possible, even probable, that the soft parts of Silurian *Lingula* differed radically from those of living ones. However, the genus in modern usage is apt to be a much smaller group than it was when Darwin wrote (see later discussion on " taxonomic inflation ").

THE SPECIES PROBLEM

If our first quotation shows Darwin in general accord with present-day zoological opinion, the second and third quotations express aspects of his thought which are at variance both with the views of his best contemporaries and with the views of leading authorities today. It is curious but understandable that Darwin should have been led into error over the very species problem with which his book is expressly concerned. If we remember that the argument constantly brought against early evolutionists was that species are something essentially and qualitatively different from varieties—nobody seems to have denied that varieties might have evolved—then we may understand why Darwin came to overemphasize the purely quantitative side of species differences. An author well versed in the Hegelian dialectic might have pointed out that quantitative differences pushed beyond a certain point come to appear as qualitative ones; however Darwin, whose liberal education did not include much German philosophy, felt compelled to deny that species-differences differed in anything more than degree from varietal ones. In his discussion of hybridism and sterility in chapter 9 he does write:

> It may be urged, as an overwhelming argument, that there must be some essential difference between species and varieties, in as

> much as the latter, however much they differ from each other in external appearance, cross with perfect fertility, and yield perfectly fertile offspring. With some exceptions, presently to be given, I freely admit that this is the rule.

But Darwin then goes on to accuse his opponents of arguing in a circle when they insist that forms which do not give fertile hybrids must belong to different species. Without saying so, he implies that in his view forms which are intersterile may properly be assigned to one and the same species. The reader is recommended to refer to the article by Dobzhansky,[1] in which Darwin's views on the species problem are examined in the light of modern genetical knowledge.

If we turn from the *Origin* to the *Monograph of the Cirripedia* we find Darwin showing no more manifest uncertainty in the discrimination of species than did his contemporaries in similar works. He appears to argue throughout on the assumption that whether forms A and B are different species is an objective and ascertainable fact. Certainly he drew attention to the great difficulty of distinguishing species in certain genera of barnacles, and to the utter confusion of the specific nomenclature in those genera at the time, but he seemed inclined to attribute this state of affairs to ignorance and superficial study only. By careful dissection of specimens and adequate study of the more recondite internal characters Darwin was able to clear up a good deal of this confusion; and he implies that if earlier systematists had done this, instead of relying purely on the very variable features of the external shell, the confusion need never have arisen. If this work represents the "experience" mentioned in our third quotation we can only say that its author has successfully hidden the doubts and uncertainties which must have beset him during its composition.

Modern systematists might smile rather grimly at our third quotation. The logic of the suggestion that, by eliminating the idea that there is any special objective basis for species, you will make for greater agreement and uniformity in their delimitation, is not immediately obvious today. Perhaps it is best explained as an example of nineteenth-century social idealism: if species discrimination is merely a matter of convenience and

[1] [Chapter 2.—ED.]

convention, then surely enlightened and rational men will very soon settle it amicably? Whatever the basis of Darwin's curious prediction, it has been strikingly falsified by the later development of the subject. Having suggested that the species was merely one arbitrary level of difference, he was no doubt logical in the last sentence of the third quotation, but here again he has been proved very wrong by modern practice. Species which are perfectly "good" in every other respect, yet show only the slightest of visible differences, are now well known in many groups of animals, notably the fruit-flies of the genus *Drosophila* (for instance, *D. pseudoobscura* and *D. persimilis*), the birds (the chiff-chaff and the willow warbler), the mosquitoes, the moths (the grey dagger and the dark dagger) and many others.

DARWIN'S INFLUENCE ON CLASSIFICATION

The reader may well ask whether, irrespective of the eventual fate of Darwin's predictions, his teachings have had any direct effect on the later development of classificatory practice. The answer appears to be, very little before the end of the nineteenth century, a certain amount in the earlier part of the present century (at least as far as species are concerned), but very little again today. To understand how this came about, we shall need to consider further that "considerable revolution in natural history" referred to in our third quotation.

That the hundred years since the writing of the *Origin* have seen a major transformation in natural history will be apparent to anyone who will take the trouble to study a few volumes of a suitable zoological or botanical journal for the 1850s and then to compare them with the corresponding volumes of a century later. A suitable example would be the *Proceedings of the Zoological Society of London.* The first feature of the volumes of this journal for the 1850s that is likely to strike the modern reader is the rich provision of beautiful pictures; on turning to the text he will find that the bulk of it is made up of systematic and descriptive papers dealing with all sorts of new animals from all parts of the world; a flood of exciting novelties was still pouring in from the exertions of people like Alfred Russel Wallace (Darwin's associate in the first exposition of the theory of Natural Selection) in the East Indies and H. W. Bates in the Amazon basin. By reading these systematic papers, with their fluent Latin diagnoses of new species and genera, their frequent reference to specimens " in the cabinet of Mr. X " or " shot at Dehra Dun by Col. Y ", and from the more discursive writings on the habits of animals, the travel experiences of Society members and so on, he will build up a picture of the men who composed the Zoological Society in those days. There were the regular officers, so often stationed in India, the medically qualified gentlemen of wide interests who contributed most of the anatomical papers, the leisured gentry and well-to-do professional men whose pride was in the rare and choice contents of their cabinets, and the hunting and shooting men with a more than ordinarily serious interest in their quarry. There was also, even then, the small band of professional systematists at the British Museum (and a few elsewhere, as at Oxford).

Darwin, in the introduction to the *Monograph of the Cirripedia*, admitted his heavy indebtedness to one of them who had given him the full freedom of the Museum's collection of barnacles, and in the course of the work mentions another who had shown him how to dissect barnacles. The relation of the professional systematist to the amateurs at that time must have been rather like that of the " pro " to the ordinary members of a golf club today. The papers printed in the *Proceedings* were, it seems, actually read at meetings of the Society; the direct, vivid and natural style of the non-systematic papers would have been suitable enough for oral delivery, but one cannot help feeling that many members must have found the solid systematic contributions an ordeal. The general standards of the Society were undoubtedly set by liberally-educated amateurs, among which class we must number Darwin himself. In their writings they used the pronoun " I " in a natural and unselfconscious way, and they referred to other investigators as they would to respected friends and acquaintances, or occasionally mischievous but entirely human enemies.

After such a mental sojourn among the zoologists of mid-Victorian London, many will find it a saddening experience to turn to the volumes of the 1950s. Gone are the beautiful pictures, the personalities, the entertaining accounts of travels and observations, gone are the colonels and the cabinets.

Instead, there rises before the eye a vision of drab white-coated figures in laboratories, expressing themselves in a dehumanized language of tortuous obscurity. Curiously enough, one section of the contents of a recent volume is likely to preserve a recognizable similarity to the corresponding parts of a century ago—the purely systematic papers. True, they will be very much fewer in number, usually forming a very minor part of the contents of a volume. The flood of exciting novelties has dwindled to a mere trickle. Practically all the new species will be based on specimens belonging to public institutions, principally the British Museum. The descriptions themselves will be considerably longer and more detailed, and probably not in Latin. But for all the differences, the systematists, alone among mid-Victorian zoologists, would probably find the works of their present-day successors intelligible.

The context suggests that Darwin expected the revolution to show itself particularly in systematics, whereas it is precisely this field which has not undergone any drastic change in the last century. I believe that the essence of the revolution which changed Darwin's natural history into modern biology lies in the rise to dominance of professionals at the expense of the amateurs in the Zoological Society (and in all the other kindred " learned societies "). Systematics, the only field in which professionals had already established their leadership in the 1850s, alone has not needed to be revolutionized since. In a sense it could be said that the revolution in systematics had taken place in the century which *ended* in 1858, that its initiator was Linnaeus (the bicentenary of whose first employment of the " binomial system " we also celebrate in 1958) and that Darwin himself was a product rather than a cause of it. Well before the appearance of the *Origin*, systematists were seeking for what they already called a natural system (one that corresponded to the " Plan of Creation ") which Darwin himself equated with his genealogical one, and in their discussions they used phrases like " closely related to " and " referable to the same basic type as " which needed only to be taken literally instead of more or less metaphorically to give them a Darwinian content. So it is understandable that the acceptance of the theory of evolution made no great difference to the practice of systematics, and in the decade or two after

1858 no evident disparity showed itself between the practice of those systematists who accepted Darwin's doctrine and those who did not.

In his personal character and relations, Darwin stands out in later nineteenth-century England, not as the vanguard of twentieth-century biology, but as one of his own " living fossils ", an eighteenth-century savant living in the railway age. He still preserved the conception of the integrity and supremacy of the intellect, of pure thought, characteristic of the enlightenment, and could never have imagined that his beloved natural history would be less altered by the acceptance of his revolutionary ideas than it was destined to be by the social changes which as a good liberal he supported.

In the sharpest contrast to Darwin stands the figure of T. H. Huxley, inventor and popularizer of the very word biology (Darwin would have called himself a naturalist to the end of his days), and prototype of the non-museum professional biologist. Huxley's efforts were devoted at least as much to establishing the professional status and training of biologists as to purely intellectual propaganda for Darwinism. It was not until students trained in the schools of biology fathered by Huxley began to irrupt into systematic work that the influence of Darwin's teachings about species began to make itself felt. The delay (at least in Britain) was increased by the conservative policy of the British Museum, which showed no eagerness to recruit its staff from the new academically trained biologists; the museum clung to the tradition that young systematists are best trained by some sort of apprenticeship under older ones. Thus it was not until the present century that we begin to find museum systematists seriously questioning the objectivity of species, and by then the modern development of genetics, which has now restored the category to something like its former pre-eminence, was well under way.

A difference existed in Darwin's time between the purely " museum " side of systematics (represented by most of the papers in the *Proceedings of the Zoological Society* of the day) and that larger view which saw in classification the synthesis of the whole of zoological or botanical knowledge. Darwin's *Monograph of the Cirripedia* evidently reflects the outlook of the second school, since he included in it practically everything known about barnacles at the time, quite consistently with his later reference to the use of " characters of any kind which have long been inherited ". The distinction still remains, though the school to which Darwin belonged seems to be gaining ground in the long run. Unfortunately, however, most systematists working in museums are forced, whether they like it or not, to take the narrower view of their task; the very extensive portions of the animal or plant kingdoms for which they have to take responsibility, the scanty research facilities available to them, and the difficulties of publication, all militate against the production of works on the pattern of the *Monograph of the Cirripedia.* The influence of Huxley, who was no great systematist, led to a decline in the status of taxonomy: from being the central core of natural history, as it was when Darwin wrote, it came to be looked on as a minor branch of biology which could well be left to a few unimportant people in museums. Fifty years ago, when a commemorative volume similar to the present one was published [4], there was no article on classification included in it. Among the present signs of a partial return to Darwin's attitude are the works of some of the leading American genetical workers on the fruit-flies (*Drosophila*), who have become increasingly occupied with systematics in the wide sense.

TAXONOMIC INFLATION

The devaluation which has been as much a feature of the history of our classificatory categories over the last hundred years as it has in that of the currency, had already shown itself in Darwin's time; he remarked in the *Origin* (chapter 14):

> Instances could be given among plants and insects, of a group first ranked by practical naturalists as a genus, and then raised to the rank of a subfamily or a family; and this has been done, not because further research has detected important structural differences, at first overlooked, but because numerous allied species with slightly different grades of difference, have subsequently been discovered.

This he cited as evidence that such groups " seem to be, at least at present, almost arbitrary ", and the position is essentially unchanged today. Many systematists have deplored this tendency to inflation, which seems to be as characteristic of a democratic order in the sphere of classification as it is in that of finance. Darwin himself, by promoting the Cirripedia to the status of a sub-class of Crustacea and establishing orders

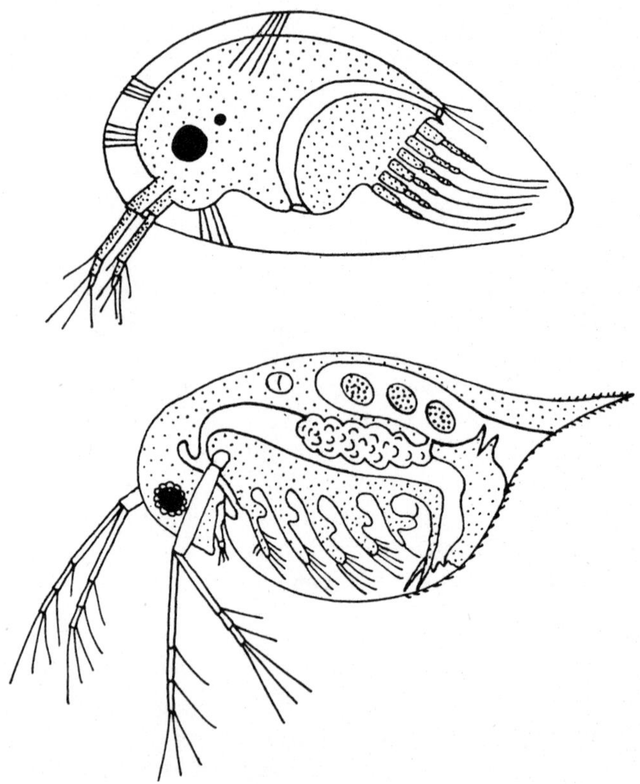

27.—Crustacea with bivalve shells. Above is the so-called Cypris larva of an Australian barnacle; the adult is a typical barnacle, living attached to rocks. Below, an adult Cladoceran, *Daphnia* sp., which is free-living. Each is enveloped in a transparent bivalve shell from which only the antennules project. It seems likely that the Cypris larva reflects the evolution of the barnacles from ancestors resembling the Cladocera, in the same way that the tadpole indicates the evolution of Amphibians from fish-like ancestors.

within it, may be held to have played his part in the process. Modern opinion would tend to regard the bivalve shell of the so-called *Cypris* stage (figure 27, top) in the development of barnacles as in some degree recapitulatory, and would hence incline to view the Cirripedia as peculiarly modified descendants of one of those "phyllopod" groups (figure 27, opposite) in which such a shell occurs in the adult stage. In a genealogical system, such a derivation would not be consistent with ranking Cirripedia as a primary sub-class of Crustacea.

There is at least one possible way of stopping "taxonomic inflation" which has not yet been tried—that of defining types of systematic categories. In each major systematic group (such as Dicotyledons, Monocotyledons, Gymnosperms, Ferns, Mammals, Birds, Reptiles, the major orders of insects) groups might be designated as types of each grade of subdivision, from the genus upwards. For example, we might define the Rodentia as the type of a mammalian order; then the question whether any other mammalian group is to be ranked as an order will mean, is it to be regarded as of the same status as Rodentia? A difficulty which would have to be faced in this particular instance is that of the position of the hare and rabbit group, the Lagomorpha. Many authorities believe these to be more closely allied by descent with various other mammalian groups, such as the elephants, than they are to the rodents proper. If they are right, then in a phylogenetic system Lagomorpha will have to be excluded from the Rodentia. Fortunately there are good hopes that this question may be settled conclusively by fossil evidence.

The selection of groups to serve as types of systematic categories in this way would need to be made with great care. A type group should be one which is well defined and as certainly natural as can be in the present state of knowledge; it should show adequate diversity at the various levels of subdivision (for example, a type genus should have well-marked subgenera and species-groups within it and, in the case of the genus, it might be desirable to establish an additional level of subdivision, between the subgenus and the species-group, as indicated in the table on page 115). Furthermore, it would be desirable for its systematic status to have been generally accorded to it for a considerable period, say, fifty years. Of course, at some stage the question of correlating the application of categories between different major groups would be certain to arise, and with it the difficulties already mentioned in connection with the genera of insects and mammals.

DARWIN AND NOMENCLATURE

Finally, we may consider Darwin in relation to the scientific naming of plants and animals, a subject which has recently developed so many pitfalls and complexities that (at least in zoology) it has come to rank as a subject in its own right, Zoological Nomenclature, with its own quasi-legal experts.

This is one aspect of natural history where reason and convenience should be the deciding factors; during the later nineteenth century the appeal to these qualities managed to produce a more satisfactory uniformity, both as between generations and between countries, in the naming of organisms than we have seen since. The twentieth century was ushered in by the promulgation of an elaborate set of International Rules for Zoological Nomenclature, designed to settle nomenclatorial disputes once and for all by appeal to objective criteria. Provided thus with a proper " objective basis ", nomenclature entered on a phase of endless controversy and instability which became a general scandal in zoology, and it is as yet too soon to say whether recent special efforts by the International Commission will end this unhappy state of affairs.

When Darwin was writing the *Monograph of the Cirripedia* there were, of course, no International Rules and no International Commission; however, a committee set up by the British Association for the Advancement of Science had published in 1842 a proposed set of rules, and Darwin explicitly adhered to these in all but a very few instances. The basic principle of the so-called " Strickland Code " was that the valid name for an animal was that first applied to it under the Linnæan binomial system, provided that the name was not pre-occupied (that is, had not previously been applied to an animal); that it was published with a proper description; and that it did not violate the rules of language and orthography. It was also generally agreed that a name which was valid by these canons might still be rejected if it were seriously misleading (as for example the name *apterus* would be if applied to an animal with wings) or if the originator of the said name had subsequently changed his mind and substituted another name for it, or if the animal had been universally known by some other name for a long time.

In dealing with his barnacles, Darwin rejected several generic names which would have been valid under the Strickland Code, on the grounds that they were misleading (for example the generic name *Trilasmus* applied to forms which Darwin showed to possess five or more valves to the shell, not three), and one on the grounds that, although it had priority, the original description was bad and nobody had used the name subsequently.

In Darwin's day almost all serious naturalists would know Latin fairly well, and many would have at least some acquaintance with Greek (Darwin's new generic names for barnacles are Greek-derived, and he explains their derivations); in such a social context the actual meaning and derivation of the names would seem to be of material importance. The fact that the International Code of 1901 did not permit the alteration of a scientific name on grounds of inappropriateness of meaning or linguistic error in its formation shows that the revolution heralded by T. H. Huxley had made great strides within fifty years.

THE FUTURE OF DARWINIAN CLASSIFICATION

The task mapped out for systematists by Darwin in our first quotation is one that we have by no means completed a hundred years later, and it seems probable that even another century will not see it brought to a finally satisfactory conclusion. During that time, the march of general human progress may well call a halt to progress on this particular line, either by bringing about the extinction of the large majority of species of animals and plants now living, or more subtly by leading to a Utopia in which no human beings are maladjusted enough to take up such unprofitable studies as plant or animal systematics. Meanwhile, those of us who value the work and traditions of our forefathers may still be guided by Darwin's precept and example. His example, in the shape of the *Monograph of the Cirripedia*, has still to be followed as far as most of the animal kingdom is concerned. Few groups of animals have yet received such comprehensive and world-wide treatment; only Europe can yet boast of anything approaching a complete coverage of its fauna by regional monographs. Even such a wealthy and technologically advanced area as North America is today not very much better provided with comprehensive works on its own fauna than it was in Darwin's lifetime.

It is probable that the future of Darwinian systematic natural history is bound up with that of serious amateur naturalists. Attempts are often made to stress the practical advantages of a fully worked out natural classification, and undoubtedly such a system would have a special value for economic zoologists and botanists, but I do not think that utilitarian motives alone could suffice to bring about its achievement. The mere building up and maintenance of adequate collections of animals and plants is itself a long and costly process, with no promise of quick or exciting rewards, and if (as would be necessary) ecological, physiological and biochemical information were also to be collected systematically over the whole range of the animal and plant kingdoms, the services of a whole army of biological specialists in well-equipped laboratories would be needed for a long time to come—and all this merely to provide the basic information for the construction of a natural system.

Given the probability that only a small proportion of the species of animals or plants are ever likely to have appreciable economic importance for man, it seems that a society dominated by utilitarian motives would not be likely to think the possible advantages of a comprehensive natural system sufficient to justify the cost of producing it. After all, modern man is able to control and exploit the plant and animal worlds quite effectively already.

If systematic zoology and botany are to progress along Darwinian lines, we shall need to have a society with many serious amateur naturalists who will play a major part in the collection of specimens and field observations, who will buy and use serious (and expensive) systematic monographs and thus make their publication possible, and who will have enough social influence to ensure that adequate financial support is maintained for zoological and botanical laboratories and museums where systematic work is carried out. Unfortunately the advanced societies of today recognize only two motives for human endeavour—economic gain and pleasure; and if the pursuit of a natural classification is not to be justified in terms of economic gain, then modern men will insist that it must come under the category of pleasure. A hundred years ago a third type of motive was socially recognized—the pursuit of virtue and piety; and in the pre-Darwin and pre-Huxley age the justification of natural history was seen in these terms. The dedicated naturalist had something of the aura of a priest or monk, as the revealer of the divine mysteries of creation, and it would have seemed irreverent to suggest that anything that was worth God's while to create was not worth man's while to study. Whether systematic natural history can continue to flourish when its practitioners are looked on merely as a rather odd variety of pleasure-seekers remains to be seen.

6

Darwin and the Fossil Record

By

Alfred Sherwood Romer

> While on board H.M.S. *Beagle*, as naturalist, I was much struck with certain facts in the distribution of the organic beings inhabiting South America, and in the geological relations of the present to the past inhabitants of that continent. These facts, as will be seen in the latter chapters of this volume, seemed to throw some light on the origin of species—that mystery of mysteries, as it has been called by one of our greatest philosophers.

Thus the opening sentences of *On the Origin of Species*. In part it was his study of living animals and plants—of the teeming life of the tropical jungles, of the sharp contrast of the animal population of South America with that of his homeland, of the curious character of island faunas—which turned Darwin's thoughts towards the problem of evolution. But equally important in arousing his interest and stimulating his thought was his acquaintance, gained during the *Beagle* voyage, with the geological past of South America and the strange nature of its palæontological record, which he was one of the first to explore.

During his Edinburgh student days Darwin had studied geology, but found it boring. At Cambridge, however, his interest was aroused; he undertook stratigraphic field work in Shropshire and, with Sedgwick, in North Wales. Under the essentially stable conditions of the mature topography of Great Britain, however, geological processes tended to be thought of as events of the past, which have little to do with the settled world of today. Far different was the picture presented along the coasts of Chile and Patagonia, where several years of the *Beagle*'s voyage were spent. The rise of the great chain of the Andes is, geologically, a relatively recent occurrence. Time after time Darwin notes in his diary the presence of raised beaches, far above the present ocean level but so fresh in appearance that the time since their formation can be estimated, at the most, in thousands rather than millions of years. Here is a major geological event that is not, so to speak, dead and buried, but belongs to the present as much as to the past. That this active mountain building is still going on was impressed on Darwin's mind by a major earthquake which took place in southern Chile during the *Beagle*'s presence there in 1834—one of the series of major earthquakes to which that region had been, and still is, subjected, and which is part of the process of Andean elevation. Obviously such revolutionary changes in the terrain must have a pronounced effect on the fauna and flora of the continent; and consideration of these effects leads inevitably to consideration of the possibility that evolutionary processes were involved.

FOSSIL MAMMALS OF SOUTH AMERICA

Today we are reasonably familiar with the general outlines of the faunal history of South America during the course of the Coenozoic Era, the Age of Mammals. Almost at the beginning of this era, some 60 to 70 millions of years ago, when the evolution of higher mammals was in its infancy, that continent was cut off from the rest of the world and remained in isolation until a time probably not more than a million or so years ago, when connections were re-established via the Isthmus of Panama. Relatively few kinds of mammals were present in South America at the time when communications were broken. During the long course of Tertiary times, when there evolved in the northern continents the majority of the mammal types familiar to us, there took place in the south a vast radiation of forms of a distinctive nature.

Descendants of certain of these "natives" still survive. They include the opossums, primitive pouched mammals little changed from Cretaceous ancestors; numerous and varied rodents belonging to a sub-order quite distinct from those common on other continents; monkeys of a type very different from those of the old world tropics; and—most distinctive of all—the armadillos, tree sloths and anteaters, which have sometimes been grouped in the order Edentata.

Even more striking were numerous forms now extinct, but abounding in South America during mid-Tertiary and often surviving into the Pleistocene. No carnivores of the familiar kinds of the old world had reached South America before the time of isolation, and (much as in Australia) pouched mammals related to the opossums developed as flesh-eaters, with forms comparable to the wolves and felids of northern continents—one even to the sabre-tooth " tigers " of North America and Eurasia.

Lacking, too, in the original fauna were any representatives of the orders to which belong the hoofed mammals of northern continents. A few primitive ungulates were present, however, and from them arose a spectacular array of hoofed forms quite unlike those of any other region; a few are illustrated in figure 28. *Toxodon*, for example, was a large and heavy beast to some extent analogous to a hornless rhinoceros or a hippopotamus; *Macrauchenia* had the appearance of an ungainly camel; other forms had a superficial resemblance to horses and proboscideans; still others could be compared to nothing imaginable short of the creatures of nightmares or of alcoholic delirium. Adding to this grotesque array, the " edentate " order produced giant types in the glyptodonts—large relatives of the armadillos with the body covered by an arched carapace of bony armour—and numerous large and clumsy ground sloths, such as the giant *Megatherium*.

At the end of the Tertiary Period, connections with North America were re-established, and the downfall of this curious and isolated fauna began. A few southern forms ventured northward. Ground sloths of several sorts and glyptodonts invaded North America, and met with temporary success, only to perish there, as in their ancestral home, by the end of the Pleistocene Ice Age. Of the adventurers to the north, only the North American porcupine, the armadillo and the opossum (apparently a re-immigrant) have persisted.

Movement along the isthmian land bridge was mostly in the opposite direction—with catastrophic results for the native types. Progressive ungulates—llamas, deer, peccaries, even proboscideans—entered from the north to compete with the native South American ungulates. Still more disastrous was the entry of advanced carnivores of the dog and cat tribes. These supplanted the marsupials as predators, and obviously found the native herbivores an easy prey, although some

28.—Some of the grotesque Pleistocene mammalian inhabitants of South America, whose bones are found in the Pampas formation. Darwin found remains of most of the forms figured here. A pointer dog is shown to give the scale. 1–2, Glyptodonts. 3, *Macrauchenia*, a rather camel-like native ungulate (Order Litopterna). 4, An extinct horse (*Hippidion*). 5, *Toxodon*, a ponderous native ungulate (Order Notoungulata). 6—7, the ground sloths *Megatherium* and *Mylodon*. (From W. B. Scott, *A History of Land Mammals in the Western Hemisphere*, courtesy the American Philosophical Society.)

lingered on through the Pleistocene; but before modern times all the more spectacular native mammals—the larger marsupials, the ground sloths, the glyptodonts and every one of the numerous and varied ungulates—had vanished.

At the time the *Beagle* sailed almost nothing was known of this story. A *Megatherium* skeleton had been found in the Argentine and sent to Madrid half a century before; mastodon teeth had been found by von Humboldt and other travellers; several writers had mentioned the presence here and there of fossil bones but had not described them. Nothing more. Darwin was, thus, one of the pioneers in the vertebrate palæontology of South America.

Unthinkingly one sometimes assumes that Darwin's scientific work on the *Beagle* voyage was done *on* the ship. This is, of course, almost exactly the reverse of the true situation. To be sure, he made some observations on marine life, and employed some time on the ship in working up his collections. But, as reference to his original diary will show, his principal occupation aboard was the enjoyment of sea-sickness. "I am now writing the memoranda of my misery for the last week"; "wretchedly out of spirits and very sick"; and "the misery is excessive" are characteristic entries. The *Beagle* was for Darwin essentially a means, although an uncomfortable one, of getting from one interesting region to another, and it was mainly on the occasions when he could leave the ship and get ashore that his scientific work was done. Fortunately the nature of the *Beagle*'s mission, that of making coastal charts, made such opportunities of frequent occurrence, for the ship often tarried for considerable periods in a given region, sailing back and forth along the shores from port to port.

The *Beagle* arrived at Monte Video, Uruguay, in July 1832; she passed the Straits of Magellan, bound for Chile, in June 1834. Thus nearly two years were spent off the coasts of Argentina, particularly Patagonia, from the La Plata to Tierra del Fuego, and Darwin had ample opportunity to explore the country. Fortunately for palæontology, although he was as much or more concerned with observation of living animals and plants, Darwin on a number of occasions met with fossil-bearing beds.

A much-simplified sketch of the geology of southern South

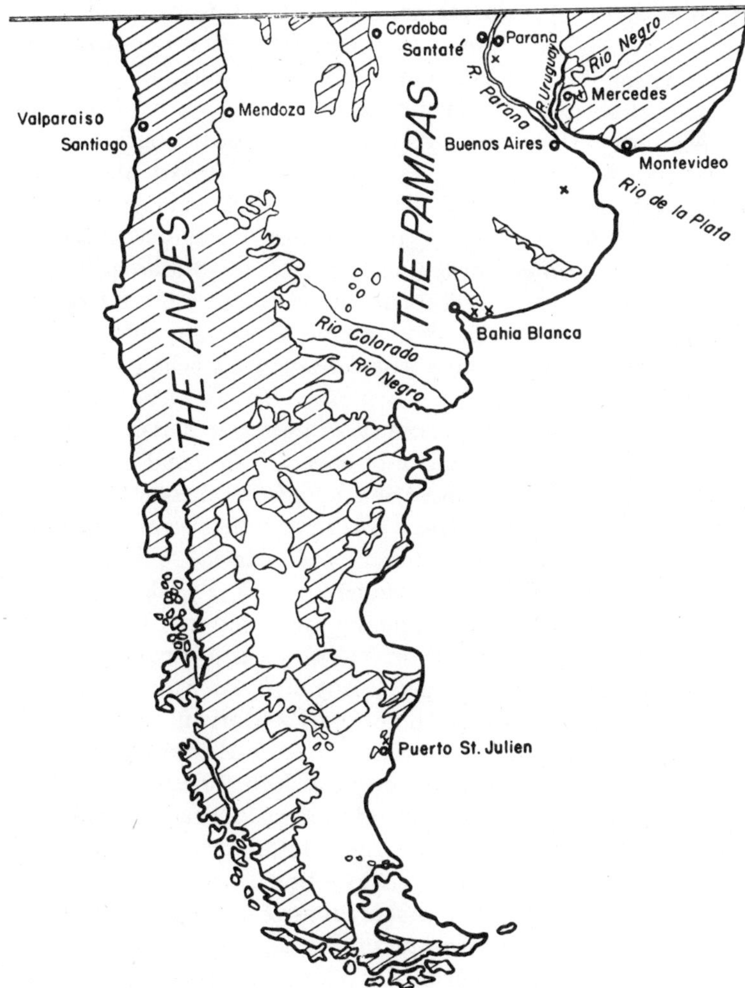

29.—A sketch map of southern South America, to show the localities where Darwin collected fossil vertebrates (marked by crosses). The white areas are the plains covered by the Pleistocene Pampas beds and, to the south in Patagonia, Tertiary fossil-bearing beds as well. Shaded are the non-fossiliferous rocks which form the core of the Andes and pre-Coenozoic sediments and volcanics which spread eastward over much of Patagonia and are present in Uruguay as well.

America is shown in figure 29. The core of the north-south running Andes and the regions adjacent to this mountain chain are formed of igneous and metamorphic rocks and pre-Tertiary geological formations. Much of Patagonia, however, consists of series of Tertiary beds, fossiliferous in various areas. In northern Patagonia and stretching far to the north through Argentina and Uruguay, lies the Pampas, a vast and nearly flat plain. This (as well as scattered areas farther south in Patagonia) is underlain by the Pampas formation, relatively recent beds contemporary in time of deposition with the Pleistocene Ice Age in the northern continents. Over this plain, exposures are few and mainly limited to bluffs along river channels and the coast, but the deposits, when found, are generally fossil bearing.

Darwin had little opportunity to visit the Tertiary beds of Patagonia, but on a number of occasions found fossils in the Pampas beds. Most of September and October of 1832 were spent by the *Beagle* in Bahia Blanca in northern Patagonia, and Darwin was ashore for much of the period. Here, Darwin found his first fossils. " September 23 . . . I walked to Punta Alta to look after fossils; and to my great joy, I found the head of some large animal, imbedded in a soft rock. It took me nearly three hours to get it out. As far as I am able to judge, it is allied to the Rhinoceros." Darwin visited Punta Alta again on a number of occasions, and this locality proved to be the most profitable one which he discovered. The animal he thought to be like a rhinoceros turned out, on later study, to be one of the large ground sloths (*Scelidotherium*); much of its body skeleton too was unearthed at Punta Alta, and in addition other giant sloth remains, a fossil horse tooth and a fragmentary lower jaw of *Toxodon*—the first discovery, I believe, of any member of the native ungulate groups. Just before leaving Bahia Blanca a brief stop was made at Monte Hermoso, farther up the bay, and some fossil rodent remains were found; the horizon here is a late Tertiary one, slightly older than the Pampas beds from which all his other material was derived. The only other Patagonian locality from which fossil vertebrates were recovered by Darwin was at Port Saint Julien, far to the south, visited a year and a half later. Here, in an outlying patch of beds of the Pampas formation, he found a partial skeleton, lacking the skull, of *Macrauchenia*, a second—but very different—native ungulate.

Darwin was able to take several extensive overland trips across the Pampas—notably across the wild region from northern Patagonia to Buenos Aires; from that city up-river to Santa Fé and Parana; and, in Uruguay, from Monte Video north-west to Mercedes on the Rio Negro, near the Rio Uruguay. On each of these journeys some fossil material was obtained, despite the paucity of exposures. Some glyptodont armour was discovered south of Buenos Aires; the cliffs of the Parana yielded further glyptodont armour, another horse tooth and one of *Toxodon*, and some poor remains of a mastodon. Still other fossils were gathered from the Rio Negro region of Uruguay, including mastodon, sloth and glyptodont specimens. Most notable was a skull of *Toxodon*, from the Sarandis River; this, although far from complete, was justifiably described in detail by Owen as the first-known cranial material of its order. That author in his account says of its antecedents merely that " it was discovered lying in the bed of the rivulet, after a sudden flood had washed down part of the bank." Darwin's note to Owen about this important specimen is worth quoting as illustrating the vicissitudes to which a fossil may be subject when found by non-scientists. " The head had been kept for a short time in a neighbouring farm house as a curiosity, but when I arrived it was lying in the yard. I bought it for the value of eighteen-pence. The people informed me that when first discovered, about two years previously, it was quite perfect, but that the boys had since knocked out the teeth and had put it on a post as a mark to throw stones at."

A letter from the distinguished geologist Charles Lyell to Richard Owen on October 26, 1836, reads:

> My dear Sir, Mrs. Lyell and I expect a few friends here on Saturday next, 29th, to an early tea party at eight o'clock, and it will give us great pleasure if you can join us.
>
> Among others you will meet Mr Charles Darwin, whom I believe you have seen, just returned from South America, where he has laboured for zoologists as well as for hammer-bearers.

What more natural than that Darwin's vertebrate fossils should be turned over for description to the energetic young comparative anatomist and palæontologist, Owen, 32 years of age at the time, currently teaching at St. Bartholomew's but about to be appointed Hunterian Professor in the Royal College of Surgeons? Owen undertook the work with his usual energy and enthusiasm. His publication on Darwin's fossils, in 111 quarto pages accompanied by 32 plates, occupies a great part of volume I of the *Zoology of the "Beagle"*. Two decades were to pass before the major rift over the question of evolution occurred between these two great scientists.

It is amusing to reflect that Owen's work on his fossils, and the light they shed on the extinct fauna of South America, appears to have been one of the two main sources from which Darwin, a one-time firm believer in special creation, derived his final belief in evolution. For in his diary for 1837 he writes: "In July opened first notebook on 'Transmutation of Species'. Had been greatly struck from about month of previous March on character of S. American fossils—and species on Galapagos Archipelago. These facts origin (especially latter) of all my views."

FOSSILS AND THE "ORIGIN"

Of exceeding interest to the palæontologist is Darwin's treatment of the fossil record in the *Origin*, and we shall discuss this at some length. Today, in an elementary book devoted to the topic of evolution, a substantial section is usually devoted to proofs of the reality of the evolutionary process—proofs derived from a consideration of the fossil record, particularly that of the vertebrates. Not so in Darwin's classic. The treatment of the subject is essentially a negative one. For the most part his argument is not that palæontology supports the evolutionary theory but, rather, that it need not be regarded as opposing it.

To understand this approach it is necessary to recall the history of palæontological work. Today, evolution and palæontology march hand in hand. Palæontology supports evolution; the truth of evolution is a basic assumption underlying all palæontological work. This, however, was not the case in earlier days.

Little work of importance was done in palæontology until the late 1700s, at which time both vertebrate and invertebrate fields began to assume importance. Intensive work in the invertebrate area arose from recognition of the fact, first clearly seen by William Smith, an English civil engineer and amateur geologist of the period, that a given set of beds tended to contain the same species of shells over vast and widely separated areas. Accurate determination of fossils could thus be of great practical use to the stratigrapher; as a result, invertebrate palæontology tended to develop not as an independent science, but as a handmaiden to the geologist—a working tool for the stratigrapher looking for oil or ores or coal. The fossil shells were rarely thought of as the remains of once-living organisms, but merely as convenient markers for the identification of successive formations, and would have been as useful had they been identifiable mineral inclusions or distinctive assortments of nuts and bolts. This point of view, incidentally, is not confined to the early days, but has continued to a considerable degree to the present—much to the disadvantage of the science. Recent years, however, have seen a considerable increase of interest in the biological aspects of their subject among invertebrate palæontologists.

With this background the invertebrate workers of Darwin's day not merely lacked interest in evolutionary ideas, but were inclined to view them with suspicion as detrimental to their work. For clear-cut stratigraphic work the species in a given formation should be stable entities, clearly distinguishable from those in the strata above and below. The idea of gradual change and of transitional forms was abhorrent.

More striking is the fact that most vertebrate palæontologists, now ardent believers in evolution, were in early days to be found in the opposition camp. Cuvier, justly venerated as the true founder of the science of vertebrate palæontology as well as of comparative anatomy, first clearly pointed out that the true road to the proper interpretation of fossil forms is to consider them as once living organisms and to interpret them in the light of our knowledge of still existing forms. But Cuvier was flatly opposed to any evolutionary interpretations, as shown by his violent opposition to Lamarck and Geoffroy St. Hilaire. For him the fossil record was one of clearly distinct periods of life, separated from one another by major geological revolutions, with the creation of an entirely new fauna after each catastrophe. Man, for Cuvier, was a geologically recent form, not part of the true fossil record; mammals are not to be found earlier than the "Tertiary" beds; the older "Secondary" rocks contain a quite distinct type of life, in which mammals were absent and reptiles were prominent; still further back were "Primary" rocks in which land animals were absent and older creations of invertebrates and fishes constituted the faunas. This non-evolutionary attitude of Cuvier was maintained, despite increasing knowledge, by most of the more prominent palæontologists of Darwin's day, notably Richard Owen, greatest of nineteenth-century English workers in vertebrate palæontology and anatomy, and Louis Agassiz, the first major student of fossil fishes.

With this to contend with, it is apparent why Darwin was thrown on the defensive in his treatment of the fossil record. He could not call on the palæontologists for support; the most he could do was to attempt appeasement, to show that it was at least possible to interpret the geological story in evolutionary terms, and that there were no insuperable objections.

Palæontological data are cited in a variety of places in the *Origin*, as, for example, in the chapters on geographical distribution. His general argument on the fossil story is, however, concentrated in chapters X and XI, "On the interpretation of the geological record" and "On the geological succession of organic beings". Of these two chapters the first is by far the more important. In it he discusses some of the arguments which might be—and were—brought against an evolutionary interpretation of the geological record, and answers most of them in a convincing fashion. The major objections, certain of which continued (although with diminishing force) to be brought against evolutionary beliefs long after the time of the first publication of Darwin's work, may be stated as follows:

First, if the evolutionary theory were true, we should expect to find many fossil species or varieties of intermediate nature—"missing links", that is, in modern popular terminology. This, said the opponents, is not the case.

Second, the extent of geological time is too brief for major evolutionary changes to have occurred.

Third, known fossils from the various periods and formations do not show a well-arranged phyletic pattern, as would be expected on evolutionary grounds, but a scattered, seemingly random, distribution of forms.

Fourth, if the history of life has been a gradual evolutionary progression, we would expect to find gradual changes between forms in the lower and upper parts of geological formations. This is not the case.

Fifth, whole groups of species appear suddenly, in an abrupt manner, in certain formations, contrary to what one would expect if evolutionary development had occurred.

Sixth, a related and more serious problem is the sudden appearance, without known antecedents, in the lowest known fossiliferous strata—the Cambrian—of a whole series of members of a variety of major animal types.

Darwin discusses these objections *seriatim* and is able, for the most part, to give convincing answers.

First, there is the absence of intermediate varieties. In an earlier section of the *Origin* Darwin had given quite satisfactory reasons for the rarity, at the present day, of types intermediate between living forms, and the same situation should hold for any given geological formation. Further, if one is looking in the fossil record for "intermediates", what should they be intermediate between? For example, says Darwin, should we look—today or in any older formation—for an intermediate between a horse and a tapir—a form which "splits the difference" between the two? It is, he says, highly improbable that either one of these has descended from the other. Rather, the two have presumably descended from a remote common ancestor; and we would be unable to recognize this remote ancestor—the true "intermediate"—without a knowledge of the lines of descent of horse and tapir from it.

Darwin could not have hit upon a happier illustration. Already in Darwin's day his erstwhile friend and later opponent, Owen, had described the skull of a small browsing animal from the English Eocene which he named *Hyracotherium* because of the resemblance of its teeth to those of the living conies, or hyraxes, of Africa and Syria. Considerably later, with the discovery in the American West of linking types, it became apparent that *Hyracotherium* was an ancestral equid, and in recent decades it has been demonstrated that *Hyracotherium* and the widely-known *Eohippus*, the " dawn horse ", are generically identical. But *Hyracotherium* is not merely a direct horse ancestor ; it is at least very close to the ancestor of the tapirs and other odd-toed ungulates. It is thus one of the " intermediates ", the supposed absence of which was argued as an objection to evolution; but just as Darwin pointed out would be the case, its nature was not recognized until further connecting links were discovered.

Second is the question whether geological time is insufficient. Our Christian ancestors were in general habituated to a chronology of the type promulgated by the learned Bishop Ussher, according to whose computations from the somewhat confused and conflicting data of the Old Testament the world is rather under 6,000 years of age. Even those of a more liberal cast of mind, who with that other, older, and more famous bishop, St. Augustine of Hippo, were willing to grant that the seven days of creation need not be taken literally, still tended to regard the lapse of time since the earth began as a relatively short period. Darwin presents the concept of the long span of geological time most persuasively. He had himself seen, in South America, geological processes proceeding in fast tempo. For his purposes here, however, he paints for his readers, instead, the picture of the English landscape and the immense amount of time necessary to effect the changes which have brought it from the conditions surely present in times past to those of the present. Strata are worn away but slowly. Even where the sea eats away a cliff on its shores, the degradation is generally slow; and, further, only a fraction of the shore-line is suffering in this fashion at any one time. But the action of the sea is only a minor process in the erosion of the land; much more important—and much slower, in general—are the breaking down of rocks by chemical and biological means and their transportation downhill and to the sea by rills and streams. Immense thicknesses of rock, it seems clear, have been completely worn away, over, surely, a vast period of time by this type of action. And equally impressive is a consideration of the sediments laid down in ancient seas and lowlands as a result of such degradation. Darwin cites an estimate of the thickness of sediments laid down since the beginning of the Palæozoic in one area or another of England as having a total thickness of over 72,000 feet—nearly fourteen miles of accumulation. A vast amount of time was surely needed to form these deposits; and since there are many gaps in the English sequence, further major additions must be added to give the total time since the fossil sequence became established in Cambrian days at the opening of the Palæozoic Era.

How long in terms of years have the various eras and periods covered? In his first edition Darwin estimates that the minimum time needed during the Tertiary to remove once overlying sediments—about 1,100 feet of them—from the Weald of Southern England must have been over 300,000,000 years, and that since the process of denudation was presumably intermittent, the actual elapsed time would have been far longer.

Darwin's figure here, it would seem, is very far above any acceptable to his contemporaries. In later editions of the book this calculation is omitted, and the only estimate of any sort for which figures in years are given is one cited from Croll to the effect that a thousand feet of sediments might be removed in about six million years. On this basis (although Darwin does not say this) the estimate for the Wealden degradation during the Tertiary might have been rather less than seven million years. This is on an order of magnitude which was more palatable, it would seem, to the geologists of the last century, and Darwin even offers a further sop to the hesitant conservative by saying of the Croll estimate that "some considerations lead to the suspicion that it may be too large", and that it might be halved or quartered.

Up to the early decades of the present century figures on the order of magnitude of the Croll calculations were those generally accepted; the time from the beginning of the Tertiary was frequently cited as about five million years, and from the beginning of the Palæozoic as about fifty millions. Such figures were based on estimates of the minimum time needed to lay down the known series of sediments, and it was agreed that in all probability some increase, although perhaps a modest one, might be needed to account for gaps in the sedimentary record. During the past quarter-century there has been a sharp upward revision of such figures due to a study of radioactive rocks present at a number of levels in the geological column and their degree of disintegration. The new figures have increased over the old by a factor of ten; the Tertiary is now commonly cited as having a duration of fifty to seventy million years, and the time elapsed since the beginning of the Palæozoic calculated to be five hundred million years or so. This is indeed a sharp jump, but even so falls far short of Darwin's original estimates. But whatever estimate one then accepted, or accepts now—the " short count ", current figures, or Darwin's original long one—the span of geological time is certainly adequate for a very considerable amount of evolutionary change to have occurred.

Third, there is the assertion that known fossils do not form a phyletic pattern, as would be expected on an evolutionary hypothesis. At the time of the first publication of the *Origin* and, to a somewhat lesser degree, at the time of publication of later editions, this objection was one of seemingly great validity. The known record of past life was a very " spotty " one; to some extent the later formations in the sedimentary sequence showed the presence of " higher " forms of life, but this could just as well be explained on the basis of separate successive creations with " improved " forms being brought forth *de novo* in the later ones. Beyond this single general trend, there was no evidence of any phylogenetic " family tree ". Even among the vertebrates, in which hard internal skeletal parts favour preservation as fossils, there was little indication at the time of any evolutionary arrangement of the animals then known. A fair assortment of fossil fishes had been described; but as Agassiz observed (even as late as in a posthumous paper in the '70s) forms of a presumed advanced type had been found in older strata than those containing any fishes of a supposedly more primitive nature—the reverse of an evolutionary sequence. Intermediates between fishes and land vertebrates were practically unknown. Plesiosaurs and ichthyosaurs, the only reptiles then adequately described, were isolated if spectacular types which shed no light on the possible evolution of reptiles. Nothing was known to connect either birds or mammals with lower groups.

Why this seeming contradiction between the known fossil record and that expected on the evolutionary hypothesis? Darwin attributes this to the poorness of palæontological collections in his day. Only a small portion of the earth had then been geologically explored, and no part explored thoroughly. Of the fossil species then described, very many were known and named from single and often broken or fragmentary specimens, rendering interpretation difficult. In addition to the inadequacies of our knowledge of fossils due to insufficient exploration, are those due to the imperfections of the geological record. Unless deposited in an area where sediments are being laid down, shells and bones decay quickly. It seems clear that at any past time, as is true today, sedimentation was taking place over only a very small portion of the earth's surface, leaving great gaps in the preserved record of life in any given area. Further, the fact that an animal was fossilized is no guarantee that it would be preserved down to modern times, for degradation of necessity goes hand in hand with sedimentation. Our knowledge of terrestrial formations in the earlier geological periods, Darwin points out, is hence extremely meagre. There are vast areas of the earth (the Canadian Shield, for example) in which no unaltered sediments at all are present today: " primordial " rocks form the surface; in consequence, we shall never obtain a record of any former inhabitant of the region. And—most discouraging fact of all—many of the connecting links between major animal phyla, according to the evolutionary hypothesis, would have been soft-bodied animals; and while Darwin's statement that " no organism wholly soft can be preserved as a fossil " is too extreme, identifiable remains of such forms are extremely rare.

Many of the reasons advanced by Darwin for the inadequacies of fossil remains—those concerned with the imperfections of the geological record—are as valid today as they were a century ago, and it is certain that we shall never be able to find and describe more than a very small fraction of the former inhabitants of the earth. But wider exploration and further exploitation of fossiliferous deposits already known have added vastly, over the intervening decades, to the number and variety of known forms. And it is of major importance that, although no " family tree ", even that of the vertebrates, is fully documented, nearly every new discovery fits into once hypothetical phylogenies. There are still, among the vertebrates, areas in which there are major lacunæ—for example, the evolution of the earliest fishes, the origin of the modern amphibian orders and of certain reptile groups—but in many portions of the "tree" the phyletic pattern is becoming increasingly clear. Beyond the middle Devonian the general pattern of fish evolution is demonstrable. The gap between fishes and land vertebrates is gradually closing through such discoveries as that of the late Devonian amphibians of Greenland. Connections between early amphibians and the reptiles are so well documented that in the case of such an animal as *Seymouria* of the early Texas Permian, it is difficult to reach a decision concerning the class to which it should be assigned. A few years after the publication of Darwin's first edition came discovery of *Archæopteryx*, filling a half-way position on the branch of the vertebrate tree leading from reptiles to advanced birds. Discoveries, first in South Africa, later in other regions, have closed much of the gap between reptiles and mammals. (Much of the early material of mammal-like reptiles from South Africa was described by Owen. In a monograph on these forms published in the 1870s he discusses the morphological similarity of these reptiles to mammals without committing himself to biological evolution; this is a masterpiece of eloquent evasion.) It is still necessary today to call attention to the obvious imperfections of the palæontological record, but the need is much less than in the last century and the approach to the fossil story can be a positive rather than a negative one.

The fourth objection was the absence of intermediate varieties in any single formation. If, as assumed by Darwin, evolution had occurred in a slow but constant fashion, we should find, in any formation, a gradual transition between species present at the time of its commencement and the differing varieties or species presumably descended from them and present at its close. This did not appear in Darwin's day to be generally the case, although much invertebrate fossil material was known from a number of formations of marine origin. Darwin admitted the strength of this argument, but gave a number of suggestions which lessen its force. It is difficult to determine how long a term of years is necessary to effect an evolutionary change from one species to a derived one; possibly many formations did not persist for a sufficient length of time for noticeable evolutionary changes to have taken place in them. Again, to witness evolutionary progress within a formation, it is necessary that the populations present at its close be descended from those found there at its initiation; but it is not at all unlikely that there may have been considerable immigration from other areas, accompanied by extinction or emigration of old residents. Darwin believed (and many geneticists today are in agreement with this conclusion) that the development of a new variety or species which may eventually supplant the parent type generally takes place in a rather restricted area, and perhaps in a relatively short time; the chance of finding a formation in which the supplanting type developed is small.

The situation may be further obscured by a factor of quite another sort—a man-made one. There is no golden rule by which a palæontologist may distinguish varieties and species. Species are founded by some workers on the basis of excessively slight differences, by others only when obvious major differences are visible; whether or not specific changes are thought to have occurred within the limits of a formation may depend as much on the working methods of the palæontologist describing the material as on the nature of the fossils themselves.[1]

Now, as in Darwin's day, there is still often no evidence of progressive evolutionary change within a formation. But over the course of the intervening century a number of detailed studies of formations have been made, ranging from the Devonian to the Tertiary, in which careful work has revealed series of finely graded changes in forms from successive levels. Some of this work was done during Darwin's lifetime. As a result, instead of being forced to state, as he did in 1859, that "geological research . . . has done scarcely anything in breaking down the distinction between species, by connecting them together by numerous, fine, intermediate varieties", he could in later editions say instead: "It has been asserted over and over again, by writers who believe in the immutability of species, that geology yields no linking forms. This assertion . . . is certainly erroneous."

The fifth difficulty is the sudden appearance of whole groups of allied species. "The abrupt manner in which whole groups of species suddenly appear in certain formations, has been

[1] [See also pages 37 and 114.—ED.]

urged by several palæontologists—for instance, by Agassiz, Pictet, and Sedgwick—as a fatal objection to the belief in the transmutation of species." Were this phenomenon a reality, says Darwin, the objection would be a serious one. But the objection is based merely on negative evidence, which experience often shows to be worthless. He points out that, for example, Agassiz had maintained that teleost fishes first appeared—and appeared then in abundance—only in the Upper Cretaceous, but that he had later discovered teleosts, in lesser variety, in the earlier Jurassic and even Triassic. Since Darwin's day many further supposed examples of the sudden appearance, full-fledged, of animal and plant groups have been found to be equally illusory. The weakness of negative evidence can be further illustrated today by cases of supposed extinction. The most familiar example is the recent discovery in the sea off the Comoro Islands of a living cœlacanth fish, *Latimeria* (figure 24, page 112), belonging to a group of which no fossils are known in beds later than the Cretaceous; this group had therefore been confidently stated to have been extinct for seventy million years. Still more striking is the very recent discovery by the *Galathea* expedition, off the west coast of Mexico, of an archaic segmented mollusc type long supposed (because of negative evidence) to have been extinct since the Ordovician—a period of perhaps 400 million years.

Finally, we have the sudden appearance of groups of allied species in the lowest known fossiliferous strata. This situation is a special case of the last, and one admitted by Darwin to be a serious difficulty for his theory. From the beginning of the Cambrian up through the rest of the geological sequence we have an abundant representation of animal life at every stage; even in Lower Cambrian formations marine invertebrates are numerous and varied. Below this, there are vast thicknesses of sediments in which the progenitors of the Cambrian forms would be expected. But we do not find them; these older beds are almost barren of evidence of life, and the general picture is reasonably consistent with the idea of a special creation at the beginning of Cambrian times.

"To the question why we do not find rich fossiliferous deposits belonging to these assumed earliest periods prior to the Cambrian system," says Darwin, "I can give no satisfactory answer." Nor can we give today any fully satisfactory answer, although some signs of pre-Cambrian life unknown to Darwin have been discovered, and although a number of palæontologists have devoted much thought to the question. Darwin advanced a hypothesis that in pre-Cambrian days the world "may have presented a different aspect, and that the older continents, formed of formations older than any known to us, exist now only as remnants in a metamorphosed condition, or lie still buried under the ocean". This hypothesis is none too convincing. Later workers have made various additional suggestions toward a solution of the problem. But even today we have not completely solved this greatest of remaining palæontological puzzles.

FOSSILS AS EVIDENCE FOR EVOLUTION

In the chapter just reviewed, Darwin was strictly on the defensive, parrying objections, based on palæontology, to the theory of evolution. In the second of the two chapters dealing with the history of life, "On the geological succession of organic beings", he advances from a negative position toward a positive one. He attempts to demonstrate that, even if in his day the fossil record could hardly be used as strong proof that evolution had occurred, it could at least be interpreted as satisfactorily, or even more satisfactorily, in evolutionary terms than under the hypothesis of special creations. But while his refutation of palæontological objections to his theory was very ably done—particularly in view of the inadequacy in his day of the fossil record—this second chapter dealing with palæontology is far from convincing, and seems to me to be perhaps the weakest link in the admirable chain of arguments that comprise the *Origin*.

Under the Darwinian theory, one would expect to find that new species would come in slowly and successively; that species of different classes would not necessarily change together, or at the same rate, or in the same degree, but that, in the long run all would undergo modification to some extent. In agreement, says Darwin, are the known facts of the fossil record. The older picture of the past history of life (that it had consisted of a few major faunas and floras, following one another in time and each separated from its successor by a major revolution with complete extinction and appearance of a total new creation) had of necessity been abandoned with increased knowledge of stratigraphy by even the most conservative of geologists and palæontologists; and by Darwin's day it was seen that there had been a long succession of faunas and floras. The content of each differed from those preceding and following, but in some cases the differences were not major; many types continue through a series of formations, their later representatives differing in a progressive fashion, though to a variable degree, from the earlier ones as we pass upward. In general animal and plant groups are of a modest and restricted nature when they first appear, later become more varied, and then frequently diminish in abundance, rapidly or slowly, and may finally become extinct. Species or groups once extinct never reappear subsequently; this is, of course, the only result to be expected under the evolutionary hypothesis, but would not necessarily be expected on the theory of special creation. The whole picture presented is, as Darwin says, in harmony with his ideas of evolutionary progression, and is much more difficult to interpret on the basis of special creation.

As noted above, the idea of catastrophic extinction had been abandoned by his day, but the reasons for the extinction of species or groups of animals and plants were obscure (and to a very considerable extent, are still obscure today). Darwin points out that acceptance of the evolutionary doctrine gives at least a partial explanation. The evolution of new forms can hardly be accomplished without the extinction of older ones. If a new species has evolved from an older one, the more successful characteristics which have been responsible for its origin through natural selection may lead to its dominance over the parent type and to that form's extinction. Further, its new and advantageous characteristics may enable it to compete successfully not merely with its own close relatives but also with members of other groups which lead similar lives, and thus cause the extinction of a quite unrelated form.

Related to the question of extinction is a phenomenon noted in the marine Tertiary record by a number of invertebrate palæontologists—the fact that successive faunal changes appear to take place simultaneously (or nearly simultaneously) over large geographical areas. This, says Darwin, agrees well with expectations from the theory of natural selection. New species presumably evolve (as said above) in a restricted area; but once evolved as dominant forms in one area, the advantages they have acquired would tend to make them dominant in a rapid fashion over broad areas, provided there were no major geographic or ecological barriers to prevent their conquest.

In a discussion of the affinities of extinct forms to each other and to living forms, Darwin overlaps part of his argument in the preceding chapter. There he had pointed out that the general pattern of distribution of known fossil forms was not in conflict with evolutionary interpretation. Here he attempts to go farther, and to demonstrate that the evolutionary interpretation better fits the facts. It aids us, he says, to understand how it is that all known forms of life, ancient and recent, make together a few grand classes, and to understand why the more ancient a form is, the more it generally differs from those now living and often tends to bring closer together two groups now quite distinct. Only on an evolutionary basis is there any necessary reason why the organic remains from any two temporally adjacent periods are more closely allied to each other than to those of other periods, or why in a sequence of successive formations the fauna of a given formation need be—as it is—intermediate in character between that which precedes and that which follows. It seemed clear to many palæontologists of Darwin's day, even if they were special creationists, that the organization of living things had, on the whole, progressed through geological time; this progression is best interpreted on the basis of evolutionary advances. And still further, in at least the later stages of the Tertiary it is known quite definitely that within the same continental areas the same structural types succeed one another without radical change; modern Australian marsupials follow allied although specifically different Pleistocene marsupials in Australia, and not elsewhere; living "edentates" succeed similar Pleistocene types in South America, but not on other continents. Such consistency is requisite only on an evolutionary basis.

Darwin thus attempts to show in this chapter that the palæontological picture is not merely compatible with an evolutionary basis, but is more readily interpretable on this basis than on one of special creation. His arguments are persuasive, but one may suspect that a special creationist would, in his day, have been able to resist them. To be sure, the creationist had been forced to abandon the idea of a few successive universal creations and destructions, but he could continue to believe in an essentially continuous series rather than continuous evolution of species. And as for the pattern of progress and differentiation which was beginning in those days to appear in the known fossil record, he could reasonably argue that in successive creations the creator—nature or a deity—might well follow a logical course, that new species created would tend to be shaped along the general lines of the old, and that the general progress seen in successive periods might be compared to the picture seen in the development of modern machines and appliances, where the " creations " of successive years are of a more and more advanced nature, even though no refrigerator or automobile gives birth to its successor.

But despite resistance, a gradual and eventually complete conversion of the palæontologists did take place. Partly this may have been due to the general reasonableness of the evolutionary point of view; in considerable measure, we may suspect, to the constantly enlarging vista of the past rendered possible by further fossil finds. This conversion was well under way during Darwin's lifetime. In the first edition of the *Origin* he remarks that " the most eminent palæontologists, namely Cuvier, Agassiz, Barrande, Pictet, Falconer, E. Forbes, etc., and all our greatest geologists, as Lyell, Murchison, Sedgwick, and so on, have unanimously, often vehemently, maintained the immutability of species." And, in favour of his own point of view, he could say only that his close friend and confidant, Lyell, " from further reflection entertains grave doubts on this subject ". Years later, in the final edition of the *Origin,* the first of these two statements remains unmodified. But for the second, he was able to substitute (no doubt with quiet satisfaction); " But Sir Charles Lyell now gives the support of his high authority to the opposite side, and most geologists and palæontologists are much shaken in their former belief."

7

Darwin and Embryology

By

GAVIN DE BEER

THE first to appreciate the bearing of the facts of embryology on the problem of evolution was, of course, Darwin, and the realization that they had such a bearing was his own discovery. It is well known that Darwin owed a general debt to Sir Charles Lyell for the principle of uniformity in the history of the earth and its inhabitants, first propounded by him in 1830: that the causes of events in the past are to be sought among natural factors known to be still in operation, without appealing to catastrophies. As T. H. Huxley later remarked [4], the principle of uniformity when applied to the realms of life was bound to lead to the theory of evolution. It is little realized, however, that on the particular issue of evolution and the significance of the evidence provided by palæontology and embryology, Darwin derived no help from Lyell at all. On the contrary, in his classic work, the *Principles of Geology*, published in 1832, Lyell rejected evolution largely because of the unacceptable nature of Jean-Baptiste de Lamarck's attempt to explain it by supposing that the organisms' "inner feelings" and unconscious strivings resulted in the satisfaction of their needs. In his great work published in 1809, Lamarck[1] had decided that there was no difference in principle between species and varieties, that species had undergone change, and that evolution had occurred. As he provided no inductive evidence in support of his views, and his explanation of the causes of evolution was not only unacceptable for animals but totally inapplicable to plants, Lyell and many others threw the baby of evolution out with the bathwater.

Lyell also rejected the progressive nature of the pageant of

[1] [References other than those to Darwin and Huxley will be found in G. de Beer [3] and a few others which are given in the bibliography of this chapter.—ED.

the fossil record " from the simplest to the most perfect forms " because fish were found in the Silurian period, at that time regarded as the earliest fossiliferous formation, and dicotyledonous flowering plants in the Carboniferous. Representatives of the highest plants and the highest animals *seemed* to have existed from the start of life.

Finally, Lyell rejected the possible parallelism which F. Tiedemann in 1816 and P. M. T. de Serres in 1824 had suggested between the successive rungs on the scale of vertebrates and the successive stages through which a mammalian fœtus passes during its embryonic development. These considerations, wrote Lyell, " lend no support whatever to the notion of a gradual transmutation of one species into another, least of all to the passage, in the course of many generations, from an animal of a more simple to one of a more complex structure ". It is salutary to ponder over these statements, for they emphasize the difficulty of Darwin's task, the relevance and accuracy of his observations, the extent of his powers of critical judgment, and the magnitude of his achievement.

Darwin was of course aware of the suggested parallelism between the scale of beings and the stages of embryonic development, but it was only a speculation on which he refused to build, and he required evidence. He found what he was looking for in the principles of comparative embryology enunciated by Karl Ernst von Baer in 1828.

EMBRYOLOGY AND EVOLUTION

The so-called " laws " of von Baer state that the younger the embryos of different members of the same group of animals are, the more closely they resemble each other; and that they become progressively different as they grow older. For Darwin the important point in this " law " was the fact of resemblance; for just as the unity of type presented by the skeleton of the forelimb of bats, horses, porpoises, and man provides evidence of affinity and community of descent, so does the resemblance between the rudiments of those structures in early embryos, and the general resemblance between early embryos of fish, birds and mammals. In his *Sketch* [1] written in 1842, Darwin used this argument to show the importance of embryology in determining the affinities of organisms.

> In early stages the wing of bat, hoof, hand, paddle are not to be distinguished. . . . This similarity at the earliest stage is remarkably shown in the course of the arteries which become greatly altered, as fœtus advances in life and assumes the widely different course and number which characterizes full-grown fish and mammals.

The basic reason for the similarity between the embryos was their " genetic affinity ", which meant that they were descended by evolution from a common ancestor.

> The less difference of fœtus—this has obvious meaning on this view: otherwise how strange that a horse, a man, a bat, should at one time of life have arteries running in a manner which is only intelligible in a fish! The natural system being on theory genealogical, we can at once see why fœtus retaining traces of ancestral form is of the highest value in classification.

Without genetic affinity there could of course be no similarity between embryos or unity of type in a group. There must, however, be a complementary reason for the greater similarity found between embryos than between adults, because adults have the same hereditary endowment as their embryos. The adults have become different from one another; why have the embryos not also become different? Darwin's answer to this question was that natural selection and preservation of variants most efficiently adapted to their environment applied to adults in their various habits of life much more than to the embryos in the uniform conditions inside their membrances, shells, and the womb of their mothers.

A further point of great importance arises here. Darwin regarded the embryo of a chick or a man as in some measure representative of the ancestral vertebrate plan of structure, and in this he was doubtless right, because in this group of animals the plan of structure is laid down very early during individual development and is retained throughout life. This applies particularly to the arrangement of the arteries, which run between the gill-pouches on each side of the throat. No question arises yet of the chick or human embryo representing the *adult* ancestral fish. The laws of von Baer state that the embryos of animals higher up the scale do not resemble the adult forms of animals lower down, but resemble their embryos. Darwin accepted this point of view in his *Sketch* of 1842:

> It is not true that one passes through the form of a lower group, though no doubt fish more nearly related to fœtal state. They pass through the same phases, but some, generally the higher groups, are further metamorphosed [1].

These words are as valid today as when they were written; embryos do not represent or " recapitulate " the *adult* stages of their ancestors, they repeat the *embryonic* stages of their ancestors, to varying degrees. Sometimes, as in the vertebrates, the embryo may indicate the general plan of structure of the ancestor; but this cannot be generalized since the trochophore larva of marine worms (annelids) can no more represent the plan of structure of the ancestral annelid than the nauplius larva can represent that of shrimps (Crustacea), because these larvæ have few organs and their bodies are not divided into many segments, whereas the ancestral adult annelids and Crustacea must have had bodies of many segments, even more numerous than they have now. Furthermore, as will be seen below, there are animals like the insects where the larval forms cannot represent the ancestral forms at all.

The next general work to appear on these subjects was a statement made by J. L. R. Agassiz in 1850 of his views on a parallelism between the palæontological succession of fossils, the systematic position of various classes of animals in the hierarchy of classification, and the stages through which animals pass in their embryonic development, in the whole animal kingdom. Agassiz wrote: " In my researches upon fossil fishes I have on several occasions alluded to the resemblance which we notice between the early stages of growth in fishes, and the lower forms of their families in the full-grown state, and also to a similar resemblance between the embryonic forms and the earliest representatives of that class in the oldest geological epochs." It is a pity that with this threefold parallelism before his eyes, Agassiz nevertheless persisted in refusing to accept evolution as the natural explanation of the facts.

On Darwin the effect of Agassiz's views is seen in his classic work, *On the Origin of Species*, where, after giving evidence which proves that resemblance between embryos is a result of their genetic affinity, he turned to the question of the ability of embryos to represent the type of their ancestors.

> As the embryonic state of each species and group of species partially shows us the structure of their less modified ancient progenitors, we can clearly see why ancient and extinct forms of life should resemble the embryos of their descendants—our existing species. Agassiz believes this to be a law of nature [2; 1st ed., p. 449].

Darwin was careful to add that this was a speculation not yet proved : " I must follow Pictet and Huxley in thinking that the truth of this doctrine is very far from proved " (1st ed., p. 338).

In 1864 Fritz Müller published a book which had important consequences, for it was the first major work by a competent biologist to proclaim adherence to Darwin's views on evolution. In this book, however, besides accepting and repeating von Baer's principle of embryonic resemblance and progressive divergence, Müller went on to suggest another possible relationship between embryonic development and evolutionary descent, in which the descendant in its development might pass through the developmental stages and beyond the final adult stage of its ancestor. This implies that, contrary to von Baer's " laws ", the descendant in its development might not diverge from the developmental course of its ancestors, but copy it rigidly to the ancestral adult stage, and then build a new adult stage as it were on its roof.

In the fourth (1866) and subsequent editions of the *Origin*, Darwin made use of Müller's suggestion saying, " it is probable that at some remote period an independent adult animal, resembling the Nauplius, existed, and subsequently produced, along several divergent lines of descent, the above-named Crustacean groups ". As already mentioned, this conclusion is unacceptable because Crustacea were certainly descended from multi-segmented ancestors like the trilobites in which the number of segments was even greater; whereas the nauplius with its three segments is, like many other larvæ, a recently evolved larval form, developing into a more or less sedentary adult, and specially adapted to secure wide dispersal in the sea;

it therefore cannot have represented any ancestral form at all. It must be stressed that this additional concept, that the embryo might represent the *adult* ancestor, is quite independent of the general proposition that resemblance between embryos indicates their affinity and descent by evolution from a common ancestral form, whatever that form may have been like in the adult state.

Furthermore, Darwin was not prepared to generalize on the past evolutionary significance of early stages of development, because " the various larval and pupal stages of insects have thus been acquired through adaptation, and not from inheritance from some ancient form." In other words, Darwin recognized that in the insects the larval forms not only did not represent the adult ancestors, but did not even reflect the ancestral plan of structure.

A FALLACY AND ITS CONSEQUENCES

Presently the subject was taken up by E. Haeckel who, in 1866, carried away by his enthusiasm for Darwin's work, elaborated Agassiz' threefold parallelism between the palæontological record, the ladder of living animals, and the stages of embryonic development, by instilling into it the principle of evolution and genetic affinity. In Haeckel's hands, the old " laws " of embryonic resemblance of von Baer became converted into the " biogenetic law " or " theory of recapitulation " which held that the stages of the embryonic development of a descendant are not only a recapitulation of a series of *adult* ancestral stages in the evolutionary history, but that the events of embryology are " mechanically caused " by those of evolution.

By these speculations, for which he provided no critical evidence, Haeckel discarded von Baer's principle of progressive divergence during development and denied his " law " that descendants did *not* represent the adult of the embryos of lower forms, but represented their embryos. Instead, Haeckel admitted only Müller's second principle, of the descendant passing through all the developmental stages of the ancestor and beyond the ancestor's adult stage.

The original unicellular protozoon-like ancestor was supposed to be represented by the egg in all animals. Basing himself on the fact that, in the embryonic development of most multicellular animals, there is a many-celled stage resembling a single-layered hollow ball (a blastula), and that this is converted into a two-layered hollow sack with an open mouth (a gastrula), Haeckel in 1874 invented corresponding hypothetical stages in evolution: the " blastæa " and the " gastræa ", which were supposed to bridge the gap between the original unicellular protozoon ancestors and the jelly-fish and sea-anemones (cœlenterates) which he regarded as the most primitive metazoa.

In addition, Haeckel claimed that the various special structures often found in young stages of development and characteristic of larvæ, had been intercalated as so-called " cænogenetic adaptations " without any past evolutionary significance at all. This view is still acceptable today, but it requires to be extended by admitting the possibility that cænogenetic characters of larvæ may have significance for *subsequent* evolution.

Seldom has an assertion like that of Haeckel's " theory of recapitulation ", facile, tidy, and plausible, widely accepted without critical examination, done so much harm to science. The puerile notion that the past evolution of a race was sufficient to explain the mechanism of the processes of embryonic development of descendants, blinded biologists for a time to the necessity of introducing experimental methods into embryological research. If evolution (phylogeny) really was the " mechanical cause " of embryology (ontogeny), there was no incentive to discover the causes of cell-division, the bringing into position of embryonic materials, the formation of organs, and the differentiation of tissues. Thanks to Wilhelm His in 1874 and to Wilhelm Roux in 1890 this fog was successfully dissipated; the progress of experimental embryology which resulted has been a scientific achievement that need not be considered here.

The other aspect of Haeckel's " theory of recapitulation ", however, took a long time to expose. At first it appeared to triumph when A. Hyatt in 1872 and L. Würtenberger in 1880 published beautiful series of fossil ammonites; these were organisms with spiral shells that were continually added to during life. They arranged them in sequences which showed that the features of the outer (and therefore adult) whorls of the shells of the descendants were represented in inner (and therefore earlier-formed) whorls of the shells of the ancestors. Here, it seemed, was proof of " recapitulation ". So seductive did this picture appear that some years were to go by before A. Pavlov in 1901 showed that, if ammonite shells are arranged in such a sequence, the *stratigraphical order of the geological succession has to be reversed.* In other words, Würtenberger's and Hyatt's series falsified the evidence and were utterly valueless. Furthermore, Pavlov discovered, not only that ammonites did not recapitulate their adult ancestors in their embryonic development, but that on the contrary they showed instances where the *adult* descendant reflected characters of the *young* ancestor.

These conclusions were independently and completely confirmed by L. F. Spath in 1938, and they are of such importance that it is worth while to consider some actual examples from series of ammonites that have been so intensively studied, both anatomically and stratigraphically, that there is no doubt about their sequence and succession. *Liparoceras cheltiense* is an ammonite found in Lower Jurassic strata, with simple plain thick whorls. Above it, and therefore later in time, there is found another ammonite of similar general form, *Androgynoceras sparsicosta*, but its shell is more slender, and its inner whorls show the presence of fine ribs. Above this another ammonite appears, *Androgynoceras henleyi*, in which only the outermost whorls are smooth. Above this again and still more recent in time, yet another ammonite, *Androgynoceras latæcosta*, appears in which all the whorls are ribbed.

The evolutionary novelty of the ribs therefore first appeared in the young stage of the ancestor *Androgynoceras sparsicosta*, and in the descendant *Androgynoceras latæcosta* it was retained and prolonged into the adult. The ribs in the young of *A. sparsicosta* therefore reflect no adult ancestral stage at all, since *L. cheltiense* had no ribs, but they are *prophetic* of the subsequent evolution into *A. latæcosta*. This situation is the exact opposite of " recapitulation ". Further examples of " prophetic fossils " have been given elsewhere by G. de Beer and W. E. Swinton in a separate publication.

Meanwhile, on theoretical as well as observational grounds,

the " theory of recapitulation " was seriously attacked by C. H. Hurst in 1893 and by A. Sedgwick in 1894, both of whom criticized its conceptual foundations and pointed out that Haeckel had neglected the evidence in formulating his theory. For all their similarities when young, embryos are always

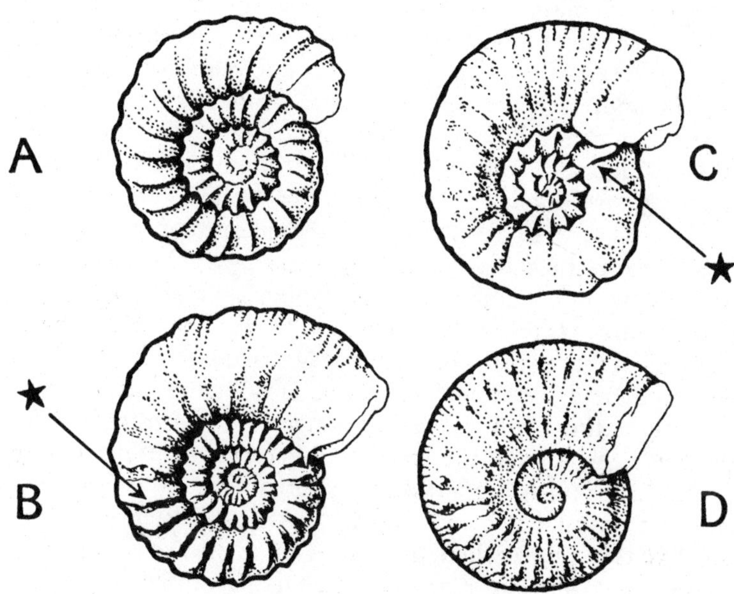

30.—Pædomorphosis in Ammonites.

Diagram showing the inception of the evolutionary novelty of ribbing in the Liparoceratidæ. In the lowest and oldest form, *Liparoceras cheltiense* (D), there are no ribs. In the succeeding form *Androgynoceras sparsicosara* (C), ribs appear on the innermost whorls representing early stages of development, as far as the whorl marked with the asterisk. In the next form, *Androgynoceras henleyi* (B), the development of ribs affects still more of the whorls as far as the asterisk, only the outermost, developed in the adult stage, being devoid of them. In the final form, *Androgynoceras latæcosta* (A), all the whorls are ribbed.

The evolutionary novelty of ribbing first appears in the young of the ancestor and becomes prolonged into the adult stage of the descendant.

distinguishable, and even the eggs of different species are distinct ; to claim that the embryo of a mammal resembles the adult fish in such a way as to represent it is a sheer misstatement of facts. The gill-pouches of the mammalian embryo do *not* resemble the gill-slits of the adult fish; they resemble the gill-pouches of the embryo fish. From this stage the fish and the mammal *diverge* in their development: the mammal does not go through and beyond the development of the fish.

Furthermore, it is in many cases quite clear that the sequence of events in embryonic development cannot have been the same as that which took place in evolution. For example, in mammals the tongue develops earlier than the teeth, but in evolution teeth appeared earlier than the tongue. In other words, the stages of embryonic development are not a pile of compressed successive ancestral adult stages.

The last champion of the " theory of recapitulation " was E. W. MacBride, who in 1914 saw in it support for his own views on evolution. If recapitulation were true, and the embryonic stages of development of the descendant represented a sequence of successive adult stages during evolution, then it must follow that evolutionary novelties made their appearance only at adult stages in the ancestors, to become subsequently compressed into earlier and earlier stages of embryonic development in their descendants. This notion of evolutionary novelties originating only at adult stages had attractions for MacBride, who saw the key to the cause of evolution in habitudinal efforts undertaken by organisms to meet their needs, as postulated by Lamarck. " A new step in evolution usually takes place when the adults of a species seek a new environment and in reaction with it have their structure modified." Fallacies are frequently linked.[1]

Darwin, with a greater respect for accuracy, had written in his *Essay* of 1844 that " variation of structure takes place at all times of life, though no doubt far less in amount and seldomer in quite mature life " [1].

It was reserved for Walter Garstang in 1922 to demonstrate the whole extent of the fallacy of the " theory of recapitulation ". He did this by providing evidence derived from an accurate comparison between developmental stages of different Crustacea that during embryonic development an organism does not recapitulate the adult stages of evolutionary history, but repeats the corresponding stages of embryonic development of its ancestors up to the point where divergence takes place. In other words, Garstang replaced on the rails of von Baer the train of thought which had been derailed by Haeckel. In

[1] [See also chapters 1 and 2.—Ed.]

addition, Garstang made two further contributions to this subject. The first was to show conclusively that the many larval forms such as trochophore, veliger, nauplius, or auricularia, are adaptations that provide for the dispersal in seas and freshwater of species whose adult powers of locomotion are limited; they cannot be regarded as ancestral adult relics at all. With as much justification could the winged seeds of sycamore or the pappus seeds of dandelions be regarded as representative of the fully grown ancestral flowering plants. They are in the same position as the insect larvæ which Darwin himself recognized as devoid of past ancestral significance.

PÆDOMORPHOSIS

Garstang later made a second contribution in 1926 which followed naturally from the first. He showed that, when considering the evolutionary history of a group, it is not obligatory to look for its possible derivation only from *adult* forms of other groups, for it may have been derived from the *young* form of another group by retention of its youthful features. To this mode of evolution, where the adult descendant resembles the youthful ancestor, Garstang applied the term *pædomorphosis*. It could never have been admitted as possible during the dogmatic sway of the " theory of recapitulation ", for it allows for the possibility that evolutionary novelties need not always have appeared in the adult stages of ancestors, but may have appeared at early stages of development of ancestors and have been retained in the adult stages of the descendants.

There are two methods by which pædomorphosis may play a part in evolution.[1] The first is by the appearance of an evolutionary novelty in the young, which thereafter affects the young and adults of all descendants. A good illustration of this is provided by the gastropod molluscs, the snails and slugs. It is quite clear, particularly since H. Lemche's discovery in 1957 of the primitive, segmented and symmetrical *Neopilina*, that the original ancestral molluscs were symmetrical [5]. It is equally clear from comparative anatomy and palæontology that the original gastropods were forms in which the body had undergone torsion through 180 degrees, with marked resulting asymmetry affecting the relations of all the organs in the body.

[1] [This topic is discussed also in chapter 5, pp. 105 *et seq.*—Ed.]

It had always been difficult to imagine how such a torsion could have been produced by gradual modification of the adults. In embryonic development the torsion takes place at an early larval stage quite suddenly, and is completed in a few hours, as a result of the mechanical instability of the shape of the larva and the contraction of the muscles developed asymmetrically on one side, as D. R. Crofts showed in 1955. At one stroke the plan of structure of all gastropods is laid down, and must so have been laid down in the original symmetrical ancestor whose larva first underwent this torsion. In that ancestor this evolutionary novelty would have been "cæno-genetic"; that is, it would have represented no stage of any of its own ancestors, but it affected all its descendants. From the moment torsion first appeared, those molluscs that possessed it diverged in their development at an early stage from those that did not. Such a mode of evolution has been called *deviation*. The only peculiarity of the gastropods in this respect is that their deviation started at a very early stage of larval development with the result that it is an ancestral larval character, torsion, that has been retained (in whole or in part) by all gastropods.

The other method by which pædomorphosis may play a part in evolution is by the retention in the adult descendant of the plan of structure of the ancestral larva; the larval arrangement displaces and becomes substituted for the ancestral adult type of structure. This may be illustrated from the evolution of the chordates. There have been numerous attempts to derive the chordates from the adult plan of structure of many groups of invertebrates, but all such attempts were far-fetched and none is satisfactory. In particular it was impossible to derive the chordate dorsal tubular nervous system from the paired, solid ventral or lateral nerve cords of any invertebrate. Garstang noticed that the structure of the auricularia larva of echino-derms such as starfishes provided a plan from which the chordates could simply be derived. The auricularia's ciliated band has exactly the same relations to the body as the neural folds of the chordate embryo; its general shape is very similar to that of the tornaria larva of the acorn-worms (*Balanoglossus*) which are undoubted chordates with dorsal neural tube and gill slits; its three pairs of body cavities are similar to those in the tornaria; and it has a band of cilia near its mouth in the same position as the endostyle of *Amphioxus* and of the ammocœte larva of the lamprey. All that is required to convert the auricularia larva of the echinoderm into a typical chordate is the perforation of gill-slits; even *Balanoglossus* has no notochord.

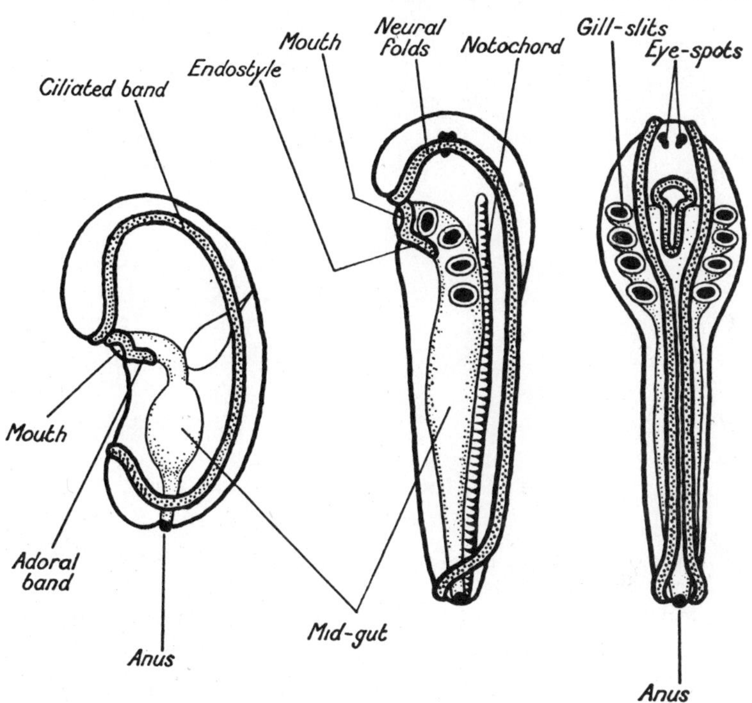

Starfish larva Two views of idealised primitive chordate

31.—The pædomorphic origin of Chordates.

Diagram showing the basic similarity between an echinoderm (starfish) larva and an idealized primitive chordate. The correspondence will be noted between the ciliated band and the neural folds and their relation to the mouth and the anus; the adoral band and the endostyle. (After Garstang.)

That the echinoderms had an affinity of some sort with chordates had been known from the facts that in both these groups the skeleton is internal and mesodermal, and that some muscles of echinoderms show a biochemical similarity to those of chordates in their creatine phosphate mechanism. Nobody would dream of deriving the chordates from adult starfish or sea-urchins; but if a strain of starfish descendants retained their larval plan of structure, elaborated it and became adult in this condition, chordates could result. From the classic analogy of the newt-like axolotl which is neotenous, that is, becomes sexually mature in the larval stage, the mode of evolution involving the retention in the adult of the larval organization of the ancestor is known as *neoteny*.

The insects are another group which may be regarded as having evolved by pædomorphosis and neoteny. That the insects have affinity with centipedes and millipedes (the myriapods) is well known; but some myriapods when young have a stage at which there are three pairs of walking legs and a short " abdomen " with rudimentary legs. In later development the myriapods add on to their hind ends the large number of segments, each with functional legs, that characterize them. The insects may have evolved from the young 6-legged stage of the myriapods by retaining this youthful form in the adult. Some primitive insects even show vestigial legs on the segments of the abdomen, thereby presenting even stronger resemblance to the young myriapods.

As another example, the evolution of man was the subject of an illuminating analysis by L. Bolk who in 1926 showed that in many morphological features, such as the absence of brow-ridges, presence of a chin, dentition, position of the foramen magnum, delayed fusion of the bones of the skull, and the axis of the vagina, adult man resembles the youthful anthropoid; and that these features could be explained by the retention into the adult stage of man of characteristics which are youthful and transient in the anthropoid.

Modern man may have been descended from ancestral man by pædomorphosis. The application of the principle of pædomorphosis to the evolution of man has been all the more fruitful since the discovery of juvenile as well as adult material, including skulls of *Australopithecus*, *Pithecanthropus*, and Neanderthal man, which has provided further examples of juvenile forms from some one of which *Homo sapiens* may have been derived. It was even shown by N. J. Spuhler in 1954 that the milk-teeth of *Australopithecus* resemble those of modern man, whereas its permanent teeth are like those of the gorilla.

Darwin himself may be said to have envisaged the possibility

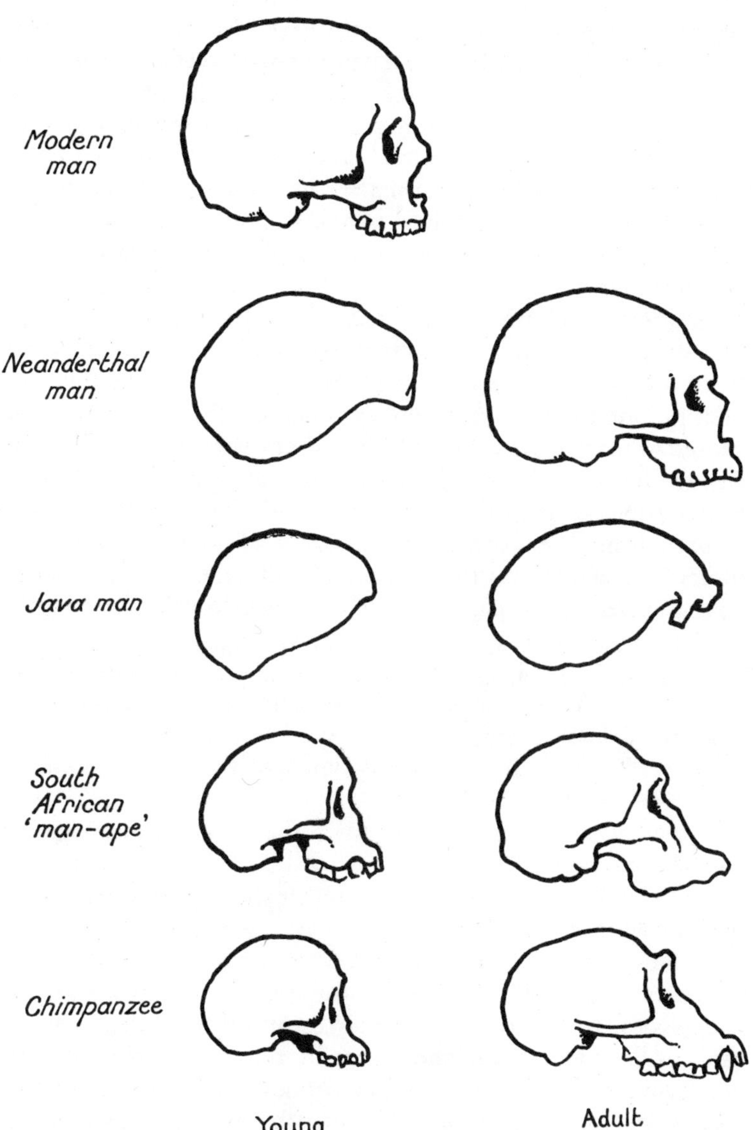

32.—The pædomorphic origin of man.

In the absence of brow-ridges, vertical position of forehead, dome-shaped brain-case, and delayed closure of sutures between the bones, modern man is derivable from the young forms of the preceding series, but not from the adult.

of pædomorphosis. In the first place, he recognized that " variations ", as he termed evolutionary novelties, need not appear only at the adult stage, and that in the offspring they tended to reappear at stages of development corresponding to those at which they originally appeared. His words were: " In a state of nature natural selection will be enabled to act on and modify organic beings at any age, by the accumulation of variations at that stage, and by their inheritance at a corresponding age."

In the second place, with that carefully controlled power of speculation of which he was a master, Darwin showed that, if the life history of a species necessitated the development of adaptations in the young, not only would these young stages represent no " ancient form ", that is, ancestral adult, but " we can see how by change of structure in the young . . . animals might come to pass through stages of development, perfectly distinct from the primordial condition of their adult progenitors." Referring to the remarkable adaptation found in the young stages of the parasitic beetle *Sitaris*, Darwin continued:

> Now, if an insect undergoing transformations like those of the *Sitaris*, were to become the progenitor of a whole new class of insects, the course of development of the new class would be widely different from that of our existing insects; and the first larval stage certainly would not represent the former condition of any adult and ancient form.

In other words, Darwin suggested the possibility of pædomorphic evolution.

On considering the various groups in the evolution of which pædomorphosis is believed to have occurred, it is to be noticed that the importance of the evolutionary result achieved differs in the various cases. Some, like the neotenous newts and the wingless birds (ratites), are specialized and degenerate. In the gastropods, insects and chordates, at the other extreme, the new groups resulting from pædomorphosis are not only of the high systematic value of Class or Phylum, but they have preserved a high capacity for further evolution as is shown by their radiation and subdivision into numerous subgroups which have been extremely successful. It would seem as though " big " evolution of the kind that produced the insects, the chordates, and man, retaining evolutionary plasticity without specialization, was associated with pædomorphosis. A. C. Hardy in 1954 called this process " the escape from specialization ".

In an unexpected manner the theory of pædomorphosis has also been able to contribute towards the solution of a problem in palæontology: the so-called imperfection of the geological record. In cases where evolution has been accomplished by pædomorphosis, the adult descendants resemble the ancestral young, and therefore the characters of the adult descendants first appeared as evolutionary novelties in the ancestral young stages. But these young stages are seldom preserved in the fossil record since they usually lack hard parts. This must have had the result that evolution proceeded as it were clandestinely, until in the descendants the previously youthful stages had been retarded into the adult and, owing to the development of hard parts, capable of fossilization. Such stages would then appear abruptly in the fossil record without visibly preserved antecedents. Gaps in the fossil record would therefore be expected in the early fossil history of groups that have evolved by pædomorphosis. These groups, including the gastropods, insects, and chordates, which are characterized by marked differences from their nearest relatives, are precisely those in whose early history the fossil record shows gaps.

EXPERIMENTAL EMBRYOLOGY

So far in this study, embryology has been considered only in its relation to evolution, since that was the aspect of the subject that presented the greatest interest to Darwin. It is, however, not the only aspect, since embryology has problems of its own, in addition to others which concern its relation to genetics.

Darwin had no possibility of expecting, let alone appreciating, the successful results of research in experimental embryology. Nevertheless, he formulated some of the relevant problems in a remarkably prophetic manner in the questions which he put to himself to explain the " several facts in embryology "; why embryos resemble each other, where in the life-history " variations " appear, and why some animals undergo " more " embryonic development than others. He answered them with the help of two principles: that evolutionary novelties appeared at some stage during development, and that they reappeared in the offspring at corresponding stages of development. These principles are at the centre of the problem of genetical control of development and of developmental physiology.

Turning now to the evidence that can be derived from experimental embryology, one of Haeckel's greatest mistakes was to regard developmental stages as static structures, whereas it is now clear that they are time-cross-sections of dynamic processes. In the course of research in the lower animals extending over half a century, Jovan Hadži in 1944 and later came to conclusions which knock away some of the main props of Haeckel's speculations. Haeckel regarded the two-layered open-mouthed invaginated sack, his " gastræa ", as the adult ancestral form of all multicellular organisms, because the cœlenterates resemble permanent gastrulæ and many higher animals pass through a gastrula stage during development. But on examining the facts, Hadži found that, even in the cœlenterates, the two-layered stage is reached by a variety of methods in different groups: by immigration of cells all round; or by immigration of cells at one point; or by folding in and invagination (which is the least frequent method). In the higher animals the so-called gastrula stage is so inconsistent that it is impossible to define the gastrula as a structure at all. As J. Pasteels showed in 1937, gastrulation is a set of processes whereby the various parts of the embryo are brought into position, and these processes vary in time, direction, and intensity. In other words the " gastrula " is not a suitable object of comparison in different groups.

But that is not all. Haeckel's " gastræa " theory rested on the assumption that the original primitive many-celled animals were hollow two-layered sacks like cœlenterates. Hadži has brought forward reasons to believe that these original primitive metazoa were not hollow sacks at all, but solid, like protozoa subdivided by internal partitions into cells, a condition now illustrated by some of the flatworms (Turbellaria Acœla). There is nothing left of the " gastræa " theory.

Many other instances can be given of processes that take place during development and cannot represent ancestral features. This is particularly evident when, as in the case of the gastrula in cœlenterates, the processes differ in related forms. In some vertebrates the spinal cord arises as a groove which becomes converted into a hollow tube; in others it arises as a solid rod. It is not legitimate to deduce from either of *these* methods what the structure of the adult ancestral spinal cord was, even if comparative anatomy and the structure of the primitive *Amphioxus* (in which it develops in yet a third manner) show that the spinal cord was hollow.

EMBRYOLOGY AND GENETICS

These questions lead on to a consideration of the evidence that modern genetics can bring to bear on the problem of the relations between embryology and evolution. It is now clear that evolution proceeds by the production of variation due to the mutation of genes, and the permutation and recombination of genes brought about by random fertilization of germ cells. The products of this variation are acted upon by selection which preserves those collections of genes, or gene-complexes, that control the development of characters which endow their possessors with advantages in the conditions of their various habitats. This applies to all stages of the life-history including embryonic development. Gene-complexes are remarkably specific for each species and they differ sufficiently from the gene-complexes of their sister-species and, even more important, of their parent species, to impose incompatibility and sterility when they are crossed. Yet the " theory of recapitulation " would require that the members of any species contain not only their own unique gene-complex, but also a library of the gene-complexes of their ancestors, all the way back to the start of life. All the successive ancestral stages were supposed to be telescoped into the life-history, and therefore into the gene-complex of the descendant.

Since the effects of all genes are known to be affected by many or all other genes in the gene-complex, and since the gene-complexes have constantly been changing by mutation and recombination of genes, it is impossible to imagine that the gene-complex of any living organism contains any unchanged " old " genes capable by themselves of controlling the development in the descendant of " stages " representing the finished products of the gene-complexes of the ancestors that were built up under very different conditions of selection.

In so far as the descendants during their embryonic development repeat any portion of the embryonic development of their ancestors, the descendants show that, in the elaboration of their own gene-complexes, they have retained certain patterns indicative of their affinity with and descent from those ancestors, although the actual genes involved in these patterns may have changed during evolution. Even in one and the same species today, genes can be substituted for one another. Some of these gene-complex patterns, such as those that control cleavage of the fertilized egg and the differentiation of the embryo, are so fundamental for the successful outcome of embryonic development that they appear to have a latitude of variation somewhat restricted by selection, although great differences can occur, such as the various methods by which gastrulation is brought about in quite closely related animals. Other features, including vestigial organs, may be dropped, although this may take a very long time. Marsupials have been viviparous at least since *Eodidelphys* lived, seventy-five million years ago. Yet there are traces in embryos of living marsupials of vestiges of the egg-tooth, used in the oviparous ancestors to enable the embryo to hatch out their egg-shells, as J. P. Hill and G. de Beer showed in 1950.

Finally, genes have been shown to exert their effects by controlling the rate of physiological processes. Variation of rate affects the time at which characters appear in embryonic development and provides the mechanism for the retardation of youthful characters involved in pædomorphosis.

In conclusion, therefore, it may be said that modern views on the relation between embryology and evolution are in substantial agreement with the conclusions which Darwin reached in his *Essay* of 1844, and with much of what he wrote in the *Origin*. If from the latter work be subtracted the views which Darwin quoted from Agassiz and Müller and which he accepted only with reserve, it will be seen how objectively grounded were his own conclusions on the significance of embryology for evolution; for they continue, as Darwin said of his principle of natural section, " to guide our speculations ", to suggest hypotheses to be tested, and to provide a framework for the interpretation of results.

8

The Study of Man's Descent

By

WILFRID LE GROS CLARK

THE publication by Charles Darwin of *The Descent of Man* in 1871 was, without doubt, an act of great moral courage. In order to avoid, if possible, the intrusion of emotional prejudice into discussions on his general theme of evolution and natural selection, he had made no more than a passing reference in *On the Origin of Species* to the central problem of man's relationship to the rest of the animal world; he merely indicated, quite incidentally, that by his studies " light would be thrown on the origin of man and his history ". As is well known, the *Origin* aroused very bitter controversies, and it is clear that many of Darwin's severest critics were largely influenced in their views by the obvious implication that man himself owed his origin to a process of evolution rather than to a special act of creation.

In the introduction to the first edition of *The Descent of Man*, Darwin specifically stated that he had for many years been collecting notes on this subject, not with the intention of publishing them, " but rather with the determination not to publish ", as he thought that by so doing he would only add to the prejudices against his general conclusions regarding the evolutionary process. It seems that he was led to change his mind as the result of several circumstances; these were: first, the publication of Haeckel's works on morphology and evolution; secondly, the comparative anatomical studies of T. H. Huxley; and thirdly, the evidence adduced by archæologists and geologists that man had a far greater antiquity than Archbishop Ussher's estimate of 4004 B.C. (his assumed date of the creation). Today we recognize in Haeckel's works many crudities and much far-reaching speculation, while Darwin's statement that Huxley " has conclusively shown that in every single visible character, man differs less from the higher apes than these do from the lower members of the same order of Primates" certainly requires some qualification. But the great quantity of anatomical data which these distinguished biologists had accumulated provided so much detailed evidence of man's relationship to the other Primates that Darwin felt he could rely on the strongest and most influential support when he came to focus attention on man as a product of evolution. The evidence for the antiquity of man was at that time much less impressive, or conclusive, than it is now, and by itself it did not add much to the sum of evidence in favour of man's descent from prehuman ancestors. But it did provide Darwin with a forceful argument for contesting the commonly accepted Biblical account of Adam and Eve, an account widely held even by men of high intellect. We can readily understand, therefore, that he felt the time was ripe for the publication of his notes in the form of a book, even though (as we know from other sources) he certainly felt some qualms about it. That these qualms were justified is made clear by his reference in the preface to the second edition of the book (1874) to "the fiery ordeal" through which the first edition passed.

Since Darwin's day, of course, the evidence relating to human evolution has grown to vast proportions. Comparative anatomical studies have multiplied almost indefinitely the structural details which ally man far more closely to the anthropoid ape family (Pongidæ) than to any other group of Primates, and the fossil evidence now carries back the zoological family to which man belongs (Hominidæ) not only a few thousands of years, but something near a million. No longer do we have to depend for conclusions regarding man's relationships on indirect comparisons of modern man with modern apes, for we now have available many fossil remains of their extinct precursors showing, in this or that feature, an intermediate character which appears to blur, or even to blot out completely, the contrasts seen in their successors today. We are not concerned in this essay with a detailed consideration of all the accumulated evidence. Rather shall we review briefly the nature of a few of the items of the purely morphological evidence adduced by Darwin for his thesis of man's descent, and discuss the validity of his arguments in the light

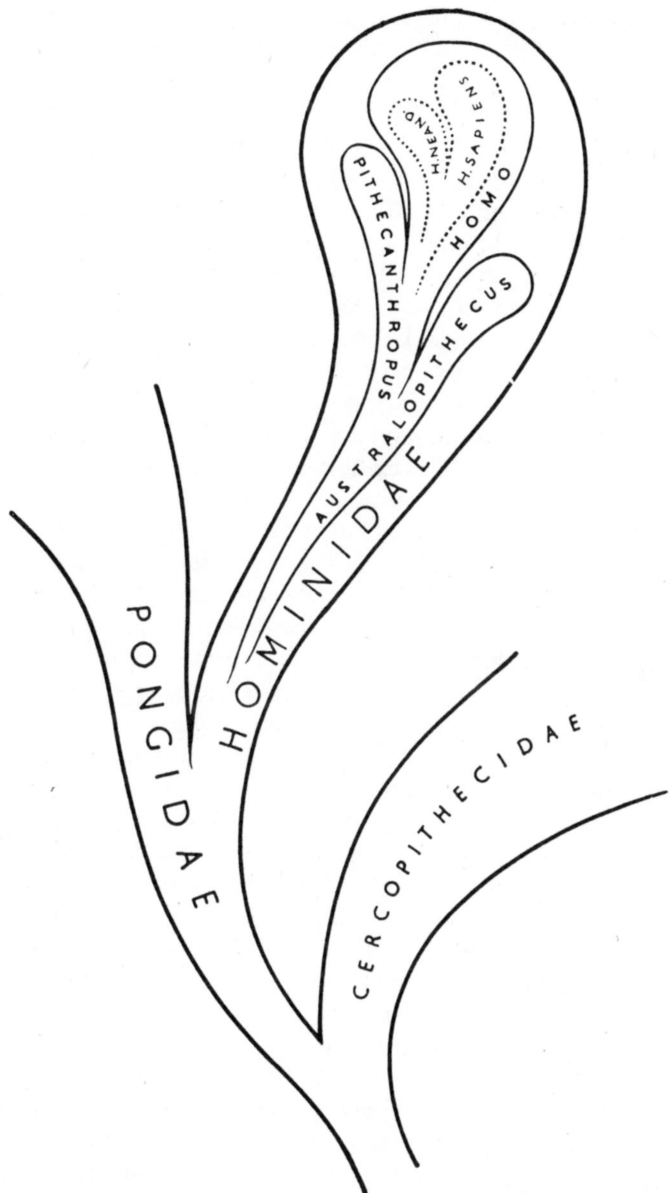

33.—The probable evolutionary relationships of the known genera of the human family (Hominidæ), and their relationship to the ape family (Pongidæ). (This figure and all subsequent figures in the chapter are reproduced by kind permission of the University of Chicago Press from *The Fossil Evidence for Human Evolution* by W. E. Le Gros Clark.)

of knowledge which has accrued over the last eighty years or more.

UNITY OF ANATOMICAL DESIGN

Even the pre-Darwinian and anti-evolutionary biologists recognized, of course, that the human body is constructed on the same general plan as that of other mammals, particularly monkeys and anthropoid apes. But if they gave any serious thought to this seemingly odd circumstance, they regarded such similarities as no more than evidence of a general plan in the mind of the Creator, a basic design to which mankind had perforce to conform like any other living creature in order to fulfil the natural requirements of life on this earth. Darwin examined more searchingly this problem of identical design, and particularly drew attention to the fact that it involved such small and apparently trivial details of structure, in creatures so diverse in their habits of life, as to be difficult of rational explanation except on the supposition that they were inherited from common ancestors before diversity of function had developed. The common basic plan in the structure of the fore-limb in birds, bats, seals and terrestrial animals is a well-recognized example of this phenomenon. So far as man is concerned the detailed correspondence which his anatomical structure shows with that of the lower Primates is perhaps even closer than Darwin in his time had reason to suppose. Not only is it true, as he remarked, that " all the bones in his skeleton can be compared with the corresponding bones in a monkey ", it is actually quite difficult to distinguish some of the bones of a human skeleton from those of a chimpanzee. This particularly applies to the long bones of the upper limb, for example the upper arm bone or humerus. It is true that certain muscular markings and ridges are commonly more strongly developed in the chimpanzee, but this is not always so. In the case of the lower limb bones there are obvious differences in the details of their structure related to functional differences, but even here the resemblances are very striking; the skeletal elements are all strictly equivalent in their basic architecture even though they have become differently modified here and there in their shape and proportions.

Still more striking is the resemblance between certain

individual teeth of the hominid dentition and the corresponding teeth of apes. For example, the molar teeth and the incisors may be well-nigh indistinguishable if comparisons are made with fossil apes in which the dentition had not yet acquired the specializations characteristic of the modern anthropoid apes. The question naturally arises—how far is such a degree of resemblance valid evidence of natural affinity in Darwin's sense? It could be argued that the identity or close similarity of structural details does not necessarily indicate a genetic relationship; they might be adaptive features acquired (or "conferred" by some creative agency) independently to meet similar functional needs. But let us look more closely at this question, with particular reference to the complicated patterns in which the individual cusps of molar teeth are often arranged. The fact is that functional needs are by no means always precisely the same in creatures known to be closely related: they may differ considerably in their normal diet even though they show identical or closely similar cusp patterns. Conversely, different types of animal which are distinguished by corresponding differences in the cusp pattern may yet feed on much the same kind of diet. For example, the different genera of the small lower Primates are readily distinguishable simply by an inspection of the teeth even though some of them have similar habits of life.

In order to emphasize further the validity of the argument that close similarity of anatomical design betokens an evolutionary relationship, reference may be made to what has been termed "the opportunism of evolution". This phrase implies that the evolutionary process by no means requires the development of a character in such a form that it is the absolute acme of perfection, and the only possible form for the function which it is required to serve; there are functions which can be adequately performed by making use of a structural modification in various forms. The salient point in Darwin's exposition of natural selection is that advantage is taken of such heritable variations as occur at random, and as a result of this process something is eventually built up which is good enough for the organism, so far as survival value is concerned. Thus different groups of animals may independently develop structures which are functionally equivalent but which differ considerably in their basic pattern, and the chances are remote that two types which are not closely related will in this way acquire precisely similar patterns. An example which has been used to illustrate the opportunism of evolution is the remarkable variety of shapes of horns among different genera of African antelopes, even among those of comparable size. So far as can be determined the horns serve the same general purpose in most of these genera, but they are by no means identical in shape. The differences are explicable on the basis of independent evolutionary development in each genus. Conversely, an identity of horn shape in different species within the same genus must be taken to indicate a close relationship for the reason that the probabilities are all against the *independent* development, by the accumulation of a sequence of random variations, of identical horns in quite unrelated types.

This principle of opportunism is, of course, equally applicable to the evolution of the cusp patterns of teeth. For example, the basic pattern of the lower molar teeth in hominids is made up of a characteristic group of five cusps with a complex of intervening grooves and associated crests, all disposed in quite a well-defined pattern, and they form an uneven grinding surface by which the food is triturated against the corresponding upper teeth. But it is not to be supposed that the lower molars would be any less efficient for performing their elementary grinding function if they were composed of an equivalent series of similar cusps grouped together in a somewhat different pattern. Yet the basic pattern in hominid molars is identical with that of the large apes, so much so that (as already mentioned) it may be difficult to distinguish the two. The identity of pattern must, therefore, be based on inheritance from a common ancestor, or at least (what amounts to much the same thing) the inheritance of a mechanism conferring a common potentiality to develop the same pattern. This evidence of the dentition is of particular importance because comparative studies over many years have now made it abundantly clear that great reliance can be placed on dental anatomy for the assessment of the systematic affinities of mammals. There are occasional aberrations which can be readily recognized as aberrations, but apart from such cases groups which are known from other evidence to be closely related are found to possess a similar type of dentition with similar cusp patterns. The fact that within a common natural group the dietary habits may vary considerably (although the dentition is constructed on the same basic morphological pattern) is an interesting example of the extent to which, in the process of evolution, the " same " (i.e. " homologous ") structure can be put to varied uses with a functional efficiency which is evidently quite adequate.

A similar phenomenon is made very evident by a comparative study of the muscular system. It is a common source of astonishment to the student of anatomy, when he has made himself acquainted with the muscles of the human body, to find on dissecting one of the lower mammals the same muscles showing the same general disposition and attachments. They may show differences in relative size or complexity, and the precise extent of their attachments may vary in relation to different requirements of leverage, but most can be immediately identified as corresponding to human muscles. The similarity of the muscular system of man to that of the large apes is particularly close. At one time, it was commonly stated that at least one muscle, the peroneus tertius of the foot, is peculiar to man, but more extended studies have shown that, while it is usually absent in the quadrupedal monkeys, it is present with a frequency of 5 per cent in the chimpanzee and 18 per cent in the gorilla, and it is also sometimes absent in man. The function of this muscle is to evert or lift up the outside border of the foot, an action which is related to the position of the sole of the foot in the erect posture. Hence its occasional presence in the anthropoid apes, which are capable of temporarily assuming this posture, is of particular interest. It is also interesting to note that the development of the muscle in man is controlled by a single genetical factor.

In the course of evolutionary change, homologous muscles may come to assume different functions by comparatively slight skeletal modifications, which lead to an alteration in their relationship to the joints whose movements they control. An example of such a functional shift is seen in the gluteus maximus, one of the muscles of the buttock. In the apes and monkeys this muscle is placed on the lateral aspect of the hip joint, so that in contraction it abducts the thigh, that is, draws it outwards. In the evolutionary development of the erect posture and gait characteristic of the Hominidæ, the muscle becomes converted into an extensor of the thigh and, as such, serves an important function in the movements of walking upright. The transformation is brought about by the backward bending of the upper part of the hip bone, the ilium, which gives attachment to the origin of the muscle, so that the latter now comes to be placed more behind than to the side of the hip joint. This mechanical adaptation, which seems to have been a significant factor in the evolution of the Hominidæ, was thus achieved, not by the appearance of a completely new muscular element, but by making use of one which was already a component of the muscular pattern of lower Primates, a pattern which evidently represents a common heritage. Similar transpositions of function are seen in the small foot muscles. In the sole of the human foot exactly those muscles are found which are used for the mobile functions of the ape foot, even though in man this mobility has been lost ; the muscles have abandoned their original functions, but they retain their identity.

It would be possible to enumerate many other examples of structural similarities between man and apes, in some cases extending to the finest details of microscopic anatomy. Data of this kind have accumulated in great quantity since Darwin's time. Altogether they serve to establish unequivocally his thesis of a unity of design between the two groups, and thus provide *prima facie* evidence of a genetical affinity.

Today the main arguments relating to this kind of evidence from comparative anatomy centre on the degree of the affinity which can be inferred from it. No one doubts that, in general, the anatomy of the human body approximates to that of the anthropoid apes more closely than it does to the quadrupedal monkeys. Huxley made the rather different statement that, in his anatomical structure, man differs less from the higher apes than the latter do from the quadrupedal monkeys. This proposition was subsequently labelled by Haeckel the " pithecometra thesis ", an unfortunate term since it might be taken to imply that the degrees of structural affinity between man, anthropoid apes and lower Primates are capable of expression in strictly quantitative terms. But though it is possible to express quantitatively individual measurements considered as isolated abstractions, or even combinations of measurements by the complicated statistical technique of multivariate analysis, it is not possible, with present knowledge, to give a quantitative value to different morphological characters according to their relative importance for the assessment of *degrees* of evolutionary relationship.

It is of considerable importance to recognize the difficulties which are inherent in attempts to apply quantitative methods to problems of this sort, for uncritical applications of these methods have on occasion led to very misleading statements. For example, in making comparisons of one zoological group with another, the *principle of taxonomic relevance* is of essential importance. Some anatomical characters have a much greater relative value than others for the estimation of genetical affinities, and it is of course those characters which are relevant to such an enquiry. As a rather crude illustration, we may refer to the Tasmanian wolf, *Thylacinus*. It would be easy to select a number of dimensional characters of the skull of this animal in which they correspond far more closely to an ordinary dog's skull than they do to (say) a cat's skull. But these particular characters have no taxonomic relevance for comparisons of this sort, since other distinguishing, and much more fundamental, features show at once that the Tasmanian wolf belongs to a totally different group of mammals, the marsupials; and, in fact, the dog and cat are far more closely related to each other than either is to the Tasmanian wolf. This extreme example is cited in order to emphasize a principle which is sometimes overlooked when attempting to assess the affinities of less distantly related types.[1]

Each natural evolving group of animals is defined (on the basis of data mainly derived from comparative anatomy and the study of the fossil record) by the progressive evolutionary development of a certain pattern of morphological characters which its members possess in common, and which are found to be sufficiently distinctive and consistent to distinguish its members from those of other related groups. It was the initial development of this particular pattern which, in a sense, was responsible for the evolution of the natural group which it comes to characterize, and it is this pattern, therefore, which

[1] [See also chapter 5.—ED.]

is relevant in the assessment of affinities. The final vindication of the validity of such morphological criteria no doubt depends on the accumulation of data from the fossil record, and it is perhaps one of the most remarkable sources of confirmation of Darwin's deductions from comparative anatomy that in so many cases the affinities which have been inferred from such indirect evidence have been substantiated by the discovery of the postulated evolutionary links. This is no less true of the higher Primates, including the Hominidæ, than it is with many other groups of mammals.

Reference should be made to the evolutionary phenomena of convergence and parallelism, for it is well known that these can lead to structural similarities which, taken by themselves, may be misleading. The term convergence is applied to the occasional tendency for distantly related types to simulate one another in general proportions or in the development of analogous adaptations in response to similar functional needs. But such similarities are easily distinguishable by a detailed comparative study of the animal as a whole. For example, the resemblance in general appearance, and even in a number of morphological features, of the Tasmanian wolf to a dog does not obscure the fact that in fundamental details of their anatomical construction they belong to quite different mammalian groups. On the other hand, the potentialities of parallelism have been grossly overestimated by some anatomists, who have sought to discount the structural resemblances between man and the anthropoid apes by attributing most (if not all) of them to this process. But to attribute all those similarities which form component elements of a highly complicated total morphological pattern to long-standing parallelism reduces to an absurdity the morphological criteria employed in the assessment of genetical affinities. There is, in fact, no evidence that such a degree of parallelism ever does occur, and there is a good deal of indirect evidence that it does not do so.

The whole question of convergence and parallelism has been discussed in considerable detail in recent years, and particularly by G. G. Simpson, a palæontologist whose long record of experience well qualifies him to adjudicate on it [7]. He points out that the whole basis of parallelism depends on an *initial* similarity of structure and the inheritance of a common potentiality for reproducing similar (or homologous) mutations, and that, this being so, the initial similarity and the homology of mutations themselves imply an evolutionary relationship. Thus " closeness of parallelism tends to be proportional to closeness of affinity ". It is important to recognize this principle, for it has been overlooked (and still is overlooked) by some students of human evolution. Perhaps the most extreme hypothesis of human evolution which has been advanced was that which involved the derivation of *Homo sapiens*, by a long line of evolution presumed to be entirely independent of the evolution of monkeys and apes (and stretching back for some fifty million years), from a tiny creature like the modern tarsier which still inhabits the forests of Borneo. On this hypothesis, every one of the structural resemblances peculiarly common to man and the anthropoid apes was held to be the result of convergence—a conclusion which today seems so fantastic that we may well wonder that it was ever countenanced. But lesser degrees of the misapplication of the principles of convergence and parallelism are by no means infrequent, and indeed appear to be mainly responsible for the diverse conclusions drawn by different anatomists from the same data of comparative anatomy. The fact is that by arbitrarily assuming convergence and parallelism for selected structural resemblances (and incidentally taking no account of the statistical improbabilities at once made evident by a consideration of the genetical factors involved) it is possible for anyone to draw just those conclusions regarding systematic affinities between man and apes which conform most nicely with personal predilections; and the arbitrary nature of such lines of argument is clearly demonstrated by their inherent inconsistencies. It may not always be possible to exclude finally the factor of convergence as an explanation of a similarity in certain individual structural features, but it is not permissible to dismiss a complicated pattern of resemblances (with a complicated genetical basis) as merely the expression of convergence without presenting convincing evidence in support of such an " interpretation ". On the basis of the extensive data of comparative anatomy and palæontology which have now accumulated over many years, it is justifiable to assume that groups of animals which show a preponderance of structural resemblances in taxonomically relevant characters are genetically related to a corresponding degree, unless flagrant discrepancies exist in some basic features such as could be explained only by a long period of independent evolution.

Darwin insisted that, in the assessment of affinities, reliance must not be placed on isolated anatomical characters: " a classification founded on any single character or organ . . . is almost sure to prove unsatisfactory ". Again, " numerous points of resemblance are of much more importance than the amount of similarity or dissimilarity in a few points ". In spite of this warning, it is too often the case that anatomists are tempted to overemphasize the importance of one particular system or one particular organ as an indication of relationship (usually the system or organ which is the subject of their own special study). Reference was made above to the " total morphological pattern ", by which is meant the integrated combination of unitary characters which together make up the complete functional design of a given anatomical structure. It is important to emphasize this concept of *pattern* because the assessment of systematic affinities must be based on the total pattern which characters present in combination, and not on the comparison of individual characters, one by one, treated as isolated abstractions. A fundamental similarity in total morphological pattern may be superficially obscured by a sharp contrast in some individual features, and this has sometimes misled the systematist. But it needs to be recognized that a single gene mutation can lead to quite an abrupt and discontinuous structural change in individual features; such a sharp contrast, superimposed on a fundamentally similar morphological pattern, may thus be the result of an occurrence which, geologically speaking, is quite recent.

We may sum up the implication of the morphological similarities by saying that the intensive studies of comparative anatomy over the last half century make it even more clear than it was in Darwin's time that the Hominidæ are more closely related to the anthropoid ape family (Pongidæ) than to any other group of Primates. This conclusion is expressed in the now generally accepted scheme of classification, which includes both the Hominidæ and Pongidæ in a common superfamily, Hominoidea. The assumption follows that these two families have ultimately been derived from a common ancestral stock by an evolutionary process of divergent modification.

The validity of the morphological criteria on which this assumption is based received remarkable confirmation in 1904, when Nuttall published his classic work on Blood Immunity and Blood Relationship [4]. He studied in great detail in many animal species the precipitin reaction of the blood, an immunity reaction which can be quantitatively assessed. Briefly, a species A which is immunized against the blood of a species B by repeated injections of small quantities of blood serum of the latter produces specific antibodies, or antigens. Then, if the serum of A is mixed with the serum of B in a test-tube, a precipitate is formed. Moreover, a similar reaction is found to occur if the serum of A is mixed with the serum of species known to be closely related to B, and the intensity of the reaction parallels the closeness of the relationship. On the other hand, no reaction is produced if the sera of unrelated or very distantly related species are mixed. It would be difficult to overestimate the importance of the precipitin reaction for the purpose of a natural classification, because, first, it is determined by the biochemical constitution of the protein in the blood—a most fundamental property of the whole organism ; secondly, it provides a means of testing the validity of conclusions based on comparative anatomy ; thirdly, it is an objective test ; and fourthly, the reaction can be expressed in quantitative terms. Lastly, there is good reason to suppose that it may overcome one of the complicating factors that always requires to be considered by the systematist who relies only on morphological criteria—the factor of convergence. In a recent review on systematic serology [2], Boyden states (on the basis of an extensive series of comparisons) that : "As far as serological convergence is concerned, we have as yet no proven case for protein antigens. As further work is undertaken, they may appear, but it is unlikely that they will be frequent." With these considerations in mind, it may be noted that the precipitin reaction indicates a close relationship between man and the large anthropoid apes, a less close relationship with the quadrupedal Old World monkeys, and no more than a distant relationship with the lower Primates. In other words, this serological test confirms precisely the relationships which had already been inferred by Darwin, Huxley, and their contemporaries from purely anatomical data. It has provided the strongest validation of the morphological criteria which are used for assessing affinities, and the final justification for placing the Hominidæ and Pongidæ together in a common taxonomic group, the Hominoidea.

DISTINCTIVE CHARACTERS OF MAN

The opposition to Darwin's thesis of the evolutionary origin of man naturally led his critics to search for anatomical characters in which the human body could be said to be " unique ", thus providing arguments for removing man in any system of classification as far as possible from other mammals (especially the apes). In some cases, indeed, these arguments were pushed to an extreme of absurdity, which today we are apt to find rather astonishing. The celebrated wrangle over the comparatively unimportant feature of the brain called the " hippocampus minor " is a relevant example. This feature, present in the human brain, was said to be absent in the ape's brain, and its absence was assumed to be of great significance. Huxley's easy demonstration that, in fact, it is present also in the chimpanzee's brain was hailed as a great victory for the evolutionists—an interesting commentary on the emotional atmosphere of such discussions at the time, and amusingly satirized by Charles Kingsley in *The Water Babies*.

In the second edition of *The Descent of Man* there is included an appendix by Huxley on the comparative anatomy of the human and ape brain, in which he shows that the two are constructed on the same total morphological pattern and that the difference between the two groups is not greater than those which occur between the large anthropoid apes and the quadrupedal monkeys. These conclusions have certainly stood the test of time, for today it can be safely asserted that, in its gross and also its microscopical structure, the brain of a man has not been found to show any qualitative differences from that of a gorilla: the differences appear to be quantitative only. As we now know, even the quantitative difference is a good deal smaller than was commonly supposed. Thus in gorillas the largest cranial capacity so far recorded is 685 c.c., while the lowest capacity recorded for a human being of "normal" intelligence is less than 900 c.c. The difference between these extremes is therefore no more than about 200 c.c., and the difference in the volume of actual brain tissue is presumably even less, since the cranial capacity includes also the space occupied by the membranes of the brain, the cerebrospinal fluid and blood vessels. It is exceedingly difficult to suppose that such a small addition of brain substance in man could, by itself, account for his vastly superior mental powers, and it seems necessary to assume that the latter depend on some complexity of functional organization which is not reflected in structural organization so far as it has been possible to define this anatomically.

If this is so, obviously it is very hazardous to draw any inferences regarding the mental processes of extinct hominids simply by a consideration of their cranial capacity. Darwin remarks that "no one supposes that the intellect . . . of any two men can be accurately gauged by the cubic content of their skulls", and repeated statistical studies have since shown that this statement still holds true for *Homo sapiens*. To what dimensional limits it is equally true for extinct types is simply not known (except for the very limited inferences which can be drawn from the complexity of associated cultures found with fossil remains, etc.). But the gradations now recognized in the cranial capacity ranging from man to ape are close enough to obliterate any sharp distinctions based on this character alone.

It is opportune here to make reference to one of the most curious, yet insistent, sources of confusion in discussions on human relationships, coming even from authorities who might be expected to be more discriminating; that is, the variable and careless use of the colloquial term "man". Darwin wrote: "In a series of forms graduating insensibly from some ape-like creature to man as he now exists it would be impossible to fix on any definite point when the term 'man' ought to be used. But," he says, "this is a matter of very little importance." In the sense which he intended, he was of course perfectly right. But he could hardly have foreseen the sort of

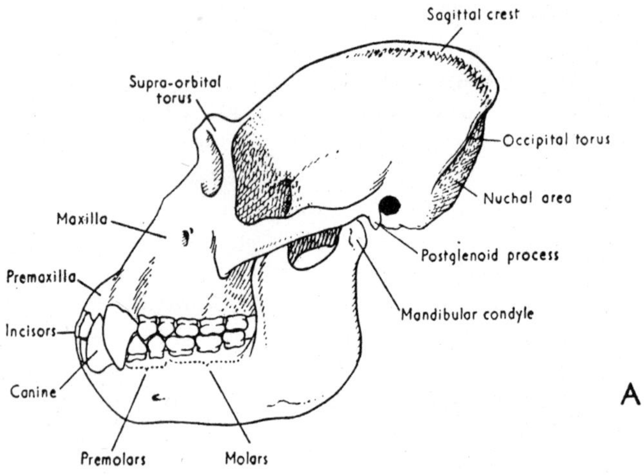

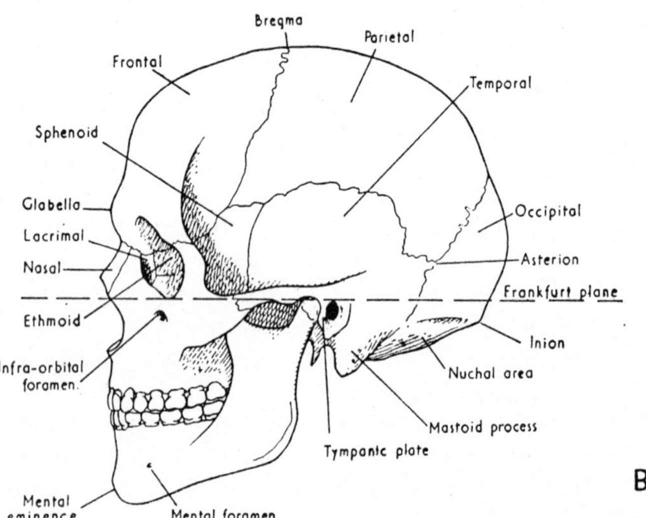

34.—The skull of a male gorilla (above) and of a modern man, to show some of the important anatomical landmarks.

sterile argumentation about the significance of primitive fossil types which has often arisen from the misuse of the colloquial term.

Unfortunately, even today it is by no means uncommon to

read a paper in which (say) the skull of a problematical fossil Primate is compared with a series of European skulls, with the result that a statement is made that in this or that character the fossil type resembles apes rather than "man". All that the author is entitled to say, of course, is that, in those particular characters, it resembles apes rather than the *modern European*

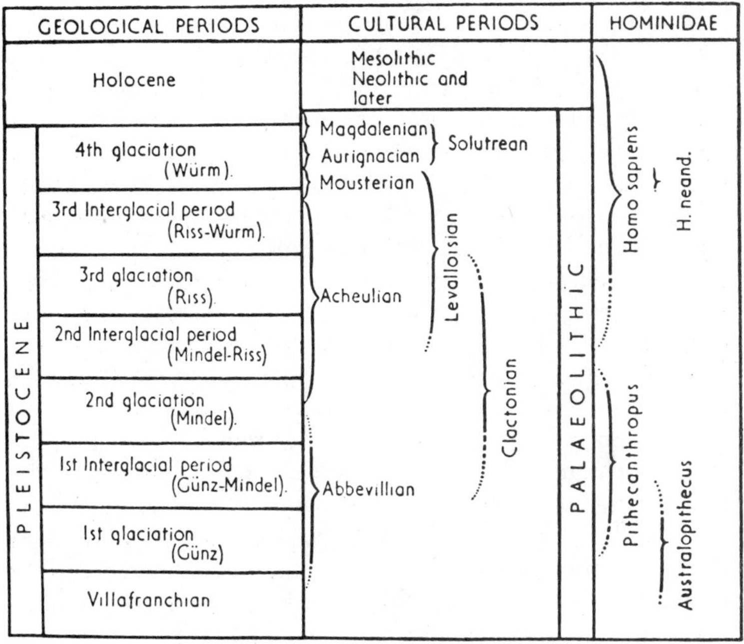

35.—The cultures of the old stone age, and the different human types, in relation to the series of ice ages of the past half a million to a million years.

variety of Homo sapiens; for the term "man" must presumably be taken to include not only all the varieties of today, but also a whole range of extinct and primitive types, such as Neanderthal man and *Pithecanthropus* (Java man), which are themselves, in certain features, also more like apes than the modern European. It may seem surprising that colloquial and unscientific phraseology is still employed among anatomists and anthropologists to the extent that it leads to mis-statements of such a kind, but it is well to beware of this fallacy, particularly when reading expositions of a controversial type.

Except where their meaning is unequivocal, there is now a strong case for eliminating altogether the colloquial terms " man " and " human " in scientific discussions dealing with the evolution or affinities of our own species, if only to exclude as far as possible the emotional prejudices which so often unconsciously obtrude themselves, even in the scientific mind.

In any event, a clear distinction must be made between these terms and the taxonomic term Hominidæ. The latter refers to the zoological family of which *Homo sapiens* is a terminal product, in fact the only terminal product which survives today. In modern taxonomy a zoological family is taken to include the whole sequence of the evolutionary radiation to which it refers (and its various ramifications) from the time when the sequence became initially segregated from collateral radiations having their origin in the same common ancestry. Thus the family Hominidæ includes not only *Homo*, but all those representatives of the evolutionary sequence which finally led to the development of this genus from the time when the sequence first became segregated from the evolutionary sequence of the anthropoid apes. So far as the brain is concerned, therefore, its large size in *Homo* may be taken as a diagnostic character of this genus, but not of the Hominidæ as a whole. For it is perfectly clear that, in the initial stages of the evolution of the hominid sequence (and almost certainly for a long period of time after its segregation as a separate adaptive radiation), the brain could hardly have been bigger than that of the large anthropoid apes. Indeed, the fossil evidence so far available indicates fairly clearly that the hominid brain did not begin to undergo the expansion characteristic of *Homo* until the latter part of lower Pleistocene at the earliest, that is, only about half a million years ago; and that the increase in its size then proceeded with great rapidity.

These points are particularly important because the discovery in South Africa of exceedingly primitive representatives of the Hominidæ, the South African " man-apes " of the genus *Australopithecus*, at first encountered a good deal of scepticism, the argument being that hominids are characterized by a large brain and therefore these extinct creatures could not be hominids. But this is to confuse the taxonomic term Hominidæ (or its adjectival form hominid) with *Homo*. It also involves a

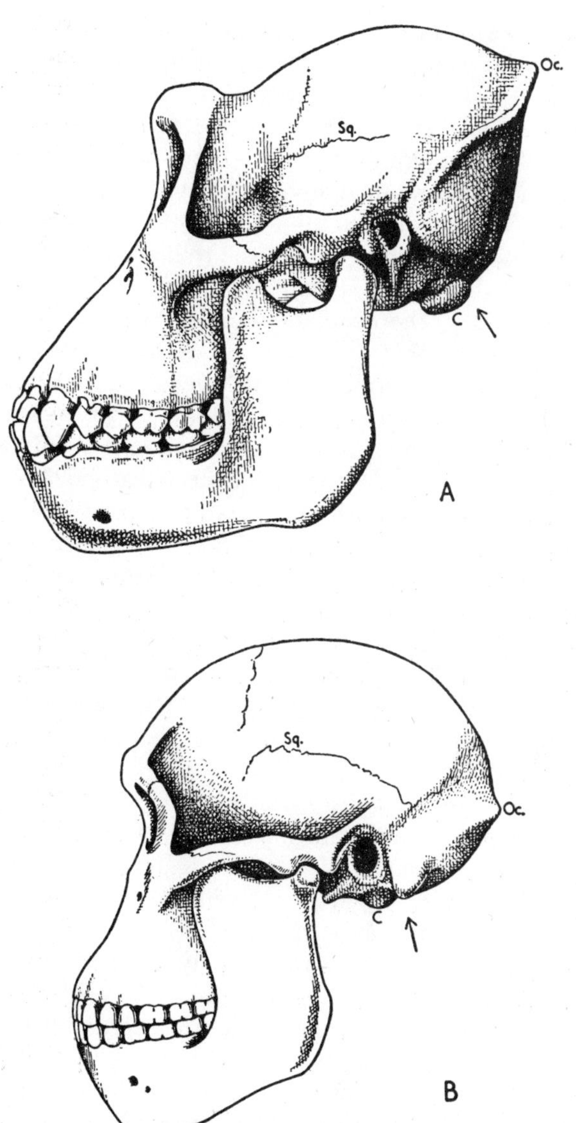

36.—The skull of a female gorilla (above) compared with that of a South African "man-ape" (*Australopithecus*) reconstructed partly by reference to other specimens. The arrow shows the position of the foramen magnum, and indicates the more completely upright stance which must have been displayed by the "man-apes".

fundamental misconception of the definition of familial groups of zoological taxonomy: such groups, in so far as they are taken to represent major adaptive radiations following different evolutionary trends, can be defined only in terms of these trends. The family Equidæ (horse family) includes not only the modern one-toed horse but the whole evolutionary sequence of this group right back to the small *Eohippus* of Eocene times, that is, long before the reduction of the toes had reached the extreme of the genus *Equus*.

Similarly, since the term Hominidæ includes the earliest representatives of this adaptive radiation, it is not to be confined to those terminal products of evolution characterized by a fully expanded brain. The fact that the australopithecine fossils of South Africa represent early hominids is fully attested by the clear evidence that these extinct creatures (in spite of the primitive characters which they still retained in the size of the brain and jaws) had already advanced a considerable way in the direction followed by the hominid sequence of evolution and quite opposite to the direction which was followed by the anthropoid ape sequence [3]. For example, as G. G. Simpson has pointed out, in almost every important feature of the dentition, it conforms to the hominid pattern (and, as we have seen, dental morphology has proved to be of exceptional value for taxonomic determination).

The discovery of *Australopithecus* (as also the earlier discovery of *Pithecanthropus* in Java) has narrowed very considerably the structural gap between man and ape which existed in Darwin's time, particularly in regard to brain size. Thus it is now clear that only in the *later* phases of their evolution can the Hominidæ be said to contrast strongly with the anthropoid apes by the possession of a large brain. Darwin, recognizing that an individual anatomical feature in a common zoological group may sometimes show an unusual degree of development without indicating a corresponding degree of taxonomic distinction, remarks " we must bear in mind the comparative insignificance for classification of the great development of the brain in Man ". Had he been acquainted with the gradations in brain size which are now known to exist between the Hominidæ and the anthropoid apes, he would certainly have felt this statement to be well justified.

The confusion of ideas to which the loose and uncritical use of the colloquial term " man " can lead is particularly well shown in those lists of anatomical features which are from time to time enumerated with the intention of demonstrating his uniqueness in the animal world. For example, it has been

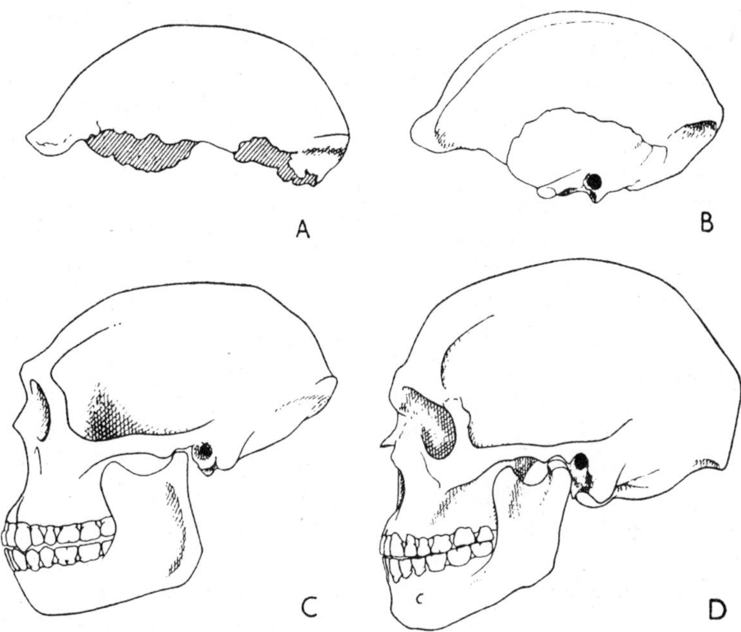

37.—Java man (*Pithecanthropus*) and modern man (Australian aborigine). The two upper drawings show two incomplete skulls; the left hand one below, a reconstruction of a third skull. The chin, reduced prognathism and larger cranium of the modern skull (lower right) are quite evident and make a contrast with Java man.

claimed that " man " is distinguished from all other Primates by such characters as the structure of the genital organs, the prominence of the calf muscles, the red lips, the shape of the female breast, the comparative nakedness of the body, and so forth. It does not always seem to be realized that features of this kind may be no more than distinctions at the specific or generic level—characteristic of the species *Homo sapiens*, and perhaps of the whole genus *Homo*. But, of course, we have no idea whether they were also characteristic of extinct hominids such as Neanderthal man, *Pithecanthropus* and *Australopithecus*. In other words, it is not justifiable to assume that they are characters which distinguish the Hominidæ (as a family) from the Pongidæ. Certainly *Homo sapiens* shows a combination of distinctive features in which he contrasts with other mammals, but this is also true of any other species of mammal. Indeed, it may be said that, anatomically, *Homo sapiens* is unique among mammals only in the sense that every mammalian species is unique among mammals. In this connection, we may refer to another remark of Darwin's (quoted from Mivart) that man " is but one of several exceptional groups of Primates ".

If confusion has been introduced into discussions on hominid evolution by limiting comparisons on one side to *Homo sapiens*, it has sometimes been still further increased by restricting comparisons on the other to the *modern* apes, that is to say, to the specialized terminal products of pongid evolution. For example, it has been argued that the contrasts between the structure and proportions of the limbs of man and apes are so great as to preclude any suggestion that the former could have been derived from the latter by a process of evolution; apes, it was argued, are in these respects far too specialized to provide a basis for the evolutionary development of man. Such arguments have involved the curious assumption that the extinct apes of earlier geological time must have shown aberrant specializations in limb structure comparable with those characteristic of the modern anthropoid apes. In fact, it is now known from the discovery of fossil apes in East Africa that in the early part of the Miocene age there existed primitive anthropoid apes with limbs of a remarkably generalized type (even though they provide evidence of *incipient* changes in the direction of later pongid evolution). It seems not a little remarkable that modern theorists sometimes still overlook the warning which Darwin gave in 1871: " we must not fall into the error of supposing that the early progenitors of the whole Simian stock, including Man, was identical with, or even closely resembled, any existing ape or monkey."

To some it will be surprising that, if we take into account all the known hominids (living and extinct) and all the known pongids (living and extinct), duly recognizing that our information on some of the fossil types is still far from complete, there are only two major distinctive anatomical characters by which the two groups are rather sharply differentiated—the dentition, and the structural adaptations to posture and gait. We have already noted that some of the individual teeth of the dentition may be difficult to distinguish in the two families. Others, however, show a strong contrast, and consideration of the terminal products of their evolution makes it apparent that in these dental characters the Pongidæ and Hominidæ have

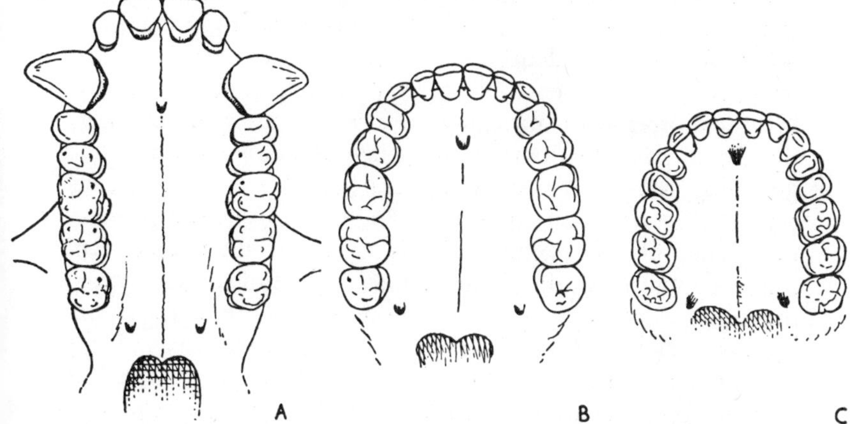

[*By courtesy of the Trustees of the British Museum*]

38.—Palates and teeth of a male gorilla (left), "man-ape" (centre) and Australian aboriginal (right). The middle drawing (of *Australopithecus*) is a composite.

followed different trends which are opposite in direction. Thus, in the evolution of the apes the tendency has been for a progressive enlargement of the incisors (with an associated widening of the front of the jaws), the development of sharp-pointed and powerful, interlocking canines which show a pronounced sexual dimorphism, an enhancement of the sectorial (or cutting-edged) shape of the front lower premolar tooth, and so forth. On the other hand, in the evolutionary development of the Hominidæ the incisors and canine teeth have undergone a progressive reduction, the canines assume a spatulate form approximating in morphology and functions to the incisors and lying in uninterrupted series with them, and the front lower premolar becomes bicuspid and non-sectorial. These and other distinctive features are pronounced in all the known types, and already in the most primitive hominid so far discovered, *Australopithecus*, the total morphological pattern of the dentition (as already noted) conforms fundamentally to that found in more advanced types of the Hominidæ such as *Pithecanthropus*. Thus, so far as dental morphology is concerned, there still remains a distinct gap between the known hominids and the known pongids (even though this gap is by no means so large as might appear by comparing, say, a modern European with a modern gorilla).

Although the palæontological evidence now available indicates that the differential development of the pongid and hominid types of dentition occurred quite early in their evolutionary divergence, it is probable that the structural modifications for a contrasting posture and gait were far more significant as factors determining their initial segregation into diverging adaptive radiations. Indeed, it is now generally agreed that the most important single factor in the primary emergence of the Hominidæ as a separate and independent line of evolutionary development was related to the specialized function of erect bipedal locomotion, and that the other evolutionary trends characteristic of the family were secondarily conditioned thereby. If this is so, it is to be expected that the structural modifications requisite for these functions would be developed in the early hominids long before gross changes affecting other parts of the body, for example, the expansion of the brain and the reduction of the jaws, had become established. The fossil evidence has now confirmed this inference. The mechanical adaptations associated with the development of erect bipedalism affect many parts of the body. They are manifested most obtrusively in the anatomy of the pelvis and lower limb, but they are also reflected in the construction of the base of the skull and the vertebral column. As with the dentition, in the total morphological pattern of its major (and functionally important) features the pelvis of the extinct genus *Australopithecus* conforms to that of the Hominidæ and contrasts abruptly with the anthropoid apes. Since the distinctive features of the hominid type of pelvis are certainly related to the mechanical requirements of posture and gait, it may be inferred that *Australopithecus* had acquired the erect posture and gait so characteristic of the human family. For,

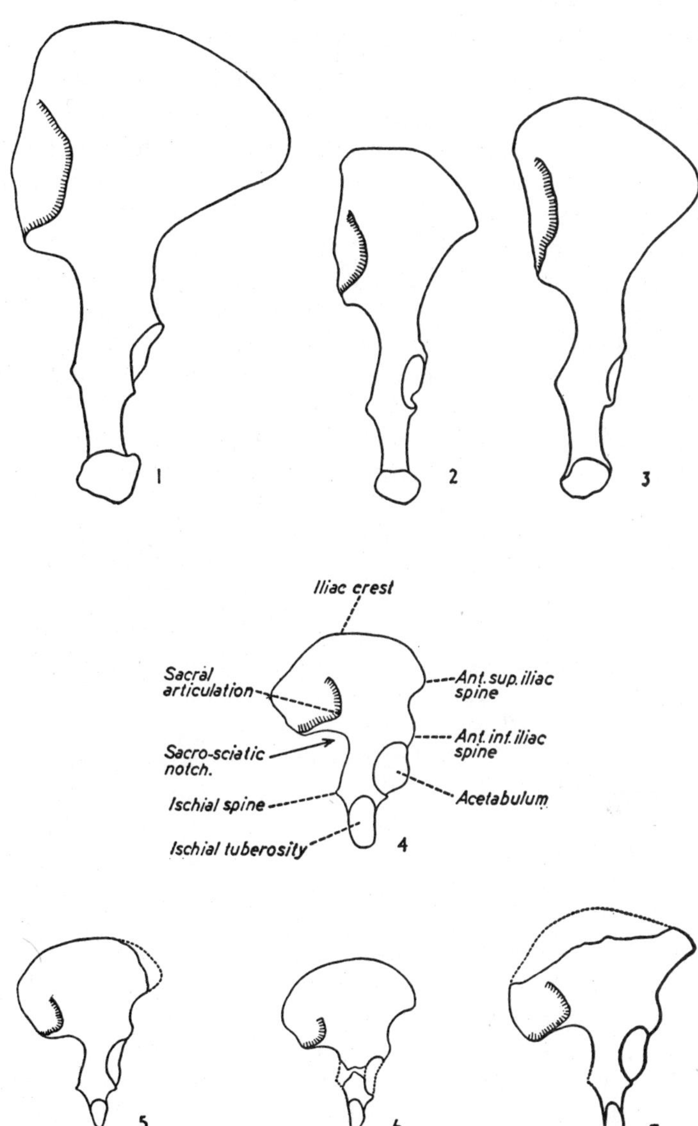

39.—Lateral view of three specimens of australopithecine pelvic bone (5, from Sterkfontein; 6, from Makapansgat; 7, from Swartkrans) compared with those of *Homo sapiens* (4) and the modern large apes (1, gorilla; 2, chimpanzee; 3, orang). The position of the ventral margin area of contact with the sacrum on the inner aspect of the bone has been indicated. Note that in its fundamental characters the australopithecine bone conforms to the hominid pattern. On the other hand, it contrasts strongly with the apes. (This has been confirmed by comparison with 87 pelvic bones of modern apes.)

as recently emphasized by A. H. Schultz (who has made special studies of the pelvis in the higher Primates), " there can be no longer any doubt on the basis of the formation of their hip bones " that these South African creatures of the Early Pleistocene were bipedal, erect Primates [6]. Such an inference receives further corroboration from the associated changes which are also evident in the base of the skull, the lumbar region of the vertebral column, and the lower extremity of the thigh bone. There are various indications, however, that (as of course would be expected in hominids of such a primitive type) the erect posture had not developed in this genus to the degree of perfection found in *Homo sapiens* of today.

The most serious time gap in the fossil record of both the Pongidæ and the Hominidæ now appears to cover the Pliocene period of geological chronology. Unfortunately, fossiliferous deposits of Pliocene age which are likely to yield evidence of this phase of evolution of the higher Primates are rarely to be found, particularly in Africa where, from consideration of the existing fossil record, we might hope to find important transitional types. It seems probable that, when such deposits are discovered, the earlier stages in the evolutionary development of the hominid dentition and the hominid type of pelvis will be demonstrated.

EMBRYOLOGY

The great pioneer of comparative embryology, K. E. von Baer, had published his classic studies on the development of animals as early as 1828, and in *On the Origin of Species* Darwin referred to his observations on the astonishing similarity of the embryos of vertebrates which, when fully developed, are very different indeed. This is expressed in von Baer's formal proposition that during its development an animal departs more and more from the structural organization of other animals.[1] In *The Descent of Man* Darwin referred to the observation that the proposition applies as much to the human embryo as it does to other mammals. Thus a human embryo of about three weeks is very similar to the embryo of a dog or monkey at a corresponding stage of development. The early development of the human embryo also involves a series of

[1] [See chapter 7.—ED.]

transformations during which one type of organization characteristic of lower vertebrates is exchanged for a more advanced type of organization. Such transformations appear sometimes to be almost dramatic, and they seem to have no rational explanation unless we suppose they have reference to evolutionary history. The suggestion that in a modified form ontogeny (embryonic development of the individual) repeats phylogeny (evolution of the type) has been termed " recapitulation ". But it must be emphasized that this does not mean that the successive stages in the embryonic development of an individual in any way represent the *mature* forms of successive stages of its evolutionary history. It means only that, broadly speaking, in his ontogeny the developmental stages through which man passes reproduce the *embryonic* form of certain ancestral types [1]. Thus, it is well known that in the early human embryo a foundation of gill arches is laid down in the neck region, similar to that which, in fishes, finally leads to the establishment of functional gills. But, as in other mammalian embryos, the elements of these gill arches become completely reorganized so as to form, not the gills for which it seems certain they were originally intended in past evolutionary history, but quite different structures such as the framework of the larynx and its muscles, the facial musculature, and so forth. This transformation of the gill arches involves a most remarkable rearrangement of skeletal elements, muscles, nerves and blood vessels, and it seems impossible to explain such a profound modification unless it is supposed that the embryonic basis of the gill arches has been inherited from a remote ancestor of fish-like form. Many other examples of similar kinds of change or replacement could be enumerated, such as the withdrawal of a projecting tail into the floor of the pelvis to form the coccyx, the sequence of events which leads to the partitioning of the heart, and the reorganization of certain groups of nerve cells in the brain. It may be argued that each stage in these transformations provides the essential preliminary for the next stage and that, such being the case, it is not logically necessary to postulate an evolutionary basis. It is true, for example, that certain elements of the gill arch system appear to function as " organizers " during development, determining and controlling the subsequent development of adjacent regions. But the most satisfying explanation of this phenomenon is still to be found in evolutionary terms, i.e. that, since during the course of phylogeny each stage in the sequence of evolutionary development is determined by the preceding stage and itself determines the subsequent stage, the same mechanism still persists as a requisite basis for the embryological development of the individual. Nor are the early stages in a sequence of transformations to be explained by reference to the immediate requirements of the embryo at that time. Darwin makes a point of this when he remarks (in the *Origin of Species*):

> We cannot, for instance, suppose that in the embryos of the vertebrates the peculiar loop-like course of the arteries near the branchial slits are related to similar conditions in the young mammal which is nourished in the womb of its mother, in the egg of the bird which is hatched in a nest, and in the spawn of a frog under water,

and again

> there is no obvious reason why, for instance, the wing of a bat, or the fin of a porpoise, should not have been sketched out with all the parts in proper proportion, as soon as any structure became visible in the embryo.

Perhaps a more striking example is provided by the development of the definitive arterial pattern of the limbs in the human embryo; in the latter the arteries are at first disposed in a pattern similar to that of lower vertebrates, but this subsequently becomes rearranged in a new pattern which is presumed to be functionally more suitable for limbs of human structure. But there is no evidence that the embryonic pattern is determined by the mechanical requirements of the circulation at that particular stage of development, and it happens from time to time that the pattern persists as an abnormality into adult life without any concomitant abnormality of muscles and other associated structures. The fact that in the development of the human embryo the various tissues and organs often do not proceed directly to the patterns of organization adapted to the functional requirements of the mature individual, but follow (as it were) a circuitous route which leads them through stages of development characteristic of lower forms of life, and which may involve the " scrapping " of temporary structures so that the latter can be replaced by structures of a very different pattern—all this provides formidable evidence for the thesis of the evolutionary origin of man. Much more evidence from the study of comparative embryology since Darwin's time has added support to his affirmation that " community in embryonic structure reveals community in descent ".

Apart from the broad generalizations regarding man's place in the animal world which can be drawn from embryological data, there are certain other points of interest in this field of study which have reference to his more immediate relationship to the higher Primates. One of the distinctions between the Hominidæ and Pongidæ which has often been cited for its taxonomic importance is that of body proportions, in particular, the linear dimensions of the limbs. For example, in association with the erect posture, the lower limbs in man are relatively long, whereas in the apes the arboreal mode of life has led to a relative lengthening of the upper limbs. Although such contrasts in body proportions certainly exist between modern men and the modern apes, they have been rather over emphasized by some writers, and (as already noted) there is reason to suppose that in fossil representatives of the two families the contrasts were by no means obtrusive. This general subject has been systematically investigated over many years by A. H. Schultz [5], and the results of his studies have shown, first, that variability in adult limb proportions is actually much greater than commonly supposed, even in the modern anthropoid apes, and, secondly, that there is a much closer resemblance between man and apes in early developmental stages than in adult life. This last observation is clearly relevant to the problem of changing proportions of limbs during evolution, for there is no theoretical difficulty in the supposition that the relative lengths of the limbs in adult forms could be considerably modified by a single mutation of a gene determining their rate of growth during development. Of all the structural modifications which may occur during evolution, those involving the relative dimensions of the main parts of the body are the least difficult to understand in terms of the underlying genetical mechanism.

While in general it may be said that man and the apes diverge in their structural organization more and more during successive stages of their embryonic development, this must not be taken to mean that any feature which shows itself at an early stage must have a corresponding evolutionary antiquity. It is desirable to draw attention to this fallacy,[1] for it has occasionally been assumed from von Baer's propositions (or, rather, from the rather crude derivation from these propositions which Haeckel elaborated as his "biogenetic law") that, since certain hominid characteristics appear very early in human development, therefore the hominid sequence of evolution must have diverged from the pongid sequence at an incredibly remote time. But there are many examples which make it clear that a phylogenetic novelty can sometimes display itself in the very initial stages of ontogeny. The embryo of a Manx cat can be distinguished from the embryo of an ordinary cat during the process of vertebral segmentation in the first few days of development. But it is not to be inferred therefrom that the evolutionary independence of Manx cats and ordinary cats extends back to the initial stages of vertebrate evolution!

RUDIMENTS

Darwin naturally laid great stress on the importance of vestigial structures as evidence for evolution, for, as he pointed out, the common explanations offered by the older naturalists that such vestiges had been created "for the sake of symmetry", or in order to " complete the scheme of nature ", were really no explanations at all. In fact, such rudiments are entirely meaningless unless they are recognized to be relics of some organ or structure which was fully developed in a past ancestry. It may be (and has been) argued that, because it is not possible to ascribe a specific function to a rudimentary structure, it does not necessarily follow that it is completely functionless; indeed, the fact that a relic persists at all has been assumed to imply that it must have some selective value, however slight, or that its development must be in some way

[1] [Also discussed in chapter 7.—Ed.]

correlated with that of another feature which does have a selective advantage. But, in either case, the rudimentary or vestigial nature of a structure which no longer serves its original function is not in question.

Vestigial structures are so well known in the human body that there is no need to enumerate them here. Darwin himself drew attention to certain muscles (like those of the external ear) which are normally present in man and which clearly represent remnants of a more extensive and functional musculature in lower mammals, and also to such features as the nodular remnants of the tail in the coccyx, the lanugo covering of hair in the human fœtus, and so forth. These examples could be multiplied to make an impressive list. No one today doubts

40.—A diagram illustrating the longer and more robust root of the canine tooth in our own species, compared with the roots of neighbouring teeth; this reflects the evolutionary history of the canine, since in our ancestors this was a much larger tooth.

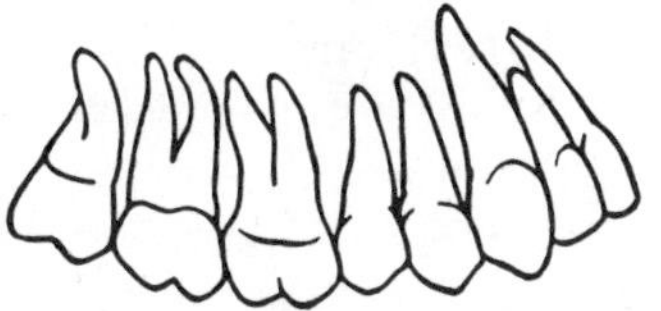

that they are properly regarded as evidence of a past ancestry: we can only be impressed with the remarkable conservatism of morphological elements whereby they persist in modified and apparently functionless form for, it may be, thousands or even millions of years after they have ceased to serve what was originally their main function.

In this connection, reference may be made to the canine tooth because of the abrupt contrast which this shows in modern man as compared with modern apes—so abrupt, indeed, that some have argued that the human type of canine tooth, small and spatulate, could not have been derived by evolutionary modification from a large and sharp-pointed tooth of the ape type. But (apart from the fossil evidence) there is direct morphological evidence in the canine tooth of modern *Homo sapiens* that it has undergone a secondary reduction, and thus in some sense is a rudimentary structure. For example, the newly erupted and unworn canine (particularly the milk canine) may project quite markedly beyond the level of the adjacent teeth and may also occasionally be furnished with a very sharp point. Again, the permanent canine is provided with an unusually robust root and the latter is also longer than that of the immediately adjacent teeth. Such features are difficult, and indeed impossible, to explain on a purely functional basis, for in modern man the canines have no special functions to perform. But they do become intelligible if we suppose that they had special functions in the past similar to those which they serve in the anthropoid apes.

POSTLUDE

We have seen that the morphological evidence advanced by Darwin in *The Descent of Man*, supported by the vast accumulation of knowledge since the publication of this book in 1871, provides a valid basis for the assumption of a natural relationship between man and other Primates, and that this relationship is particularly close with the anthropoid apes. It is also the case that predictions implied in the conclusions drawn from this evidence have been verified in a most remarkable way by the discovery of the actual remains of extinct hominids representing, or closely approximating to, some of the intermediate phases which had been postulated (particularly in the series *Australopithecus—Pithecanthropus—Homo*). In this essay, only the most cursory reference has been made to this palæontological evidence, though it is of quite crucial importance. But we have been concerned rather with the nature of the evidence on which Darwin himself relied, and he could only refer to " the absence of fossil remains seeming to connect man with his ape-like progenitors ". Opinions today still vary on the *degree* of the relationship between the Hominidæ and the Pongidæ, and also on the approximate date in geological time when the two families became separated in the course of their evolutionary divergence. These differences of opinion are in some cases due to an evident misunderstanding of fundamental evolutionary principles, but they are partly due to the fact that the fossil record of hominid and pongid evolution, though year by year it becomes more ample, is still too incomplete. It is well to recognize this fact, for until the documentary evidence of the fossil record has been accumulated in greater quantity, it is not to be expected that there will be general agreement on the details of hominid evolution. Yet, even though there are still conspicuous gaps in the record, it is important, indeed essential, to offer the most probable interpretation of such evidence as is available. Such an interpretation may not be the only possible one, but only by the construction and deliberate formulation of hypotheses is the opportunity given of putting them to the test by further observation and discovery.

At the beginning of this essay we observed that Darwin must have required considerable moral courage to publish *The Descent of Man* at the time he did. Almost ninety years afterwards, though from a somewhat different aspect, the study of human origins still requires a certain degree of moral courage—courage to advance interpretations some of which are quite likely to be proved erroneous by the accession of further evidence, and courage freely to acknowledge errors of interpretation when these come to be exposed.

9

The " Expression of the Emotions "

By

S. A. Barnett

> No emotion [wrote Darwin] is stronger than maternal love; but a mother may feel the deepest love for her helpless infant, and yet not show it by any outward sign; or only by slight caressing movements, with a gentle smile and tender eyes. But let any one intentionally injure her infant, and see what a change! how she starts up with threatening aspect, how her eyes sparkle and her face reddens, how her bosom heaves, nostrils dilate, and heart beats. [7]

Scientists rarely write in this manner today. One of the reasons for this is the much more physiological, even mechanistic, manner in which Darwin's subject is now studied—as may be seen if we examine another famous book in the same field. W. B. Cannon's *Bodily Changes in Pain, Hunger, Fear and Rage* was first published in 1915, forty-three years after *The Expression of the Emotions.* Here is Cannon's conclusion on the biological or functional significance of the processes he describes:

> Every one of the visceral changes that have been noted—the cessation of processes in the alimentary canal . . .; the shifting of blood from the abdominal organs to the organs immediately essential to muscular exertion; the increased vigor of contraction of the heart; the discharge of extra blood corpuscles from the spleen; the deeper respiration; the dilation of the bronchioles; . . . the mobilizing of sugar in the circulation—these changes are *directly serviceable in making the organism more effective in the violent display of energy which fear or rage or pain may involve.* [5]

The work of Cannon and his colleagues was an appropriate response to Darwin's last sentence in *The Expression of the Emotions*: " the philosophy of our subject . . . deserves still further attention, especially from any able physiologist." What further advances can we claim now, in the 1950s?

QUESTIONS OF METHOD

Lord Morley said that an educated man is one who knows when a proposition has been proved. In this sense biologists and psychologists today are better educated than Charles Darwin. We have a clearer notion of the precautions to take in assembling and assessing evidence; we owe to mathematicians routine methods of designing experiments, and of displaying numerical results so that their significance, if any, can be readily discerned. Darwin took many of his examples of the behaviour of "native races" from casual observations made by men who were chiefly concerned in representing the British government or the English church oversea: for instance, Mr. Bulmer, "a missionary in a remote part of Victoria", wrote:

> With Europeans hardly anything excites laughter so easily as mimicry; and it is rather curious to find the same fact with the savages of Australia, who constitute one of the most distinct races in the world. [7]

We are now more conscious of the dangers of the anecdotal method. We no longer believe, as did some of Darwin's contemporaries (and others later), that male gorillas abduct women for their harems, or that wild rats resist our attempts to kill them by means of a highly developed intelligence: the first belief has given way before accurate observation and reporting, the second has been disproved by experiment. Our writings are more precise and rigorous, more likely to lead to correct prediction, and usually a good deal less readable. We consequently have the techniques needed to solve clearly definable problems, such as discovering the best among several methods of feeding pigs or killing mice. This is a real advance: it enables people of only modest ability to do useful scientific work. On the other hand, it does not at all help us to perceive a truly novel scientific development or to make an original synthesis of disparate facts or theories.

It is indeed salutary to reflect on what Darwin achieved without these methodological amenities. In the lack of rigour of his method and writing he reflected—as he inevitably did in so many other ways—the period in which he worked; and in

The Expression of the Emotions we see this nineteenth-century background equally clearly in the Lamarckian assumption which runs through the whole book. "That some physical change", writes Darwin, "is produced in the nerve-cells or nerves which are habitually used can hardly be doubted, *for otherwise it is impossible to understand how the tendency to certain acquired movements is inherited*" (my emphasis). Today, no biologist would write in these terms.[1]

This apparently paradoxical Lamarckism was a facet of Darwin's most important contribution to the study of behaviour: by giving this study an evolutionary foundation, he helped to make it rational. A century ago there was no science of ethology, as we call it now. Accounts of animal behaviour—often myths—had principally a moral significance; although stockbreeders, hunters and others had a real understanding of the natural history of the species they knew, this sort of learning rarely made its way into scientific journals. For the learned as for the peasant, the wonders of nature still represented the creativeness of God, and the natural order was divinely determined. But, once organic evolution was accepted, a new significance was given to the exquisite adaptation of bee to flower or of gull to flight: these, and everything like them, were products of the blind yet rational operation of natural selection. The complex skills of instinctive behaviour in insects or the flexible intelligence of mammals owed their existence to their "selective value": they were open to objective discussion, as products of a "mechanism" or a process which, though slow, could be studied in action. Darwin wrote:

> It will be universally admitted that instincts are as important as corporeal structure for the welfare of each species, under its present conditions of life. Under changed conditions of life, it is at least possible that slight modifications of instinct might be profitable to a species; and if it can be shown that instincts do vary ever so little, then I can see no difficulty in natural selection preserving and continually accumulating variations of instinct to any extent that may be profitable. It is thus, I believe, that all the most complex and wonderful instincts have originated. [6]

[1] [See chapters 1 and 3 for discussions of this topic.—Ed.]

This is not to say that Darwin achieved at one stroke that state of total objectivity which is sometimes represented as a main aim of ethologists today. Far from it. "Although we can hardly account for the shape of the mouth during laughter", he writes,

> which leads to wrinkles being formed beneath the eyes, nor for the peculiar reiterated sound of laughter, nor for the quivering of the jaws, nevertheless we may infer that all these effects are due to some common cause. For they are all characteristic and expressive of a pleased state of mind in various kinds of monkeys. [7]

41.—A baboon (*Cynopithecus niger*) "in a placid condition" (left) and "pleased by being caressed". (All the illustrations in this chapter are from *The Expression of the Emotions.*)

This may itself lead to a slight quivering of the jaws in the modern reader. But Darwin was not uncritical. He gives some admirable examples of the pitfalls of anthropomorphism, including a lifelike drawing (shown above) of an Anubis baboon wearing a ferocious expression with a caption which informs us that the animal is "pleased by being caressed". As he remarks, "this expression would never be recognized by a stranger as one of pleasure".

EMOTION

Darwin, however, took it for granted that terms like *love*, *fear* and *desire* can usefully be employed to describe the behaviour of animals—or at least of mammals—generally. He accepted the colloquial use of the word emotion. In doing so, he assumed (by implication) that other species have feelings like our own; he also grouped the feelings of both pleasure and unpleasure under one head. Since his time it has gradually been found more convenient to describe animal behaviour, not in terms of feelings of which we are directly aware only in ourselves, but in terms of the activities which can be seen and recorded by any observer; we may also try to describe the internal processes which bring these activities about. Thus today it is unusual for ethologists to speak of emotions, though they continue to study the various types of behaviour which Darwin described as expressive of emotion. If the word emotion were to be used in the scientific study of animal behaviour, its meaning would have to be shifted from the familiar, subjective one: it would have to be used to refer, not to feelings, but to internal changes which could be studied physiologically.

Etymologically, " emotion " implies " that which moves or motivates ". In this it has much in common with the term " drive ", as this is now used by some ethologists: W. H. Thorpe has defined " drive " as " the complex of internal states and stimuli leading to a given behaviour " [13]. This definition is objective: it assumes the existence of processes in the animal's organs which are causally related to its behaviour, but it says nothing about feelings or thoughts. Drive in this sense is related only to the unpleasant emotions—or at least to the unfavourable conditions which, as we shall see, impel an animal or a man to action; and it is this aspect of Darwin's subject which will concern us here.

Drive in the sense given is expressible quantitatively. A female rat in œstrus will cross a barrier to reach a male more often than when she is not in œstrus. This example may suggest that " drive " is in some way related to " instinct ". Today this word is often employed by ethologists to name a kind of behaviour and not, as might be expected, any internal process or agent. However, the term is not used uniformly, and many writers on behaviour prefer to avoid it altogether, in order to prevent confusion. Darwin's own definition nevertheless explains sufficiently clearly what sort of actions he refers to by the term, and it is quite adequate for use at the present stage of this argument:

> An action, which we ourselves should require experience to enable us to perform, when performed by an animal, more especially by a young one, without any experience, and when performed by many individuals in the same way, without their knowing for what purpose it is performed, is usually said to be instinctive. [6]

Thus Darwin, in writing of instinctive actions, refers to stereotyped sequences of behaviour which seem to develop in a certain way, namely, without practice, imitation or any process which could be called " learning ". Today these sequences are often called fixed action-patterns.

What connection, then, has instinctive behaviour, in this sense, with emotion? An example of a stereotyped act may be given by continuing the account of the female rat in œstrus. On her encounter with a male she may nose him ; she will then probably run a short distance away, and pause; on his following and mounting, she will adopt a specific posture, with raised coccyx, which facilitates intromission; her head will also vibrate. The behaviour described is found in females of all varieties of *Rattus norvegicus*, including the wild type, the uncommon melanic form and the several laboratory strains regardless, as far as we know, of individual experience. A further characteristic feature is that the (tame) female's posture with raised coccyx may be evoked (when she is in œstrus) by gently palpating her flanks with the fingers. The response is made, in fact, to a rather specific stimulus, in this case a tactile one. The response of the female, then, in its uniformity and in the way in which it is produced by internal and external agencies, is a typical instinctive sequence or behaviour pattern.

In man sexual behaviour is accompanied by the experience of violent emotion, and it is natural to attribute similar feelings to other animals. This attribution seems further justified by the persistence and energy displayed by animals in reaching a partner of the opposite sex. Nevertheless, it may be suggested that, on the contrary, an animal whose nervous system is organized in advance to produce standardized responses may be regarded as needing no " emotional incitement or prompting " to perform such responses: the behaviour occurs " automatically " as soon as the appropriate stimulus situation turns up. But in an animal in which behaviour depends largely on a gradual, adaptive learning process in each individual, there must be such internal incitement or prompting to ensure that the learning takes place [16]. This would apply above all to our own species.

This hypothesis illustrates some of the difficulties which face us when we try to interpret animal behaviour in terms of human concepts such as emotion. How can we decide whether even so closely related an animal as a rat is actuated by emotion? This question can be put more rigorously if we ask exactly what we mean by emotion, or by what criterion we identify it.

THE CRITERIA OF EMOTION

One sort of objective criterion is clearly set out in chapter 3 of *The Expression of the Emotions*: " certain actions ", writes Darwin, " which we recognize as expressive of certain states of the mind, are the direct result of the constitution of the nervous system ", and this may bring about responses such as a general trembling of the muscles, changes in the secretion of the digestive organs, alterations of the heart-beat, and vasomotor disturbances, for instance blushing. We concern ourselves, in fact, with observations of behaviour and of physiology. As we have seen, this type of enquiry was later pursued especially by W. B. Cannon and his associates at Harvard, and Cannon explained the changes observed in these states as enabling the animal to cope with sudden emergencies. In this way he developed an important Darwinian thesis: Darwin held that the structures involved in the expression of emotions had been " acquired through variation and natural selection "; for Cannon the whole constellation of processes occurring in a mammal in a state of pain or danger is biologically advantageous.

These processes are very diverse. On the one hand we have,

for instance, increased heart rate, a rise in the sugar content of the blood and the discharge of red blood cells from the spleen: all these increase the capacity of the body for violent exercise, but none has a direct effect on the appearance of the animal. On the other hand we have visible, audible or olfactible changes: these include the adoption of particular postures or facial expressions, raising of the hair or feathers, reddening of the skin (in man and some other primates),

42.—A "dog approaching another dog with hostile intentions". The raised hair is one of the signs of stimulation of the sympathetic nervous system, emphasized by W. B. Cannon in *Bodily Changes in Pain, Hunger, Fear and Rage*. See text.

trembling, the utterance of cries and increased sweating. These, like the members of the first group, reflect the activation of the autonomic (or "involuntary") nervous system; but their biological significance is often different, since they may influence the behaviour of other individuals of the same or different species. If members of the same species are influenced, the changes may be called "social signals".

The importance of these signals, sometimes called also social releasers or releasing stimuli, has come to be much more fully understood since 1935, when K. Z. Lorenz published an important paper on social behaviour in birds [9]. It is usual for members of the same animal species to associate together, if only at the time when the eggs of the female are inseminated by the male. (Some aquatic species, of course, discharge their eggs and sperm promiscuously into the water.) But social

43.—" Cat terrified at a dog "—another example of the sort of response studied by Cannon (compare figure 42).

interactions, even apart from courtship, are very widespread. N. Tinbergen has been particularly associated with the study of releasing stimuli in social behaviour [14, 15]. If two animals approach each other, typical responses are first, avoidance or flight or, second, an attempt on the part of one to eat the other. Neither of these is biologically appropriate when an animal encounters a potential mate or, say, its own offspring

Hence, to ensure mating, animals have signalling devices which bring them together—such as the courtship ceremonies found in many vertebrates, and even in invertebrates from insects to cuttlefish. The ceremonies also prevent attempts at mating between different species, and ensure the accurate timing of insemination. Similar arrangements prevent predatory animals from attacking their own young, or their mates (at least, until mating has been consummated).

Examples are found in the mating of some spiders. The female is predatory and larger than the male. The male's freedom from attack depends on his performing a stereotyped series of acts characteristic of his species: in species with good eyes these may involve the adoption of specific postures, or in others the female's web may first be agitated in a particular way. When insemination is completed, the male is now merely a small animal, and as such may be caught and eaten by the female. He has no further signals to prevent this. Evidently, there is little or no selective advantage in the survival of the male once he has mated.

The use of the word "signal", derived from human behaviour, might suggest that the use of these signs is deliberate on the part of the animals. In fact, many observations show that it is "automatic", and evoked by other specific signs. For instance, a bird may utter its alarm call at the sight of a hawk (or of a cardboard model of a hawk) even though there is no other bird to hear it. In general, animals apart from man do *not* adjust their signals so as to ensure that they are received by other individuals to whom they might be thought to be addressed.

Nevertheless, in man the importance of indicators of emotion is obvious. As Darwin said:

> The power of communication between members of the same tribe by means of language has been of paramount importance in the development of man; and the force of language is much aided by the expressive movements of the face and body. We perceive this at once when we converse on any important subject with any person whose face is concealed. [7]

In the attempt to study emotion objectively we have then been obliged to turn entirely away from the *feelings* which we all experience, and which we call emotional. Instead of trying to give a scientific account of these very complex subjective effects, we have resorted to the more modest, but sufficiently difficult, task of studying overt behaviour in a variety of species, together with the observable internal processes which accompany the behaviour. The type of behaviour which seems to reflect emotion is that in which the activities are violent, especially those arising from a threat to the life of the animal or in connection with reproduction or the care of young. However, there is no definable boundary between an animal moving in a leisurely way from its lair to easily accessible food on the one hand, and the same animal frantically seeking and devouring food in times of scarcity, on the other. In fact, in attempting the scientific study of emotion we are involved in enquiring into the springs of all behaviour, and not only one particular group of activities. We are involved in studying what is now commonly called motivation.

MOTIVATION

In essaying such a study we observe that the *expression* of a particular emotion, or internal state, may vary in intensity from zero to a maximum at which a trifling stimulus can touch off an explosion of activity. We are therefore obliged to ask how this quantitative change takes place.

We have already seen that the internal state of an animal varies in such a way as to make its behaviour alter. This holds for stereotyped, " innate " activities, as in the example of coitus in the rat, and also for behaviour developed as a result of individual experience, for example, finding the way to a source of food. The important fact is that a given situation, or set of stimuli, does not always evoke the same response. This is familiar enough: a hungry animal eats, while a full one sleeps; a pregnant female builds a nest, while a female in œstrus runs or copulates.

Recent studies by zoologists have been especially concerned with the motivation of stereotyped acts. A simplified account of the concept of instinctive behaviour, recently developed, is as follows. There is first an internal change in the animal: this may be autonomous, as in the œstrous cycle of some mammals; or it may be induced by external agencies, for instance the longer days which induce the breeding state in some birds. Second, the internal change alters the behaviour: often, at this stage, the animal moves over a relatively wide area, either apparently at random or in a particular direction which may be dictated by some external influence such as light or temperature; this is the phase of appetitive behaviour, " the variable introductory phase of an instinctive behaviour pattern or sequence " [13]. Third, the animal reaches a situation in which a consummatory act, or series of such acts, can be performed: these acts constitute much of the truly rigid, stereotyped behaviour; examples, studied especially in birds and fish, are coitus, nest-building and " fighting ". The performance of a consummatory act is usually followed by an abrupt or gradual decline in " drive " or " motivation ", as measured by the amount of stimulus required to induce the animal to perform the act again.

There has been much discussion on how this change in responsiveness is brought about. At first it was assumed that the performance of a specific act could cause a " discharge " or " tension " or " nervous energy ", but it was soon realized that such expressions are merely similes and do not describe what actually goes on in the nervous system. It seems more reasonable to suggest that the performance of a consummatory act is accompanied or followed by a corresponding input to the central nervous system, and that this puts a stop to the output and so to a continuance of the consummatory act.

This emphasis on the importance of the nervous input is reinforced by observation of the effects of straightforward sensory changes: a cold skin, for instance, may lead to appetitive behaviour which ceases when a warmer place is reached (the cat near the fire); or it may evoke a consummatory performance, such as increased nest-building in a mouse. In any case, it is the attainment of a particular sensory input, or consummatory *situation*, which puts an end to the activity.

The same principle seems to apply to certain kinds of behaviour which at first sight resemble appetitive movements, but are carried out when all obvious biological needs are satisfied. These movements are *exploratory*. The exploratory behaviour of an animal such as a rat leads to the consummatory situation in which the whole of an area around the nest has been investigated and " learned ". The " stimulus " which evokes this behaviour is the perception of novelty: an unfamiliar object or area induces investigation. The behaviour gives the impression of " inquisitiveness " and, in its tendency to seek novelty, of an attempt to avoid boredom. This behaviour has a quite obvious biological function: it enables the animal to learn all the important features of the topography of the area, and regularly to reinforce that learning [2]. Exploration and the learning it promotes are widespread in the animal kingdom.

Some learning—often of this topographical kind—is a feature of all complex behaviour, however stereotyped it may seem, and in some vertebrates and molluscs there is a great reduction in stereotypy, and a great deal of the animal's activity is adapted to particular conditions.

In general, whether the behaviour consists mainly of stereotyped action patterns or is principally learnt, it is such that it brings the animal into certain positions, or creates in the animal certain states, which we may call consummatory. Once the consummatory state is reached, the activity ceases, and any " emotion " which has led to the activity subsides.

Since I have used the word " stereotyped " (a figure of speech taken from the printing trade), I must add that learned behaviour patterns often become exceedingly rigid in an adult, and so simulate innate behaviour patterns. Learned patterns, however, may vary from one individual to another. In our own species they may be fairly uniform within a single culture, while other cultures are found with quite different patterns; the form the behaviour takes is determined by the social experiences of the individual.

ABNORMAL BEHAVIOUR

Since we know that both " emotions " and, consequently, their expression, are likely to cease when consummation is achieved, an important question is, what happens when a consummatory act is prevented or, more generally, when a consummatory state is *not* attained? One might suppose that appetitive or exploratory behaviour then would continue, and often this is indeed what happens. But sometimes there are more complex consequences. The subject of frustration has—

naturally—been studied more in our own species than any other, and we find J. H. Masserman describing its effects like this:

> When, in a given milieu, two or more motivations come into conflict in the sense that their accustomed consummatory patterns are practically or wholly incompatible, kinetic tension (anxiety) mounts and behavior becomes hesitant, vacillating, erratic and poorly adaptive (i.e. " neurotic ") or excessively substitutive or symbolic (i.e. " psychotic "). [10]

44.—A dog " in a humble and affectionate frame of mind ".

Ever since I. P. Pavlov induced severely disturbed behaviour in a dog, by facing it with an insoluble problem, it has been known that mammals other than man could display behaviour analogous to neurotic behaviour in ourselves. The insoluble problem, in such experiments, is usually whether to approach an object or to withdraw from it, since the animal is unable to distinguish whether the object is a source of food, or of punishment. An animal in this state, to quote Masserman again,

> showed erratic salivation, ceased to feed, howled and struggled against the harness and became restless and uncoöperative in further training.

As well as these behavioural changes, there are alterations of internal function: the pulse rate and blood pressure rise, sugar and oxygen in the blood increase and there are other indications of excitation of the sympathetic nervous system, such as trembling, erection of the hair and dilated pupils. In fact, the changes which Cannon studied, and described as enabling an animal to deal with an emergency, occur too in " experimental neurosis ". They are also observed in severe anxiety states in our own species, and Masserman has no hesitation in using the term anxiety, not only for (subjective) states reported by human beings, but also for states studied behaviourally or physiologically in other mammals.

Clearly, we can at most only venture a surmise on the feelings of these other species; while, if we ignore the philosophical pitfalls and adopt a commonsense attitude on human communication, we can usually assume that, if a person says he feels anxious, we know what he means. We are not, however, obliged to rely only on speech, for our knowledge of the effects of frustration, and of other emotional states, in man. We have seen that anxiety is accompanied by physiological changes which can be objectively studied and even measured; but there are other conditions in which similar bodily changes are easily detected, but in which no anxiety or comparable emotion is acknowledged. These are the so-called psychosomatic disorders, in which emotional (" psychic ") states influence " physical " processes. The distinctive feature of these illnesses is that they come to the attention of patient and physician through some malfunctioning of an organ, for instance the lungs in asthma or the stomach in gastric ulcer: the patient often reports no emotional symptoms at all. However, it is sometimes possible, by prolonged enquiry, to uncover anxiety or some other unpleasant emotion; and the somatic condition may then be cured.

There is therefore good reason for regarding the " psychosomatic " disorders, and the " psychoneuroses " such as anxiety states, as a single group of illnesses [1]: in all these disorders the primary malfunction is in the central nervous system and is not directly detectable with existing equipment. (Of course, not *all* central nervous disorder produces illnesses of these kinds.) Further, these conditions all arise from frustration due to a conflict between alternative courses of action. A further justification for grouping them together is the phenomenon of "syndrome displacement": sometimes, when one group of signs or symptoms is cured, another group replaces it; and the alternative states may be either "psychosomatic" or "psychoneurotic". When this happens it is assumed that the patient is suffering from some "emotional" or "nervous" disorder which is concealed, perhaps, both from himself and his physicians; and that the observed signs and subjective symptoms are the different forms in which the damaged condition of the central nervous system manifests itself.

This account of "psychosomatic" disorders is oversimplified in at least one important respect: it leaves out the fact that the "somatic" conditions are of two kinds. First, there are those of which the individual is unaware: for instance, one's blood pressure may be raised; if this is due to anger, verbal expression of the anger may be accompanied by a return to the normal level. Second, the somatic condition may be specifically related to the emotional state which brings it about—although the connection is not evident to the patient: for instance, a man reluctant (unconsciously afraid) to return to work where he is persecuted by a superior, may prolong his sick leave by developing a painful blistering of his feet. Other such effects have become widely familiar through the rather unreliable source of the cinema: for example, hysterical blindness or paralysis are conditions (exceedingly rare) in which the defect symbolizes the patient's emotional problems.

UNCONSCIOUS MOTIVATION

We saw above that there are powerful influences (sometimes called "drives") of unknown nature which tend to bring an animal into favourable situations, and that these favourable situations are often reached as a result of the behaviour which we call expressive of emotion. We now see that, when a favourable situation is not attained, the mode of expression of the emotion may alter, sometimes to the detriment of the animal. The important thing seems to be that expression takes place somehow. This conclusion is further supported by other aspects of psychopathology, as we shall further see below.

The preceding argument has used ideas derived from psychoanalysis. (By psychoanalysis I mean the psychological theories which we owe to Sigmund Freud, not his method of therapy.) The psychoanalytic method is inapplicable to species other than man, since it depends on verbal communication. This is in a way paradoxical, since a central feature of psychoanalysis is the emphasis on unacknowledged (unconscious) motivations, emotions and so on. Using the statements made by patients as his starting-point, Freud formulated an exceedingly complex system of concepts and explanations, all in terms of a " mind " which consisted largely of " unconscious wishes ", " repressed fears " and so forth. This system developed quite separately from the contemporary growths of psychology and physiology, and its seemingly bizarre terminology (even apart from its other disconcerting features) tended to erect a barrier between it and the orthodox sciences of behaviour. Nevertheless, it is now widely agreed that Freud made a contribution of the first importance to our knowledge of ourselves, especially in the sphere of motivation—a contribution comparable to that of Darwin in biology; certainly his ideas are today generally used in social anthropology, educational psychology and many other fields, as well as psychological medicine.

Even though Freud's method is of no use for other species, it might be hoped that, if his theories are sound, they could suggest something of value for ethology as well as psychology. Perhaps they have done so already. A notable example of an hypothesis from psychoanalysis proving fruitful in the laboratory is the demonstration by O. H. Mowrer of analogues of the phenomena of regression and reaction-formation in rats: both types of behaviour are among those to which Freud first drew attention.

Mowrer put rats in cages in which they were subjected to mild electric shock from a floor grid; they could reduce the shock by adopting a particular posture, and this they soon learned to do (" habit A "). The rats were then put in a similar cage in which the shock could be completely switched off by depressing a pedal, and this response too was learned (" habit B "). Once this second response had been established, it was arranged that the rats would receive a shock from the pedal on depressing it. When this happened, the rats discarded habit B and returned to habit A. By contrast, control rats, which had never learned habit A but which had learned habit B, did *not* discard habit B in the new circumstances. This then is a simple analogue of regression in man—a return to an earlier learned pattern in frustrating conditions.

Even more interesting are the following observations:

> After habit B was well established [the rats] customarily sat very near the pedal after pressing it, in readiness for the next . . . shock from the grill. When, however, shock was put on the pedal, their behavior was noticeably altered. After discovering that the pedal was charged, these animals would frequently *retreat* from the pedal end of the apparatus soon after they began to feel the grill shock, i.e. as soon as they began to have an impulse to press the pedal. In effect, they were thus *running away from the pedal because they wanted to go toward and touch it;* they were, in other words, manifesting a simple but presumably genuine type of reaction-formation. [12]

In man, reaction-formation is said to occur when an impulse, say of hatred, is suppressed, and an excessive tenderness or kindness is displayed instead: the nature of the response is shown by its compulsive and exaggerated character; it is not a "natural" or "normal" kindness, but a defence against the opposite feelings (which arouse guilt or fear). In general, then, reaction-formation consists of doing the opposite of what is, to some extent, desired.

Mowrer, like most other writers on the subject, holds that a full validation of Freudian hypotheses must depend on experiments on human beings. (Such experiments have been attempted, though not yet on a large scale; they have sometimes confirmed the hypotheses they set out to test [8].) For our present purpose the interest of Mowrer's and similar experiments is in the complexity of animal motivation which they reveal.

We must not, however, digress further on the problem of the extent to which "abnormal" behaviour in animals may properly be compared with our own neuroses. Anyone interested should consult the valuable review by E. R. Hilgard in *Psychoanalysis as Science* [8]. Our present concern is with the effects of expressing emotions or feelings, and in particular those which are, or have been, unconscious or at least not acknowledged.

Sometimes the mere expression of an emotion becomes important in itself, regardless of any change in the objective situation it may bring about. For example, a small boy was subjected to severe " discipline " and prevented from expressing any aggression at all against adults or other children. He resorted to pinching himself, banging his head and pulling out his own hair. He recovered when taken to an environment in which excessive restraints were removed, after going through a stage in which violent hostility was directed against the adults around him. Another example is that of a man whose carefully accumulated savings were dissipated by his wife. At first the man displayed only bitter self-recrimination; but sharp words from his wife later evoked an outburst of rage directed against her. After this the man became quite cheerful and was able to plan new ways of saving. [10]

In both these examples the normal expression of an emotion was at first blocked; but when the inhibition was eventually released there was a general improvement in the state of the individual concerned. We see, then, once again, that where the normal expression of an emotion is prevented, some outlet is nevertheless found. Moreover, in these examples we see further that the expression of an emotion sometimes has a favourable effect and becomes an end in itself. The idea of the cathartic value of the release of feelings was familiar to Aristotle but has been much extended by psychoanalysis; (in particular, much attention has been paid to the origins of the *resistances* to expressing the emotions). It seems probable that there is a general principle here which has an application in fields of study much wider than that of human psychopathology.

DISPLACEMENT AND DISPLACEMENT BEHAVIOUR

In studies of animal behaviour the phenomenon of *displacement behaviour* seems to provide a parallel to the observations on release of feelings in human beings [1, 3, 15]. Displacement activities are most clearly seen in animals with a varied repertoire of stereotyped behaviour patterns, all of them familiar to the observer. These patterns may include aspects of feeding behaviour, courtship, coitus, brooding or suckling, preening or grooming, nest-building and fighting. All have a clear significance in the life of the animal, but at times they *seem to be performed at an inappropriate moment*, or " out of context ". This may happen if an animal is faced with a situation to which there are two possible responses, for example, to fight or to run

45.—" Cat in an affectionate frame of mind." Today this behaviour might be described in terms of " cutaneous stimulation ".

away; paradoxically, faced with this " choice ", it performs a third, and seemingly irrelevant, act instead, for instance grooming (in a rat) or nest-building (in a herring gull). There seems to be an obvious parallel here with the effects of " internal conflict " in man, and it seems that there is a case for regarding psychoneurotic behaviour, psychosomatic disorder and displacement behaviour as a single group of effects all with a common underlying physiology [1]. The crucial problem is how to investigate the physiology, and so to test the hypothesis.

Displacement behaviour and allied activities, as defined by ethologists, are far more complex in form and origin than I have so far indicated. Sometimes, an animal is prevented from directing a particular activity, such as a display, towards the appropriate " object ", for instance a potential mate, or a rival; it may then turn its attention to an alternative object. Margaret Bastock and her colleagues have called this a " redirection activity " and draw attention to the parallel with the " object displacement " of psychoanalysts [4]. An unpaired male black-headed gull in his territory, visited by a female, is stimulated to attack, to escape, and to make sexual advances. In this complex situation the male not only performs various acts directed at the female, but also attacks other birds or even men which he had previously ignored. A human being may fall in love with a statue, like Pygmalion, or even be sexually stimulated by articles of clothing and much else.

In these two instances the *behaviour* is directed towards a substitute object. By contrast, in syndrome displacement (mentioned above), it is the output from the nervous system which varies, rather than the object of attention. This leads to altered behaviour (as in hysteria) or to pathological changes in organs (for instance an ulcerated stomach) which are affected by the nervous discharge.

We must not be misled by the coincidence of the word " displacement " in three different contexts. But we can see that there is a whole range of phenomena, all suggesting that excitation of the central nervous system may express itself, or is obliged to express itself, in any one of a variety of motor outputs: that is, of behaviour patterns, or of responses of the involuntary musculature or of glands as in the psychosomatic disorders. (The patterns may, of course, be learned performances: they are not confined to instinctive behaviour.)

This notion was quite familiar to Darwin, and indeed to Herbert Spencer before him. Darwin wrote:

As Mr. Herbert Spencer remarks, it may be received as an "unquestionable truth that, at any moment, the existing quantity of liberated nerve-force, which in an inscrutable way produces in us the state we call feeling, *must* expend itself in some direction—*must* generate an equivalent manifestation of force somewhere"; so that, when the cerebro-spinal system is highly excited and nerve force is liberated in excess, it may be expended in intense sensations, active thought, violent movements, or increased activity of the glands. [7]

46.—" Chimpanzee disappointed and sulky."

Nowadays, we are a little shy of using terms like " nerve force ", and the neurophysiologists have not yet provided us with any adequate alternative. Nevertheless, if we are to attempt a comprehensive enquiry into motivation and emotion, it seems possible that in the study of substitutive behaviour, if we look hard enough, there will be found some unifying principle.

CONCLUSIONS

Darwin was primarily a naturalist of extraordinary gifts, imaginative, with insatiable curiosity and immense enthusiasm, thorough, critical and inexhaustible. He applied these gift to observing not only animals and plants in their usual surroundings, but also domestic animals and human beings of all sorts, including his own children. With all this, he combined an appreciation of the need to incorporate nerve physiology in any full account of the " senses, affections, passions " (as well as the rest of behaviour).

> When the sensorium is strongly excited [he wrote] nerve force is generated in excess, and is transmitted in certain definite directions, depending on the connection of the nerve cells, and partly on habit: or the supply of nerve-force may, as it appears, be interrupted. Effects are thus produced which we recognize as expressive. This . . . principle may, for the sake of brevity, be called that of the direct action of the nervous system. [7]

Accordingly he allied direct observation, as of the erection of the hair in fear and rage, with references to the functioning of the " vegetative " nervous system. Darwin would have required no convincing of the validity of the so-called psychosomatic approach in medicine. " When our minds are much affected," he writes, " so are the movements of our bodies; but here another principle besides habit, namely, the undirected overflow of nerve-force, partially comes into play." He quotes from Henry VIII:

> Some strange commotion
> Is in his brain: he bites his lip and starts;
> Stops on a sudden, looks upon the ground,
> Then, lays his finger on his temple; straight,
> Springs out into fast gait; then, stops again,
> Strikes his breast hard; and anon, he casts
> His eye against the moon; in most strange postures
> Have we seen him set himself.

In his unitary method of studying behaviour, Darwin was in advance of many of his successors. His references to " nerve force " may sound crude to a physiologist, but not more so than some of the recent expressions of zoologists who study behaviour in the field. This reflects the fragmentary way in which behaviour has been studied since Darwin. The growth of " behaviourism ", which owed much to Pavlov's study of conditional reflexes, has been largely separate from the study of nerve physiology on the one hand, and quite divorced from animal natural history on the other. Now that we are returning, with some success, to the observation of whole animals in complete environments, we are beginning to realize anew the value of Darwin's synoptic method.

The need for such a method is very obvious if we are concerned to study the "expression of the emotions". This study, to be complete, must take in both behaviour, objectively reported, and physiology; we have seen that the study of bodily changes in various "normal" emotional states has made some progress, and that something has been learnt also of the "abnormal" or undesirable bodily effects in man of conflicting emotions—in the psychosomatic disorders.

We can now see more clearly that a great deal of behaviour is activated by the *absence* of some specific sensory condition or consummatory state. It is this which gives animal behaviour generally its "purposive" appearance: as a result of the long, slow process of natural selection, every species either has an elaborate repertoire of responses which tend to put each individual in a favourable position for survival or reproduction; or it has the ability to learn such favourable responses. Activity is "switched off" when the consummatory state is reached. What Darwin described as the expression of (unpleasant) emotion occurs when the attainment of a consummatory state is, at least temporarily, prevented.

The internal changes involved mobilize the animal for special exertion. The external changes, which constitute the "expression" of the animal's state, may be merely by-products of this mobilization; but, during evolution, these outward signs have often acquired a social significance. Dilatation of the pupils (the "blazing eyes" of the novelist) or raised hair are consequences of the presence of extra adrenalin in the blood and of the action of the sympathetic nervous system—aspects of the preparation of the animal for emergency; but they may also serve to frighten off a rival.

Not all these responses are relevant or appropriate to the situation. Especially when an animal is thwarted, we find the diverse and unexpected "displacement" activities. In man, analogues of these activities have enormous importance in psychopathology. Much of the "expression of the emotions" which we see both in ourselves and in other species has this "displaced" character. It seems to reflect a fundamental characteristic of all nervous systems—one given many names but as yet not identified with any specific activity of the central nervous system. Whatever the name, it is *as if* some source of energy must have an outlet; and, if the "normal" outlet is blocked, another is used.

It is salutary to reflect that none of these ideas is wholly new. Still, we may hope that, three-quarters of a century after Darwin made his main contribution to the sciences of behaviour, a real advance is beginning to be made.

10

Sexual Selection

By

J. MAYNARD SMITH

THE members of most animal species are of two kinds, males, which produce spermatozoa, and females, which produce eggs. Sometimes there is little difference between the two sexes other than that immediately concerned with the kinds of sex cells they produce; this is so, for example, of many of the species of fish which shed their sex cells into the water, where fertilization takes place. But often the two sexes differ also in size, in coloration, in organs of sense or of locomotion, or in behaviour. Darwin saw that the origin and evolution of these differences called for an explanation.

In his theory of natural selection, he had argued that the characteristic features of species had evolved because individuals possessing them were better fitted to survive, or were more fertile, than were individuals lacking them. Natural selection of this kind can also explain the evolution of many sexual characteristics; a female mammal with an efficient placenta, or a male emperor penguin which incubates the egg while the female departs for the sea, are each likely to contribute more to future generations than are individuals which lack these properties. But there are other sexual characteristics which are not so easily explained in this way. Why, for example, should men have beards, or should cock chaffinches have pink breasts and sing in spring? Darwin suggested that such features evolved because they enabled their possessors either to mate more often, or to mate with the more fertile members of their species. This was his theory of sexual selection [2].

Sexual selection can be subdivided according to whether the mating of a female with a particular male depends in part on female choice, or is determined only by a conflict between males. " Choice " is a subjective term, so I will give examples to explain the sense in which I am using it. Male red deer are polygamous; a single stag collects a harem of hinds, with whom he mates, driving off other males who try to do likewise. There is no reason to think that differences between the behaviour of hinds determine to which stag they will attach themselves; in fact, the hinds seem to be passive and indifferent to a change in ownership. In contrast, the males of most passerine birds occupy a territory in spring, and each male is later joined in his territory by a female. Females which differ either genetically or in their past experience will tend to settle in different kinds of territory, or perhaps with different kinds of male. When such differences between females help to determine with which male they will mate it seems reasonable to speak of female choice.

Now it is easy to see how sexual selection can have evolutionary consequences in a polygamous species. If the larger stags with better developed antlers are also the more successful in collecting harems, and if they transmit these characteristics to their numerous male offspring, then this would account for the evolution of greater size and of antlers in male deer. That polygamy favours sexual selection is confirmed by the fact that in those mammalian orders (primates and artiodactyls) in which polygamy is common, the males are usually larger, more brightly coloured, or better equipped with offensive weapons than the females. Carnivores on the other hand are usually monogamous, and there are seldom any striking differences between the two sexes, other than those directly associated with reproduction. The seals provide an exception; they are polygamous, and males are much larger than females. But there are some awkward facts. Horses are polygamous, or at least domestic horses which have run wild are so, but there is relatively little difference between the sexes. In red deer, a few stags never develop antlers. These "hummels" are often larger than other stags and earlier in coming into rut, perhaps because they do not have to face the metabolic strain of growing antlers. They are successful in maintaining harems, and can hold their own against other stags. Therefore the advantages of having antlers are not so obvious as might appear at first sight, and one can only assume that on balance it must pay to have them.

But many species of mammals, and most birds, are monogamous, at least for a single breeding season, and there are about equal numbers of the two sexes. This raises difficulties, which were recognized by Darwin, for the theory of sexual selection. In such a species all or almost all mature individuals can find a mate. Why then should those individuals with particularly well-developed sexual characteristics leave more progeny than others less well endowed?

There are two possible ways out of this difficulty. First, even in such species, not all individuals in fact mate. This is true of our own species, and D. Lack estimated that only about 80 per cent of male robins in a particular region found mates [7]. If the sexual characteristics of an individual influence the probability of that individual finding a mate, then Darwinian sexual selection can be effective. But Darwin himself thought that sexual selection worked in a different way in monogamous species. His ideas will be explained by giving an entirely hypothetical example from the human species.

Let us suppose that in a particular monogamous society women prefer to marry red-haired men. In practice such a preference, if it existed in our own species, would probably be due to cultural influences, but we will assume that in this instance the preference is genetically determined. Two questions can be asked. First, in what circumstances will the frequency of red-haired men in the population increase? Clearly the red hair must be inherited, and not, for example, dyed, but this is not sufficient. A red-haired man will have a wider range of choice in his selection of a wife, but he will not therefore have more children, unless the wife he actually chooses is more fertile than the average. The second question is, in what circumstances could a genetically determined preference on the part of women for red-haired men evolve? It would evolve if women who married red-haired men had more children than they would have done had they married someone else, and this would be so if red-haired men were more fertile, or in other ways fitter as parents, than other men. It follows that, in a monogamous species with an equal number of the two sexes, sexual selection will have evolutionary consequences (either in the appearance of striking sexual adornment or in genetically determined preferences for individuals with such adornment) only if those individuals which have characteristics which make them successful in the competition for mates are also fitter than the average as parents.

Darwin thought that a correlation of this kind would in fact exist. He argued that the most healthy and vigorous individuals in a population would be the most successful both in sexual competition and as parents. Some evidence which supports his opinion will be given later in this essay, although it does not concern a monogamous species.

Darwin's ideas on sexual selection have received rather little attention from later biologists. There are several reasons for this. First, the process is necessarily difficult to demonstrate, since it must be shown both that one kind of individual is more successful in sexual competition than others and also that such individuals leave more offspring in consequence. As we have seen, in monogamous species the latter does not necessarily follow from the former. In polygamous species, although it is clear that a male which collects a large harem is likely to leave more offspring than one which does not, yet, before one can be sure that this fact will have evolutionary consequences, it would be necessary also to show that particular males owe their success at least in part to their genetical constitution and not merely to their age.

SEXUAL CHARACTERISTICS AS "SIGNALS"

Another reason for the comparative neglect of sexual selection is that two other processes have been suggested which might account for the evolution of sexual characteristics. In the first place, they may act as "signals" eliciting the appropriate response from a sexual partner, or, in species in which both parents care for the young, keeping the partners together until the young are reared.[1] An example of the former kind of signal is afforded by territorial birds, in which the usual response of an individual in its territory to another member of its own species is to perform a "threat" display, which often causes the intruder to depart. If the display is ineffective it may be followed by a physical assault on the intruder. Such a series of responses would clearly be inappropriate if per-

[1] [See also chapter 9, pp. 213 *et seq*. Ed.]

formed by an unmated male towards a female entering its territory. N. Tinbergen and M. Moynihan have shown that in gulls, in which the plumage of the two sexes is alike, this eventuality is avoided by the performance of " appeasement ceremonies " [12]. In the threat posture of Herring Gulls and of most closely related species the head is held high with the beak pointing downwards. In courtship the head is lowered and the beak held horizontal, thus presenting an appearance as different as possible to that presented in the threat posture. The " exception which proves the rule " is afforded by the Black-headed Gull, in whose threat posture the head is lowered with the beak pointing horizontally towards an antagonist. In courtship, the head is raised and turned away, with the beak pointing downwards.

In his study of Great Crested Grebes, J. S. Huxley found that courtship movements continued to be made while the young were being raised, long after pair formation and mating [5]. He suggested that the effect of these movements was to keep the pair together. Similar " bond-forming " behaviour has since been described in many other species of birds.

In these examples, patterns of behaviour have evidently evolved because they enable animals to elicit appropriate responses from members of their own species. A second kind of selection pressure influencing the evolution of courtship behaviour and of associated structures has been stressed by Th. Dobzhansky [3]. Animals which mate with members of another species usually contribute little to future generations, because species hybrids are usually infertile or inviable. There will therefore be selection in favour of individuals whose behaviour and powers of discrimination ensure that they will mate with a member of their own species. In species in which the males are potentially polygamous, discrimination is likely to depend on the females, since only they are sterilized by interspecific matings.

K. F. Koopman has been able to show experimentally that selection of this kind can be effective [6]. He studied two species of fruit-flies, *Drosophila pseudoobscura* and *D. persimilis*; these, although they apparently do not interbreed in nature, hybridize freely in the laboratory, particularly at low temperatures. The hybrid males are sterile, and the hybrid females, although fertile, produce inviable offspring. Koopman kept the two species together in a population cage, removing in each generation all the hybrid flies, so that only those individuals which mated with members of their own species contributed to future generations. The frequency of interspecific matings could be determined from the number of hybrid flies present. In three parallel experiments, the proportion of hybrids in the first generation varied from 22 to 49 per cent. After five generations it had fallen to below 5 per cent in each case. Selection had in some way altered the behaviour or powers of discrimination of the flies. In this experiment, Koopman repeated in the laboratory a process of a kind which must have occurred many times in nature.

Is selection of these two kinds sufficient to account for the evolution of all sexual characteristics? My own view is that it is not, and that Darwinian sexual selection has also played a part. The latter process would be expected to produce results of rather a different kind. In the cases described by Tinbergen, Huxley and Koopman, all that is required is that an animal should be able to make a signal which guides the activities of another animal in one of several possible directions. There is no element of competition between the signals given by different members of the same species. Provided, for example, that a female gull is able to elicit a sexual rather than an aggressive response from an unmated male, or that the song of a cock blackbird is sufficiently different from that of other species to be recognizable, there will be no selection in favour of further elaboration of these signals, except on the grounds suggested by Darwin.

A human analogy may make this distinction clearer. Notices in public places reading "LADIES" or "GENTLEMEN", or signposts reading "LONDON", "BRIGHTON" or "NETHER WALLOP", are simple and undemonstrative in design. People reading them know what they want, and wish only to be shown how to obtain it. In contrast, advertisements urging people to smoke a particular brand of cigarette, to drink coca cola, or to go to Brighton for their summer holidays, are large, brightly coloured, and designed to be as arresting as possible. If, as often seems to be the case, the sexual ornaments and displays of animals tend to resemble advertisements rather than notices, it is likely that they have evolved because they must compete for attention. This is most obviously the case in sexual selection as conceived by Darwin, and in the competition between flowers for the attention of pollinating insects.

47.—The analogies between advertisements and the competitive signals of animals on the one hand (above) and between signboards and non-competitive animal signals on the other (below).

I want now to discuss two cases in which I think sexual selection is important; in one, only a conflict between males may be involved, but in the other there is evidence that female choice plays an important part.

In early spring the activities of male passerine birds are

devoted mainly to establishing a territory, by singing and displaying at intruders. A bird which fails to establish a territory does not breed. A bird which cannot establish itself in a habitat favourable for its species, and is obliged to occupy a territory in a less favourable one, is likely to leave fewer offspring. There will therefore be selection in favour of aggressive behaviour. But this is not the whole story. L. Howard describes a conflict between two Great Tits for a particularly favoured territory which included part of her garden in which she was in the habit of feeding the tits [4]. In this conflict the loser was seriously injured. Thus there are circumstances in which it may be wise to run away and live to fight another day; that is, there will be selection in favour of individuals which respond to the displays of others by retreating. This contradiction between selection in favour of aggression and timidity is reflected in the behaviour of individuals. As Tinbergen has shown, the tendencies to advance and to retreat may be so nicely balanced in an individual displaying near the boundary of its territory that it may perform apparently inappropriate " displacement activities ", for example preening or pecking at the ground, characteristic of a state of inner conflict [11].

Once it is recognized that there is selection in favour of timidity, that is, of retreat in response to a threat display from another individual, then it is clear that there will also be selection in favour of an exaggeration of the threat display itself, since the more exaggerated the display the greater the chance that it will cause an antagonist to retreat. Natural selection, so far from favouring a policy of " speak softly, but carry a big stick ", would appear to encourage one of " shout your head off, but run away if the other chap isn't impressed ".

It seems likely, then, that the threat displays of birds have evolved because they enable individual males to establish breeding territories in a favourable habitat, and that the "timid" response of one male to the threat display of another has evolved because the timid male thereby avoids physical injury.

FEMALE CHOICE

A male which establishes a favourable territory is in any case likely to leave more offspring. Darwin thought that such a male would derive an additional advantage : he would acquire as a mate one of the healthiest and most fertile females, since such females would be the earliest to be ready to breed. It is known that a female does seek out a male in a territory, and that she does not necessarily pair with the first unpaired male she comes across. It is therefore reasonable to think of the females " choosing " in the sense I defined earlier. But we do not know whether their choice is determined by the characteristics of the male, or of the territory which he occupies ; we do not know whether they marry for love or money. Nor is there any evidence that a male established in a particularly favourable territory is also likely to obtain as a mate a female more fertile than the average, although it may be true. Thus in this example the importance of male conflict is fairly clear, but the role of female choice is obscure.

The first clear evidence of female choice in an animal other than man was found by J. M. Rendel in the fruit-fly *Drosophila subobscura* [10]. In this, as in other species of *Drosophila*, there is a sex-linked recessive mutant gene causing the body to be yellow rather than dark grey. Rendel found that yellow females would mate equally readily with yellow or with normal males, but that normal females would rarely accept a yellow male, although such a male will court a normal female, and may attempt, unsuccessfully, to mount her. In this species we do not know what it is about yellow males which causes normal females to reject them. The problem has been pursued further by Margaret Bastock in another species, *D. melanogaster* [1]. Yellow *melanogaster* males have a lower mating success with normal females than have normal males, although the difference is less striking than it is in *subobscura*. During courtship, *melanogaster* males vibrate their wings, providing important stimuli received by the females' antennæ. Bastock found that yellow males vibrate less frequently and in shorter bouts. Since the yellow males are physically capable of long bouts of vibration, she concludes that they have a lower average level of sexual excitation; colloquially, they do not desire to mate as strongly as do normal males. While I accept this conclusion, I find it a puzzling one. In *subobscura*, Rendel found that yellow males will attempt to mount unwilling females, and I can confirm this observation, which seems inconsistent with the idea that the males do not want to mate. It is possible that the causes of the failure of yellow males are different in the two species.

Recently I have studied an example of female choice in *subobscura* which does not involve any visible genetical marker, and in which it has been possible both to show that the choice is effective in ensuring that a female is more likely to mate with a more fertile male, and to suggest a mechanism whereby the choice is exercised [8]. If virgin outbred females are paired with outbred males, mating usually takes place within fifteen minutes, and takes place within an hour in over 90 per cent of cases. If similar females are paired with inbred males, mating occurs within an hour in only about half the cases. A female which has once been inseminated will rarely mate again, even if she runs out of stored sperm, so that her future fertility depends on her first mating. Females inseminated by outbred males laid an average of over 1,000 eggs which hatched, whereas similar females inseminated by inbred males, although they laid the same total number of eggs, averaged only 264 hatching eggs. The reason for this difference is that inbred males produce fewer sperm, and that many of the sperm they do produce are in some way inadequate.

Thus females are more likely to mate with outbred than with inbred males, and if they do mate with an outbred male they leave about four times as many offspring. It does not follow that the difference between the likelihood of the two kinds of mating depends on female choice. It could be that the inbred males attempt to mate but are rejected, in which case one can reasonably speak of female choice; or it could be that inbred males do not try to mate, in which case female choice is not responsible. We therefore want to know whether inbred males do not mate with the females because they cannot or because they will not. I am convinced of the validity of this distinction, although I find it very difficult to define what I mean by it. In this particular case, however, I hope that the following account of the courtship of inbred and of outbred males will show that it is reasonable to say that the inbred males do not mate because they cannot, and that the females are choosing one kind of male in preference to the other.

When a female and an outbred male are placed together in a glass tube in the light, each fly may for a few minutes behave as if the other were not present. Then, usually rather suddenly, the male appears to notice the presence of the female; he turns towards and approaches her, flicking his wings and often protruding his proboscis. When he is close to the female, he circles round so as to face her, head to head. What happens next depends on the female. If she is too young to be ready to mate, she will either back away or turn

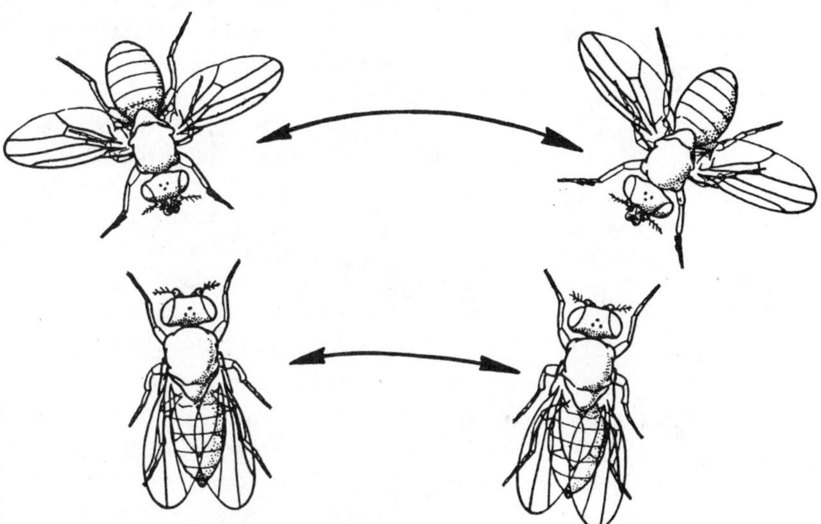

48.—The courtship dance of the fruit-fly, *Drosophila subobscura.* The male (above) sidesteps so as to keep facing the female; if he is successful in this, the female will usually accept him.

from the male, who may manœuvre so as to face her again, whereupon she will again turn away. If the female has already been inseminated, she also may turn or back away, but, if the male persists, she will make a characteristic signal by extruding her ovipositor towards the male. This signal, if repeated a few times, has the effect of stopping the male's courtship activities, of " discouraging " him. Finally, if the female is a virgin and old enough to mate, her reaction to a male approaching her head to head is to perform a characteristic side-stepping dance, moving very rapidly first to one side and then to the other. The male side-steps also, keeping his position facing the female. In a successful courtship, the female then stands still, slightly raising her wings to uncover her abdomen, and the male circles round and mounts. When the male is an outbred one, mating often follows the first dance to be performed, but sometimes there are several dances before the female stands still and permits the male to mount.

Sometimes, the courtship between an outbred female and an inbred male follows this course; but often it does not and mating does not take place. The inbred male in these cases will repeatedly approach the female head to head, and perform the side-stepping dance; but instead of accepting him the female turns or flies away. Watching such dances, it can be seen that an inbred male often lags behind, and fails to keep his position facing the female. A female will accept a male which succeeds in keeping his position facing her during the dance, and will reject one which lags behind.

I do not think that the failure of inbred males to keep up can be ascribed to any lack of " drive " or " motivation ". An inbred male which has danced repeatedly with the same female, but without success, may approach her directly from the side or from behind and attempt to mount. Although such attempts at " rape " always fail, they do suggest that the male is not lacking in sexual motivation. I think instead that the failure of inbred males to keep up is due to a lack of what in a human being we would call athletic ability. The spirit is willing but the flesh is weak.

To summarize, female *D. subobscura* perform a side-stepping courtship dance, and accept a male if he can keep up. In these experiments, in which only outbred and highly inbred males were used, many of the inbred males failed to mate. Females which mated with outbred males laid about four times as many eggs which hatched as did those which mated with inbred males. Thus there was an association between those characteristics of males making for mating success (probably athletic ability) and those making for fitness as a parent (the production of a large quantity of adequate sperm). It has not been shown that a similar association exists in wild populations, but it seems very likely that it would do so. Now, as was explained above, it is precisely such an association which Darwin thought to be a precondition for the effectiveness of sexual selection, at least for selection based on female choice. In *subobscura*, it is by means of performing a dance that females are able to discriminate between males; but this ability will not benefit a female unless she is more likely to choose a fertile than an infertile male.

It must be remembered that in *subobscura* the females are monogamous but the males potentially polygamous. It is therefore not surprising that choice is exercised by the females; R. Milani showed that males will attempt to copulate with a blob of wax of the right size which has been moved so as to simulate the dance of a female [9].

There is more room for the exercise of female choice in a species in which the males are polygamous, since each female has a choice of a larger number of males. But there is no reason to suppose that female choice could not be effective in a monogamous species, although there is little evidence that it is so. Monogamy is in the main confined to those vertebrate species in which the activities of both parents are necessary, either in feeding or protecting the young, or, in birds, in building a nest or in incubating the eggs. Such species are not particularly convenient for studies of the kind which have been made on *Drosophila*. But until such studies have been made it will be impossible to evaluate Darwin's ideas on sexual selection in monogamous species.

SEXUAL SELECTION IN MAN

Of all species our own is, for most people, the most immediately interesting, although it must be remembered that monogamy is a relatively recent and still far from universal custom. Historical changes in marriage customs and in the standards of human beauty, and the influence of social class in determining who shall marry whom, all make the study of sexual selection in man a complex one, which I am not competent to discuss. Darwin thought that there was not in the mind of man " any universal standard of beauty with respect to the human body ". He thought rather that each people tended to admire those characteristics which were peculiar to their own race, and which marked them off from others. Consequently sexual selection would tend to exaggerate any racial differences which already existed. He concluded that " of all the causes which have led to the differences in external appearance between the races of man, and to some extent between man and the lower animals, sexual selection has been the most effective ".

Darwin may well have been correct, although it is becoming increasingly difficult to discover what conditions were like during the main period of human evolution. Often, conquered or technically backward peoples have abandoned their own standards of taste in favour of those of their conquerors, just as they have abandoned their own gods for those of Christianity or of Islam. But sometimes the desire for national independence has led such people to cling to their own gods, and to their own traditions and standards of taste. It seems likely that in future racial intermarriage and the diffusion of culture will lead to a gradual obliteration of racial differences. But during the long period in which racial differences were being accentuated, it seems quite possible that sexual selection played the role which Darwin ascribed to it.

11

Darwin and Coral Reefs

By

C. M. Yonge

DARWIN's personal experiences on coral reefs were confined to the period spent in tropical seas in the Pacific and especially the Indian Ocean during the voyage of the *Beagle* from 1831 to 1836. His contributions to the study of reefs consist of observations made on the few reefs he visited, of which the atoll of Cocos-Keeling is far and away the most important, and the theory involving widespread subsidence in coral reef seas which he put forward to account for barrier reefs and atolls, and which represents deductions made from his own observations and still more those of others. Although, as he himself states, the entire theory was worked out deductively while he was still on the west coast of South America and so before he had personally encountered the coral reefs and islands of the Pacific and Indian Oceans, yet some account of his subsequent observations on these reefs should reasonably precede description and discussion of the theory with which his name will always be associated. These personal observations are contained in his book, *The Structure and Distribution of Coral Reefs*, first published in 1842 and, less formally, in his *Journal of Researches* which appeared three years later and in *Charles Darwin's Diary of the Voyage of H.M.S. Beagle*, edited by his grand-daughter, Lady Barlow, and published in 1933.

To quote the opening words of the *Journal:* " After having been twice driven back by heavy south-western gales, Her Majesty's ship *Beagle* . . . sailed from Devonport on the 29th of December, 1831." The first glimpse of corals came only some three weeks later at Quail Island in the Canaries. On this " miserable desolate spot ", quoting now from the *Diary*, there came

> the first burst of admiration at seeing corals growing on their native rock. Often, whilst at Edinburgh, have I gazed at the little pools of water left by the tide: and from the minute corals of our own shore pictured to myself those of larger growth: little did I think how exquisite their beauty is and still less did I expect my hopes of seeing them would ever be realized.

What corals he saw in rock pools in the Firth of Forth can only be conjectured: none of the few British species of stony corals (Madreporaria) would occur there; but he might have seen white or orange-coloured masses of soft coral (Alcyonacea) and would certainly have seen the pink encrustation of coralline algæ.

DARWIN'S OBSERVATIONS

But it was not until after the major work of the *Beagle*—and of her naturalist—had been completed during the greater part of four years spent in South America that Darwin had his first distant view of oceanic coral reefs. After finishing her survey of the Galapagos Islands, on October 20th, 1835, the *Beagle* began the long voyage across the Pacific and coming after some weeks to the " Low or Dangerous Archipelago ", Darwin records in the *Journal* that he

> saw several of those most curious rings of coral land, just rising above the water's edge, which have been called Lagoon Islands. A long and brilliantly-white beach is capped by a margin of green vegetation: and the strip, looking either way, rapidly narrows away in the distance, and sinks beneath the horizon. From the mast-head a wide expanse of smooth water can be seen within the ring. These low hollow coral islands bear no proportion to the vast ocean out of which they abruptly rise; and it seems wonderful that such weak invaders are not overwhelmed by the all-powerful and never-tiring waves of that great sea, miscalled the Pacific.

Later, from a height of between two and three thousand feet on Tahiti, he viewed the distant island of Eimeo which

> with the exception of one small gateway, is completely encircled by a reef. At this distance, a narrow but well-defined brilliantly-white line was alone visible, where the waves first encountered the wall of coral. The mountains rose abruptly out of the glassy expanse of the lagoon, included within the narrow white line, outside which the heaving waters of the ocean were dark-

Plate III. Seaward edge of a growing coral reef at low tide. Capricorn Islands at southern end of Great Barrier Reef. (See Page 252.)

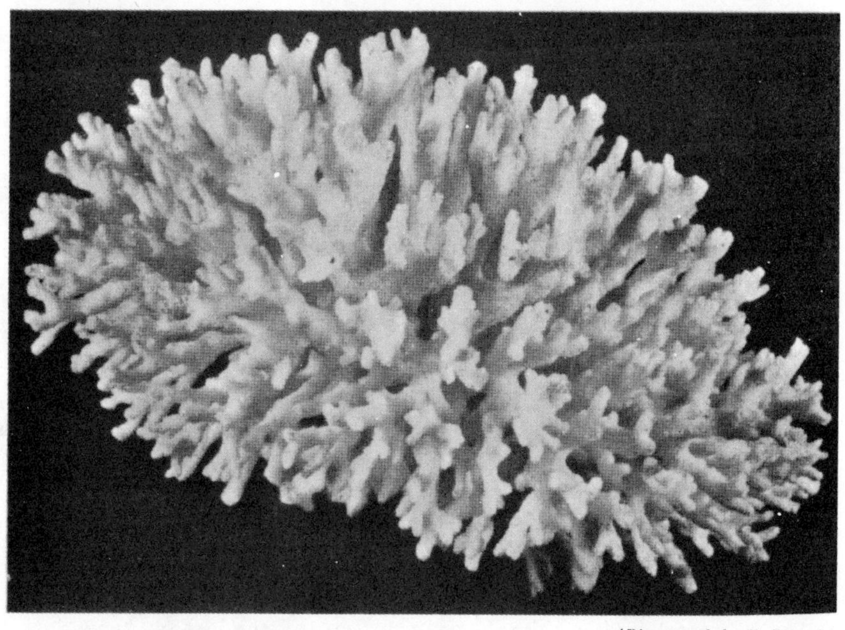

(*Photograph by F. Pittock.*)

Plate IV. Skeleton of a reef-building Coral (*Pocillopora bulbosa*). (See Page 251.)

(Photograph by the U.S. Navy.)

PLATE V. IFALUK ATOLL, WESTERN CAROLINES. PART OF THE RIM AND LAGOON ARE SHOWN: DARK AREAS ARE ISLANDS, LIGHT AREAS REEF SURFACE. THE ATOLL IS 5,500 FEET LONG. (SEE PAGE 246.)

> coloured. The view was striking: it may aptly be compared to a framed engraving, where the frame represents the breakers, the marginal paper the smooth lagoon and the drawing the island itself.

He was here describing a volcanic island surrounded by a barrier reef.

These would seem to have been Darwin's sole contacts with the coral reefs of the Pacific. Proceeding now by way of New Zealand and then New South Wales, the *Beagle* sailed around the south coast of Australia and so into the Indian Ocean where, steering more north than west, she arrived on April 1st, 1836, at the Keeling or Cocos Islands, as Darwin calls them, now usually known as Cocos-Keeling, which lie some six hundreds miles to the south-west of Sumatra. Here we must pause to view through his eyes the one coral formation he was able to study in fair detail.

> The ring-formed reef of the lagoon island [he tells us in the *Journal*] is surmounted in the greater part of its length by linear islets. On the northern or leeward side there is an opening through which vessels can pass to the anchorage within. On entering, the scene was very curious and rather pretty; its beauty, however, entirely depended on the brilliancy of the surrounding colours. The shallow, clear, and still water of the lagoon, resting in its greater part on white sand is, when illuminated by a vertical sun, of the most vivid green. This brilliant expanse, several miles in width, is on all sides divided, either by a line of snow-white breakers from the dark heaving waters of the ocean, or from the blue vault of heaven by the strips of land, crowned by the level tops of the cocoa-nut trees. As a white cloud here and there affords a pleasing contrast with the azure sky, so in the lagoon bands of living coral darken the emerald-green water.

Such was his impression of an atoll when viewed from within after the *Beagle* had entered the lagoon through the northern and leeward side of the atoll ring. Later he landed, finding a strip of land only about a quarter of a mile wide and on the lagoon side with a white coral beach, " the radiation from which under this sultry climate was very oppressive ", while on the seaward side a firm terrace of coral rock " served to break the violence of the open sea ". His account of the linear islands that bound much of the atoll ring brings vividly to mind memories of coral islands the world over.

> The solid flat of coral rock . . . breaks the first violence of the waves, which otherwise, in a day, would sweep away these islets and all their productions. The ocean and the land seem here struggling for mastery: although *terra firma* has obtained a footing, the denizens of the water think their claim at least equally good. In every part one meets hermit crabs of more than one species, carrying on their backs the shells which they have stolen from the neighbouring beach. Overhead numerous gannets, frigate-birds, and terns, rest on the trees; and the wood, from the many nests and from the smell of the atmosphere, might be called a sea-rookery.

Later he " waded over the outer flat of dead rock as far as the living mounds of coral, on which the swell of the open sea breaks. In some of the gullies and hollows there were beautiful green and other coloured fishes, and the form and tints of many of the zoophytes were admirable." Nevertheless, like many who have followed him over the surface of coral reefs, he finds that " those naturalists who have described, in well-known words, the submarine grottoes decked with a thousand beauties, have indulged in rather exuberant language ".

He accompanied Captain Fitz Roy to an island at the head of the lagoon " winding through fields of delicately branched corals " and then crossed the narrow islet to find himself on the exposed shore against which the south-east trade wind drove a

> line of furious breakers, all rounding away towards either hand. . . . The ocean throwing its waters over the broad reef appears an invincible, all-powerful enemy; yet we see it resisted, and even conquered by means which at first appear most weak and inefficient. It is not that the ocean spares the rock of coral; the great fragments scattered over the reef, and heaped on the beach, whence the tall cocoa-nut springs, plainly bespeak the unrelenting power of the waves.

He perceives the sustained power of the long trade wind swell continually breaking along the margin of the atoll.

> Yet these low, insignificant coral-islets stand and are victorious: for here another power, as an antagonist, takes part in the contest . . . Thus do we see the soft and gelatinous body of a

polypus, through the agency of the vital laws, conquering the great mechanical power of the waves of an ocean which neither the art of man nor the inanimate works of nature could successfully resist.

On the morning of April 12th, 1836, the *Beagle* sailed out of the lagoon on the long journey westward to Mauritius. While still only about a mile from the shore Captain Fitz Roy sounded with a line about one and a half miles long and found no bottom. Later soundings have shown that the depth is much less than this, but to Darwin the atoll islands appeared as ringing the saucer-shaped summit of an exceptionally steep-sided submarine mountain. This he felt to be a mystery of building far surpassing that of the pyramids which are, he wrote, so utterly insignificant " when compared to these mountains of stone accumulated by the agency of various minute and tender animals! This is a wonder which does not at first strike the eye of the body, but, after reflection, the eye of reason."

The final view of coral reefs was probably at Mauritius where the *Beagle* remained from April 29th to May 9th. Little is said on the subject in the *Journal* or the *Diary*: Darwin and his fellows were all now preoccupied with the prospect of returning home after their long absence. As he writes in a letter to his sister, Caroline, from Mauritius, " there is no country which has now any attraction for us, without it is seen right astern, and the more distant and indistinct the better. We are all utterly home-sick." Nevertheless, as appears from the pages of *The Structure and Distribution of Coral Reefs*, he was far from inactive. He visited the reef which fringes the western, and sheltered, shores which slope gently so that even at its greatest distance of some three miles from the land the reef rises from relatively shallow water. The outer edge of the reef he found to be " tolerably well defined " and a little higher than any other region and to consist chiefly of large branched species of *Madrepora* (now *Acropora*). Off these reefs he sounded " with the wide bell-shaped lead which Captain Fitz Roy used at Keeling ". Beyond the irregular border of branched *Acropora* he found that the water deepened gradually, reaching a depth of some twenty fathoms at a distance of between one-half and three-quarters of a mile from the reef. Only below eight fathoms did the arming of the lead bring up some sand, but living coral persisted down to twenty fathoms where it was almost completely replaced by sand. To the landward side of the reef margin, Darwin describes a typical boat channel with a sandy bottom and some clusters of living coral.

He noted also how the reef was breached in front of the mouths of rivers and streams, although many were dry for much of the year. Being unable to visit the eastern side of the island he had to rely on descriptions of the reef which is there exposed to a heavy surf. He was told that it had " a hard smooth surface, very slightly inclined inwards, just covered at low-tide, and traversed by gullies; it appears to be quite similar in structure to the reefs of the barrier and atoll classes."

Thus in the course of his voyage around the world Charles Darwin had viewed atolls and barrier reefs from afar in the Pacific; he had examined one atoll with some care in the Indian Ocean where he had also viewed the fringing reef at Mauritius. He had seen living corals of many types and obtained some impression of their distribution in depth and also in relation to exposure on windward and leeward surfaces. And he had obtained some information, not all of it as it eventually turned out correct, about the submarine contours of coral reefs.

In his autobiography, Darwin speaks of the period between his return to England on October 2nd, 1836 and his marriage on January 28th, 1839 as the most active in his life. He acted as one of the honorary secretaries of the Geological Society and also saw much of Lyell who showed particular interest in his views on coral reefs. To this may be due their early publication, in the form of a short paper read before the Geological Society in 1837, entitled " On certain Areas of Elevation and Subsidence in the Pacific and Indian Oceans, as deduced from the Study of Coral-formations ". Meanwhile the book on coral reefs was begun, but slowed down during the long period of illness which followed his marriage. However, " the greater part of my time, when I could do anything ", he reports in this autobiography, " was devoted to my work on *Coral Reefs* . . . of which the last proof-sheet was corrected on May 6th, 1842. This book, although a small one, cost me twenty months of hard work, as I had to read every work on the islands of the Pacific and to consult many charts." The book was published in that year as the First Part of the Geology of the *Beagle*. Its future history may briefly be mentioned. A second and much revised edition was published by Darwin in 1874. Here some additional matter was added and reference was made to certain criticisms which had appeared, but the main argument remained unaltered. It is this edition, representing Darwin's final published views on coral reefs, which has been referred to in the preparation of this article. The third edition of *Coral Reefs* was published in 1889, when, as stated by Sir Francis Darwin in the Preface, " For all that distinguishes the present from the second edition the reader has to thank Professor Bonney. He has added occasional footnotes . . . and has given, in the form of an appendix, a careful summary of the more important memoirs published since 1874." For, as we shall see, a problem which Darwin seemed initially to have clarified beyond need of further major evidence was to become, as the century advanced, the subject of increasing argument and a major battleground of conflicting theories.

THE NATURE OF CORALS AND CORAL REEFS

The object of Darwin's book, as he explains in the Introduction, is to describe the principal kinds of reefs and, as far as possible, to explain how they have arisen. He disclaims any intention of describing the corals themselves apart from considering their distribution and " the conditions ", as he puts it, " favourable to their vigorous growth". Little, it may be noted, was then known about corals. They had been recognized as animals only since the discoveries of Peyssonnel and of John Ellis in the preceding century. Their whitened skeletons appeared in increasing numbers in the natural history museums of the world, but there was great ignorance about the animals which form them. These are, apart from the colonial form of growth which most display and the invariable capacity to form a calcareous skeleton, very closely allied to sea anemones. They are by no means the sole organisms concerned in the formation of reefs. The red calcareous algæ or nullipores which form an encrusting cement on the windward surface are of fundamental importance in the consolidation of a reef, the substance of which is composed also of the shells of many animals from giant clams and great marine snails to the small chambered tests of the ubiquitous foraminiferans. Sand and calcareous fragments of all kinds pack the interstices between the larger masses of coral skeletons and shells. And both upon and within the reef mass dwell the innumerable animals and plants which, together with the reef builders, form the marine communities known as coral reefs.

The first reefs encountered by western man were probably those in the Red Sea which were certainly known to the Greeks of Alexandria. After the rounding of the Cape of Good Hope, the wide-spread reefs of the Indian Ocean gradually came to the knowledge of Portuguese, French, Dutch and British navigators and traders. The even greater coral wealth of the Pacific was revealed largely during the three great voyages of Cook. Darwin notes how the early voyagers distinguished broadly between "atolls" or "lagoon-islands"; "barrier" or "encircling reefs"; and "fringing" or "shore reefs". ("Atoll" is perhaps the only English word derived from the language of the Maldive Islands, according to Gardiner [3]. There each governmental district, which consists of a circular reef enclosing a lagoon, is known as an "atolu".) These categories hold good today, the first of each pair of alternative names now alone being used. They are easily distinguished. Fringing reefs flank the shores of continents or rocky islands to which they are closely attached, being separated by no more than the shallow boat channel observed by Darwin at Mauritius. Barrier reefs are similarly associated with a land mass, but often far removed in distance from this by a lagoon channel up to thirty fathoms in depth. Distantly viewed by Darwin from the mountains of Tahiti, barrier reefs are at their greatest along the north-eastern shores of Australia where the Great Barrier Reef—really a vast series of individual reefs—extends for a length of over 1,200 miles roughly parallel to a coast from which it is separated at its southern end by a distance of 150 miles. Atolls have no connection with any such land mass. Early navigators were enchanted and amazed by them and they remain amongst the most astonishing sights on the globe, ring-shaped series of reefs, many of them capped with sandy islands bearing trees and often inhabited, which arise from vast depths in mid-ocean, the blue of which they separate from the calm green waters of an enclosed lagoon seldom more than thirty fathoms deep. Cocos-Keeling is a typical atoll.

MODE OF FORMATION OF REEFS

There is, as Darwin pointed out, no problem where fringing reefs are concerned. Given a fairly gently sloping shore in the tropics, corals may establish themselves offshore and grow upwards until they break the surface when growth can continue mainly in a seaward direction. Here they find conditions for most vigorous growth, because it is a feature of corals that they grow most actively against the force of seas which bring both oxygen and their planktonic animal food. Where conditions are most severe, for instance on the eastern shores of Mauritius against which break seas driven by the south-east trade winds, growth both of corals and of the cementing surface of pink coralline algæ is greatest. Darwin noted how freshwater inhibited coral growth so that a passage was left opposite to each river mouth; he also comments on the harmful effect of sand and sediment. This is actually less than he and many who came after him thought; reefs are known to have arisen from the bottom of muddy bays in the East Indies while individual corals can do much to rid themselves of falling sediment. But it remains true that coral growth is most vigorous in clear water especially where this is driven before a steady trade wind. Another factor affecting the growth of reef-building corals (not all corals, because certain types live only in deep or in cold seas) is temperature. It was first pointed out by J. D. Dana that reefs flourish only where surface temperatures never fall much below 20° C. This explains why coral reefs are effectively confined to the tropics and even there never abound on the eastern sides of oceans where cold currents lower water temperature.

Clearly the width of a fringing reef will depend, other conditions such as exposure not varying, on the slope of the coast. If this is gentle the reef will extend for a considerable distance seaward, for as much as three miles as Darwin observed in Mauritius, but where the submarine slopes are steep the reef will be narrow and Darwin quotes examples from the voyages of Cook and other Pacific navigators of fringing reefs only 50 to 100 yards wide, the sea around these islands being very deep.

Of barrier reefs, Darwin, as we have seen, had little more than a distant view. He did pay one visit to the edge of the reef around Tahiti, but although it was low tide the surf was too great for him to see the nature of the corals. It appeared to him from the statements of local chiefs that " they resemble in their rounded and branchless forms, those on the margin of Keeling atoll. The extreme verge of the reef which was visible between the breaking waves at low tide, consisted of a rounded, convex, artificial-like breakwater, entirely covered with Nulliporæ, and absolutely similar to that which I have described at Keeling atoll." In this supposition he was correct. But although the solidity of coral and algal growth permits the maintenance of a reef often far removed from land, it does not explain the initial formation of such a reef. It does not arise from shallow water because the outer slopes descend steeply to the great depths of the surrounding ocean. Moreover one cannot consider that barrier reefs are just greatly extended fringing reefs, because of the depth of water (often over 200 feet as Darwin notes) which separates them from the land mass.

But if the origin of barrier reefs presents a problem, that of atolls presents an even greater one. Atolls descend steeply to great depths around their outer margins and encircle a relatively shallow lagoon. They vary enormously in size, from a few miles in diameter to the vast size of the greatest of the Maldive atolls, eighty-eight miles long and some twenty miles wide. There is a relation, as Darwin realized to a greater extent than many of his successors, between the prevailing direction of wind and weather and the form of the reef. As he stated, " the islets appear to be first formed, and are generally of greater length on the more exposed shore. The islets, also, which are placed to leeward as regards the trade wind, are in most parts of the Pacific liable to be occasionally swept entirely away by gales, equalling hurricanes in violence, which blow in the opposite direction." A reef invariably builds firmest where it is continually exposed to heavy seas, and mechanically it is much less sound in the lee. Hence it is in these regions of apparent security that damage is greatest, when for brief periods often of hurricane weather they become the exposed instead of the sheltered surface of the atoll. Coral growth tends to be inhibited in regions along the lee where openings into the lagoon, as at Cocos-Keeling, usually occur.

Atoll lagoons represent a major problem. They are never really deep, seldom over thirty fathoms and often much shallower, although a few reach a depth of fifty fathoms. Much of the bottom may be of soft calcareous mud as Darwin described, although apparently often with much growth of the green calcareous weed, *Halimeda*. But pinnacles of coral rock, usually covered with species often distinct from those living on the exposed surface of the reefs, rise from the bottom and are much more numerous than Darwin or his contemporaries and successors realized. Their numbers have only become apparent with the advent of echo-sounding: this enables a vessel to make a complete survey of a lagoon, recording the pinnacles as it passes over them, during a period of time in which the old survey vessels could only have made a few traverses with the sounding lead, no cast with which might happen to fall upon a coral pinnacle. Conditions within an atoll lagoon resemble in many respects those found within the shelter of a barrier reef. There the cemented reef crest is followed to leeward by a zone of broken coral and boulders and then a gradually shelving sandy bottom with mushroom-topped pinnacles of living coral which may break the surface of the sea at low water of spring tides [6].

The first attempt at an explanation of the form of atolls would seem to have been given by Chamisso, the naturalist who accompanied the Russian navigator, Lieutenant O. E. Kotzebue, on a long voyage of discovery including the South Sea Islands between 1815 and 1818. Chamisso, best known as the author of that strange romance of the German romantic school, " Peter Schlemihl, or the Shadowless Man ", considered corals to be essentially creatures of the surf. He thought of them as growing outwards from a central nucleus just as do " fairy rings " of toadstools in the summer fields. In this way he explained the formation of atolls. There is an element of truth in this theory, because corals undoubtedly do grow outwards against wind and weather. But at the time when the *Beagle* sailed around the world the general view held in scientific circles was that atolls were reefs which had grown up around the crater rims of submarine volcanoes.

DARWIN'S THEORY

Before he had even seen a coral reef, Darwin emphatically rejected such a view. "The idea of a lagoon island, thirty miles in diameter being based on a submarine crater of equal dimensions, has always appeared to me to be a monstrous

49.—Part of a young colony of a coral (*Porites haddoni*). A group of polyps is shown, each with a mouth and a ring of tentacles. (From T. A. Stephenson, in *Sci. Rpts. G. Barrier Reef Expedition* 1928–29, vol. III, no. 3, 1931.)

hypothesis," he writes to his sister after visiting Cocos-Keeling. We know that the idea of subsidence had been in his mind long before he even saw a coral reef. Amongst the matters which had to be decided before his evidence was complete was the depth at which reef-building corals can live, and to this matter he devotes the fourth chapter of his book. He had here the evidence of Captain Fitz Roy's soundings off Cocos-Keeling and of his own at Mauritius; he also sought the evidence of earlier navigators and naturalists, not even neglecting reports from pearl divers at Yemen and Massaura in Arabia. He had data from Moresby's detailed surveys in the Maldives and, in the second edition of *Coral Reefs*, the extensive observations made by J. D. Dana during the course of the United States Exploring Expedition under Captain C. Wilkes in 1838–41. His final conclusions are " that in ordinary cases reef-building polypifers do not flourish at greater depths than between twenty and thirty fathoms, and rarely at above fifteen fathoms ". He admits change comes gradually with increasing depth and, with his apt gift for analogy, points out:

> If a person were to find the soil clothed with turf on the banks of a stream of water, but on going to some distance on one side of it he observed the blades of grass growing thinner and thinner with intervening patches of sand, until he entered a desert of sand, he would safely conclude, especially if changes of the same kind were noticed in other places, that the presence of water was absolutely necessary to the formation of a thick bed of turf; so may we conclude, with the same feeling of certainty, that thick beds of coral are formed only at small depths beneath the surface of the sea.

Subsequent and much more elaborately conducted submarine surveys have not substantially altered Darwin's estimate of the depth at which reef-building corals can flourish.

Another matter on which he sought evidence, although inevitably with less conclusive results, was the rate at which coral reefs grow. This is by no means the same thing as the rate at which individual coral colonies under favourable conditions may increase. Thus he was informed of a ship which had acquired a layer of coral two feet thick on her bottom during a period of twenty months spent in the Persian Gulf and he quotes other instances of spectacular growth by corals. But many adverse factors have to be borne in mind when considering reefs as a whole. There is mechanical damage by wave action and destruction by exposure to the air and by the continual action of boring organisms of many kinds, both animal and plant, which quickly reduce coral rock to calcareous fragments and finally to sand. There are also less easily estimated factors such as the quantity of planktonic food.

While certainly not fully acquainted with the manifold hazards to reef growth and maintenance, Darwin realized the complexity of the problem; he noted moreover that extensive lateral growth of a reef was possible only where " conditions were favourable to the vigorous and unopposed growth of the corals living in the different zones of depth, and that a proper basis for the extension of the reef was present. These conditions must depend on many contingencies, and a basis within the requisite depth can rarely be present in the deep oceans where coral formations most abound."

After weighing the evidence then available he concludes that " any thickness of rock composed of a singular intermixture of various kinds of corals, shells, and calcareous sediment, might be formed; but *without subsidence* the thickness would necessarily be determined by the depth at which the reef-building polypifers can exist " [my emphasis]. That is, of course, some twenty fathoms. He admits there is no precise evidence about the upward rate of growth of reefs, but adds that in areas where subsidence has probably long been in progress coral growth had maintained the reefs close to the surface; this, he considers, " is a much more important standard of comparison than any cycle of years ". Uncertainty persists about the speed at which reefs may grow. In a brief summary of the position published elsewhere [7], the existing evidence indicates that uninterrupted upward growth of corals might produce a reef some ninety feet thick in 1,000 years in the Indian and Pacific Oceans; but the same growth would take substantially longer in the Atlantic, where reef-building species are fewer and growth less vigorous. Moreover, it appears that growth is probably greater at depths of a few fathoms than it is near the surface where experiments and observations on coral growth have all been made. So the potential capacity for upward growth of corals may be correspondingly greater. There is certainly nothing in modern knowledge which affects the conclusions which Darwin drew from the limited data available to him.

Darwin's " Theory of the Formation of the Different Classes of Coral-Reefs " forms the subject matter of Chapter V in *Coral Reefs*. He begins by noting that previous naturalists have concerned themselves exclusively with the problem of atolls, "and have passed over, almost unnoticed, the scarcely less wonderful encircling barrier reefs". A theory to explain the former must, he is the first to realize, also account for the latter. (The formation of fringing reefs, it may be repeated, presented no problem to him, nor does it to us today.) Darwin begins by disposing of the crater rim theory and of the earlier, and

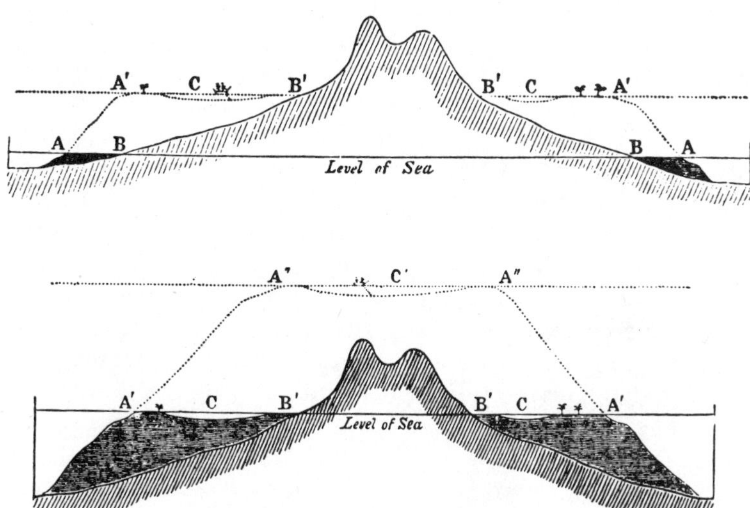

50.—Darwin's diagrams showing how, by a process of subsidence a fringing reef around a rocky island may be converted into a barrier reef (upper figure) and then, by further sinking of the land involving complete submergence, an atoll be formed (lower figure). A, A′, A″; seaward margins respectively of fringing reef, barrier reef and atoll rim. B, B′; shores of island when surrounded first by fringing and then by barrier reef. C, C′; lagoon channel between barrier reef and island and central lagoon within newly formed atoll ring respectively.

happier, theory of Chamisso. He then considers the implications of two facts, first that reef-builders live only in shallow water and second that, so far as he could determine, coral reefs and islets never rise higher than sand and fragments can be thrown by wind and waves. He is thus led to the conclusion that all reefs, including the widely scattered atolls, must rest on a basis of rock, from this they originally grew upward until their summits broke the surface, where living coral was killed by exposure and any further upward addition was due to piling up of fragments by the action of the wind and the sea.

More than this, however, was needed to explain the depths of lagoons within atolls and of lagoon channels within barrier reefs. These are usually greater than the lowest levels at which reef-builders can live. Darwin now brings in his theory of subsidence. He postulated a slow and long-continued sinking of the bed of the ocean, so that fringing reefs which bounded the shores of continents and high rocky islands would slowly be converted into barrier reefs. They would continue by their upward growth to maintain contact with the surface of the sea while the land as it sank became further and further separated from the reef by a channel of ever-increasing depth. Where in the case of an island the process was continued until all original land disappeared below sea level, only a ring of reefs would remain to indicate the former existence of that island and an atoll would come into being. The depth of the enclosed lagoon would increase as the former island mass continued to sink. Darwin's original diagrams (figure 50) in themselves are sufficient indication of his postulated changes. During the downward movement of the land the actively growing margin of the original fringing reef maintains position near the surface, to form the edge of the barrier or the atoll, while the collection of sediment within would tend to inhibit, except locally, the growth of coral within the lagoon channels and atoll lagoons. Atolls might also, in Darwin's opinion, occasionally be formed directly where coral originally fringed a shallow bank of rock or of hardened sediment which then subsided.

Surveying from his study, with the aid of all available information, the distribution of the three categories of coral reefs, Darwin produced a map of the world on which atolls and submerged annular reefs (drowned atolls in his view) were shown dark blue, barrier reefs pale blue and fringing reefs indicated in red. He also marked regions of present or historically recent volcanic activity. From this map he contended that the regions shown blue were areas of subsidence, the rate of which was less than the upward rate of coral growth. The red areas were either effectively stationary, having not subsided recently or subsided only to an insignificant extent, or had " been repeatedly upraised with new lines of reefs successively formed round them ". Such conditions are well known, most impressively perhaps in the Tongan Islands: some of these are volcanic but others consist of successive terraces of raised fringing reefs rising to heights of some hundreds of feet. In his final paragraph he points out how, in his opinion, the distribution of these two great major categories of reefs offers

> a grand and harmonious picture of the movements which the crust of the earth has undergone within a late period. We there see vast areas rising, with volcanic matter every now and then bursting forth. We see other wide spaces sinking without any volcanic outbursts; and we may feel sure that the movement has been so slow as to have allowed the corals to grow up to the surface, and so widely extended as to have buried over the broad face of the ocean every one of these mountains, above which the atolls now stand like monuments, marking the place of their burial.

In his Rede Lecture, *Charles Darwin as Geologist*, delivered at the Darwin Centennial Commemoration on June 24th, 1909, Sir Archibald Geikie says, in reference to the reception of *Coral Reefs:* " The remarkable simplicity of this explanation of phenomena that had so long been matters of dispute, together with the grandeur of the vista which the theory opened up of a stupendous geographical revolution that had been in progress since a remote antiquity, assured Darwin's views of close attention and led to their general acceptance." There is some similarity in the broad simplicity of ultimate statement between the subsidence theory of the origin of barrier reefs and atolls and the theory of evolution that appeared seventeen years later. Both represent the bringing together of evidence from many sources and reveal a mind capable of seizing upon the crucial aspects of highly complicated problems and an imagination that ranged freely through both space and time. The subsidence theory was, however, based on far less detailed evidence and its implications were less fully appreciated.

Darwin was fortunate in the immediate support of J. D. Dana, precisely four years his junior both in years and in the study of coral reefs. In the English edition of his *Corals and Coral Islands*, published in 1872, the then Professor of Geology and Mineralogy at Yale states that the cruise of the U.S.

Exploring Expedition of 1838 to 1842 followed to some extent the same course as that of the *Beagle*.

> Soon after reaching Sydney in 1839, a brief statement was found in the papers of Mr. Darwin's theory with respect to the origin of the atoll and barrier forms of reefs. The paragraph threw a flood of light over the subject, and called forth feelings of peculiar satisfaction, and of gratefulness to Mr. Darwin, which still come up afresh whenever the subject of coral islands is mentioned . . . on reaching the Feejees, six months later, in 1840, I found there similar facts on a still grander scale and of a more diversified character, so that I was afterward enabled to speak of his theory as established with more positiveness than he himself, in his philosophic caution, had been ready to adopt. His work on coral reefs appeared in 1842, when my report on the subject was already in manuscript. It showed that the conclusions on other points, which we had independently reached, were for the most part the same.

There is something here of the relationship, in scientific comradeship and generosity of spirit, later established also between Darwin and Wallace. Dana did more than give support to the subsidence theory. He added to the evidence in its favour by revealing a further implication, namely the frequent deep embayment of coastlines within barrier reefs. This is to be expected if land previously cut into deep valleys by rivers then proceeds to sink. Dana also saw many more reefs than Darwin. It is therefore fitting that the views they both upheld should often be referred to as the Darwin-Dana theory of the origin of coral reefs.

MODERN ARGUMENT AND EVIDENCE

For a quarter of a century this theory was largely unchallenged. Then it began to come under increasing attack; it was maintained for instance that atolls can be formed in regions of elevation and lagoons be hollowed out by solution of coral limestone. But the numerous opponents of the Darwin-Dana theory were far from agreeing among themselves and it is not possible here even to outline the many alternative theories that were propounded in the later decades of the nineteenth century and more recently. They have been admirably summarized by W. M. Davis, himself the leading recent supporter of the subsidence theory [2]. He lists nine alternative theories associated with the names, amongst others, of Semper, Murray, Guppy, Alexander Agassiz, Wharton and Gardiner.

The views of Murray do demand attention. As a member of the scientific staff of the *Challenger* Expedition from 1872 to 1876, he had been particularly concerned with the study of oceanic bottom deposits, and with these in mind published, on his return, views on the formation of coral reefs which were in effect precisely opposite to those of Darwin. He thought of reefs as growing upwards from pre-existing submarine platforms, probably the tops of extinct volcanoes which, if not initially near enough to the surface for reef-builders to establish themselves, were gradually raised to that level by the deposition upon them of sediment consisting largely of the skeletons of planktonic animals and plants. Once established the reef would grow towards the surface but, with the marginal corals growing with the greatest vigour, forming the beginning of the atoll ring. The more slowly growing coral in the central areas would eventually be killed by sediment and the lagoon then be formed and progressively deepened by solution.

If such was the manner in which atolls had come into being they would be assisted by elevation and certainly not demand subsidence, since this, unless very slow, might carry the reef-builders to depths greater than they could survive. There was much general support for Murray's views. Darwin himself was amongst the first to realize that there appeared a possibility of determining whether the truth lay with his theory or with that of Murray. Darwin had postulated a great thickness of coral limestone formed by continued growth upon a slowly sinking foundation; but if Murray was correct then only a relatively thin layer of coral rock, no greater than the depth to which reef-builders can live, should overlie the foundation—of rock or sediment—on which the atoll rested. The truth of the matter should therefore be determined by boring deeply into the substance of an atoll. "I wish," Darwin wrote in the year preceding his death to his friend the American oceanographer, Alexander Agassiz, "that some doubly rich millionaire would take it into his head to have borings made in some of the Pacific and Indian atolls, and bring back cores for slicing from a depth of 500 or 600 feet."

But it was the Royal Society, not the millionaire, which, fifteen years later, had such borings made on the atoll of Funafuti in the Ellice Islands. After much initial disappointment the reef was successively bored to a depth of 1,114 feet at a position some 500 feet from the reef edge on the weather side. Unfortunately this settled nothing because, although coral fragments were found throughout the length of the core, as demanded by Darwin's theory, yet, as Murray and his supporters rightly contended, the boring might well have passed through the mound of coral fragments which litter the submarine slopes of atolls and over which the rim of the atoll, by slow seaward growth, gradually extends. During the present century further borings, on the Great Barrier Reef of Australia, off Borneo, on a small island east of Okinawa and at Bikini in the Marshall Islands, have been made to depths of from 600 to 2,556 feet and in no case were the underlying foundations encountered. Success came in 1952 at Eniwetok atoll, near Bikini, where, in the course of tests made by the Atomic Energy Commission of the United States, two borings were made to the unprecedented depths of 4,222 and 4,630 feet and at the bottom of both, after passing through four-fifths of a mile of calcareous sediments, the surface of the basement rock was briefly penetrated. It consisted of volcanic olivine basalt. Study of submarine contours reveals that Eniwetok atoll consists of a vast cap of limestone—all deposited by the action of living organisms—which rests on the summit of a basaltic volcano which rises some two miles above the bed of the ocean. While the precise age of this volcano is uncertain, it does appear that at the time when it became extinct it projected high above the surface of the sea. The first 1,000 feet of the sediments were laid down in Eocene times but the contained fossils, to quote the authors of the initial report on these historic borings, " are clearly recognizable as shallow-water forms, and their occurrence in the thick Eocene section, as well as in the younger rocks above, confirms the major premise of Darwin's theory—great subsidence " [5]. Thus, under the stimulus of forces and of needs he could never have foreseen, comes successful completion of the experiment which Darwin sought, and confirmation of his views.

It may now with some confidence be stated that atolls have

been formed by subsidence as Darwin postulated. This, however, does not necessarily imply that all atolls and all barrier reefs have been formed in this manner. Other modes of origin appear possible, with a final similarity due to the same moulding action of wind and weather in trade wind seas. Moreover, by following the course of boring operations from their inception at Funafuti to culminating success at Eniwetok no mention has been made of the glacial control theory developed in the intervening period by R. A. Daly [1]. With much supporting evidence, this eminent geologist claimed that modern coral reefs have arisen on the numerous platforms cut by the lowered seas during the glacial periods when much water was locked away in the polar ice-caps. Old reefs formed during the Tertiary period could then have been cut down and also island masses themselves, especially after their protecting reefs had been killed by contemporary lowering of sea temperature.

Such shallow water platforms, perhaps some thirty fathoms deep, in tropical seas could certainly have provided bases for the establishment of reefs. Although this theory—like all other attempts to explain the origin of coral reefs—has given rise to interminable discussion, many have thought that it does contain important elements of truth. Amongst these is the Dutch geologist, Ph. H. Kuenen, who feels that in a combination of the theories of subsidence and glacial control lies the best explanation of the origin of coral reefs [4]. He thinks of the Tertiary reefs as arising during periods of slow subsidence such as those postulated by Darwin. The coral would be killed by exposure and the reefs cut down by wave action during the glacial periods of lowered sea level; finally, the sea level would rise to its present height, and the corals, re-established on the rocky margin of the platform, would form a new reef which grew upwards and outwards. On this basis an explanation is provided for the largely constant depth of lagoons and lagoon channels and for the absence of passages deeper than lagoon level through barrier reefs and atoll rims. Both have been difficult to explain on the unaided theory of subsidence.

The investigation of coral reefs is very far from completed; the problems presented will long continue to exercise the imagination of all who are interested in the structure of the globe and especially those concerned with the interaction, here so intricate, between living organisms and the diversely acting factors of the physical environment. To the full solution of this problem the contribution of Charles Darwin may eventually prove to have been the most significant.

12

Darwin as a Botanist

By

J. Heslop-Harrison

In a brief analysis published in 1899 [8], Francis Darwin distinguished two periods in his father's botanical work, one evolutionary in inclination, the other primarily physiological. The subdivision is a convenient one, for whilst it is true that there is a continuity of motive throughout all of Charles Darwin's work which transcends distinctions of subject matter, the transition from the evolutionary phase to the physiological marked a significant change of approach. The labour of the evolutionary period, which saw its climax in *On the Origin of Species* and its supporting work, *Variation of Animals and Plants under Domestication*, was essentially one of observation, compilation and deduction; that of the physiological period, from which *Fertilization of Orchids*, *Insectivorous Plants*, *Climbing Plants*, *Cross- and Self-fertilization of Plants*, *Forms of Flowers* and *The Movements of Plants* arose, was primarily one of experiment. Darwin himself regarded the experimental botanical work as a form of relaxation and recreation after the sustained intellectual effort of the evolutionary period, and there is no doubt that he found great pleasure and satisfaction in the actual handling of plants. Indeed, there were times when he felt their attraction as a temptation; whilst working on the manuscript of *Variation of Animals and Plants* we find him writing to Hooker refusing further material from Kew, saying: "... it is mere virtue which makes me not wish to examine any more orchids; for I like it far better than writing about varieties of cocks and hens and ducks!" [9, p. 270]. But it is perhaps possible to perceive a deeper motive than mere love of the material for his devotion so completely to botanical work in his later life. Darwin loathed the idea of himself as a compiler; in accepting facts second hand he felt himself continuously at the mercy of the inaccurate observer, the exaggerator and the plain fabricator, and it is no doubt this aspect of his work for the *Variation of Plants and Animals* which he found so uncongenial. What relief, then, to turn to direct observation and experiment; Nature does not lie, he wrote—and in his experimental garden and greenhouse he sought a touchstone of truth at which the conclusions of his early work, so much the product of the mind, could be put to practical test.

Notwithstanding his devotion to plants for so many years, Darwin's own estimation of himself as a botanist remained a low one; even after the publication of the *Fertilization of Orchids*, he wrote to Scott, the young head of the propagating department at the Edinburgh Royal Botanic Garden, " . . . I know only odds and ends of Botany . . . you know far more " [9, II, p. 308]. The judgment must, however, be examined in relation to the criterion upon which it was based, and doubtless to Darwin the word botanist meant what it still does in the minds of the majority—someone who knows the names of a large number of plants—a distinction to which he felt he could lay no claim. Yet were the *Origin* and the works preceding it eliminated from his achievement, Darwin would remain one of the most outstanding biologists of the nineteenth century, on the credit of his botanical work alone. Indeed, during his lifetime his fame in France, where the evolutionary tide was long resisted, rested principally upon his botanical publications, and it was in the Botanical Section that he was elected a Corresponding Member of the French Institute; although of his election he wrote to Gray: " . . . it is rather a good joke . . . as the extent of my knowledge is little more than that a daisy is a compositous plant and a pea a leguminous one " [7, III, p. 224].

Although it can hardly be said that each work marked a particular episode in Darwin's researches, the chronological sequence of the botanical publications gives some idea of how his interest in and understanding of the life of plants grew and developed. The voyage of the *Beagle* focused his attention upon the problems of geographical distribution, that " noble subject of which we as yet but dimly see the full bearing " [9, II, p. 57], and it is this aspect of botany which we see discussed in the *Origin*. The implication of the theory of natural selection is that evolution is a process of progressive adaptation, and the study of the flower structure of orchids, with its accompanying revelation of adaptations of quite astonishing intricacy served to show " . . . how natural history may be worked under the belief of the modification of species " [7, III, p. 254]. The demonstration in this work and in that of *Forms and Flowers* that the function of so many floral mechanisms was apparently to promote outbreeding strengthened Darwin's long-held conviction that cross-fertilization is in some way especially advantageous, and his eleven years of experiment on the matter culminated with the publication of *Cross- and Self-fertilization of Plants* in 1876. A more purely physiological inclination arising from an interest in the nature and causes of plant movements led him, on the one hand, to study insectivorous plants after his observation in 1860 of tentacular movements in the sundew and, on the other, to examine the nature, utility and evolution of the movements and structural adaptations of climbing plants; one research saw its culmination in the publication of *Insectivorous Plants* in 1875, and the other in the paper on *Climbing Plants* transmitted to the Linnean Society in 1865 and published ten years later as a separate volume. An hypothesis drawn from the work on climbing plants formed the central theme of the research on the general problems of plant movement carried out with his son, Francis ; research which led to the publication of *The Movements of Plants*, " a tough piece of work " [7, I, p. 98], which formed the last major effort of his career, and his last publication in book form but for *Vegetable Mould and Earthworms*.

The four major themes of the special botanical researches were thus geographical distribution, flower structure and function, breeding behaviour, and plant movements. Each field Darwin enlarged in its scope and concepts almost out of recognition, and in each his contributions have formed a ground-work upon which much subsequent advance has been founded.

THE GEOGRAPHY OF PLANTS

The two chapters in the *Origin* devoted to geographical distribution are concerned, first, to establish that " all of the individuals of a species, wherever found, are descended from common parents " [3, p. 359], and secondly, that " all the grand leading facts of geographical distribution are explicable on the theory of migration, together with subsequent modification and the multiplication of new forms " [3, p. 360]. A century later these propositions seem mild enough, but neither was without its opponents among Darwin's contemporaries, and taken together they gave a very different orientation to biogeographical thought in the decades succeeding 1859 from that prevailing during the early years of the century.

To those accepting the view that each species was a special creation, the suggestion that the same species should have been created more than once in different localities is tolerable enough; discontinuities in geographical range then cease to have any significance or to present any problem, since it is necessary only to assume independent creation in each area to explain the facts. This is essentially the position maintained by the glaciologist Louis Agassiz, of whom Darwin wrote, " . . . though he has done so much for science, he seems to me so wild and paradoxical in all his views that I cannot regard his opinions as of any value " [9, I, p. 264]. Even the distinguished systematic botanist Alphonse de Candolle early in his career had accepted a similar conception of the multiple origin of plant species, but had disabused himself of it in his *Géographie Botanique Raisonnée* of 1855.

Although few at the end of the nineteenth century remained unconvinced of the correctness of Darwin's view, the question whether the same plant species may arise independently in different localities, i.e. polytopically, is still a live one. Even accepting the Darwinian proposition that all individuals of the same species have descended from common parents wherever they may be found, a polytopic origin is still conceivable, since the ancestral stock may have been widely dispersed, and the separate populations today accepted as being part of one species may have arisen from it independently in different localities by a process of parallel evolution. We are familiar now with one kind of species formation which could certainly happen thus. This is the process of polyploidy, which involves a sudden increase in the numbers in each cell of the chromosomes, the bodies which carry the genes, the actual material of inheritance. Something like half of the flowering plants of the north temperate region have had this sort of origin, and it is quite conceivable that some of them should have arisen polytopically. There is good evidence that the *Orchis latifolia,* which Darwin knew so well at Down, is just such a case. Again, there is reason to suppose that an ecologically specialized race of a species may arise in more than one area as a result of repeated local differentiation under similar selective pressure in a specialized type of habitat, a possibility that Darwin did indeed envisage. The different populations so evolved may come to resemble each other to a degree which would cause them to be classified together, and it could be argued that such a process might lead ultimately to the polytopic emergence of a race morphologically differentiated enough to be regarded as a species. However, the probability of differentiation in disjunct populations taking a strictly parallel course to a degree establishing differences of a specific magnitude is low indeed, and there are few, if any, species for which such a case could be made out. Darwin's first biogeographical postulate must therefore be regarded as generally true, with the exception of a class of cases, those—practically restricted to plants—involving polyploidy, where a type of species origin unimagined by him is involved.

The assumption of a single centre of origin for every species defines the character of the major type of biogeographical problem, which is to establish what contemporary or historical factors determine, or have determined, existing patterns of distribution. Darwin's interpretation was that the present-day distribution of each species was to be understood in terms of its powers of migration and survival under past and present conditions. Migration he saw both as a natural outcome of the tendency for species to increase their numbers, and as a result of secular change over geological epochs which compel compensatory movements of entire species populations.

In the latter connection he stressed especially the significance of the glacial period in determining the character of the mountain floras in the north temperate zone, and even throughout the world. The central idea was that during the height of the ice age northern floras were driven southwards, only to retreat northwards again on the amelioration of

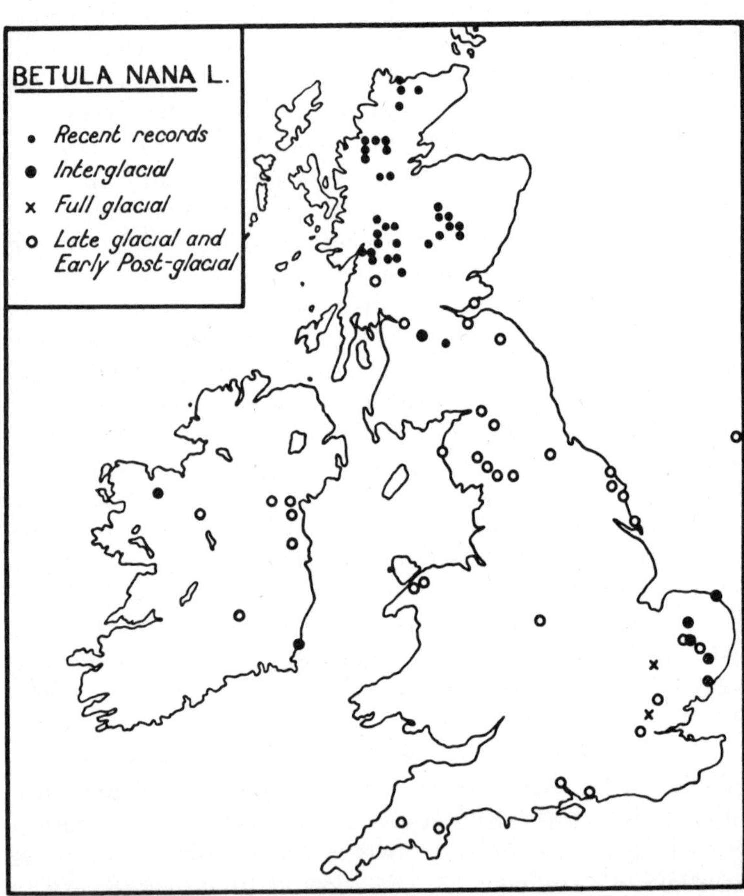

51.—Present distribution of the dwarf birch (*Betula nana*), an arctic species, in the British Isles, and sites where remains of it have been found dating from glacial times or shortly after. "By the time the cold had reached its maximum, we should have an arctic fauna and flora covering the central parts of Europe. . . . As the warmth returned, the arctic forms would retreat northward, closely followed in their retreat by the productions of the more temperate regions. And as the snow melted from the base of the mountains, the arctic forms would seize on the cleared and thawed ground, always ascending, as the warmth increased and the snow still further disappeared, higher and higher, whilst their brethren were pursuing their northern journey." (*The Origin of Species.*) (Map reproduced by courtesy of the Botanical Society of the British Isles; the records upon which it is based were assembled by Miss A. Conolly.)

climate, leaving behind a residue on southern summits where climatic conditions and absence of competition allowed survival. For this conception, Darwin gave full credit to

Edward Forbes, whose famous memoir on the distribution of the British biota was published in 1846. In his *Autobiography* there is a characteristic passage [7, I, p. 88] in which Darwin mildly regrets being thus forestalled by Forbes, since, according to a letter to Asa Gray written in 1858 [7, II, p. 136], he had worked out the theory in detail as early as 1842.

It is remarkable that this thesis of Forbes and Darwin which so neatly explains the existence in the British Isles of relict arctic-alpine plant communities on mountain summits and occasionally in other types of specialized habitat at lower levels, should have failed to gain general acceptance until comparatively recently; several botanists of the last half-century have preferred to interpret the disjunct distributions of these plants in Britain as evidence of their survival through the Ice Age, in or near their present stations. The complete vindication of the views of Forbes and Darwin has come from direct evidence which would have delighted both: the exposure of sub-fossil remains of characteristic arctic-alpine plants in low-level peat bogs and lake sediments dating from the close of the Glacial Period. It is now clear from such evidence that many of these plants were components of the tundra-like vegetation of the periglacial region, and that as the ice retreated they formed part of the first re-immigrant flora during what has been termed the Late Glacial Period. Subsequently they were extinguished throughout most of the formerly glaciated zone by the later immigration of the forest flora, leaving isolated colonies in refuges where local climate or instability of substratum inhibited the formation of closed forest. With what excited pleasure would Darwin have pored over the pages of Godwin's recent *History of the British Flora*, where the evidence for each species is reviewed in detail!

Even the evidence of the post-glacial climatic optimum he would have greeted with satisfaction, for in a letter to Asa Gray in 1858 there appears the statement, " . . . some facts have made me vaguely *suspect* that between the glacial and the present temperature there was a period of *slightly* greater warmth " [7, II, p. 136]. What some of those facts were emerges from a passage in a letter to Lyell of two years later, which so admirably sets out what is surely the most probable view of the origin and status of the group of southern plants in the flora of the south-west of England and Ireland which has been termed the Lusitanian element that it merits full quotation. After agreeing with Lyell that the survival of this Lusitanian group through the Ice Age (the view firmly held by Forbes) was improbable, he goes on to suggest

> . . . that a slightly different or more equable and humid climate might have allowed (with perhaps some extension of land) the plants in question to have grown along the entire western shores between Spain and Ireland, and that subsequently they became extinct, except at the present points under an oceanic climate [9, I, p. 461].

To this little or nothing need be added.

On the broader canvas of the world, Darwin applied the migration theory of biogeography with force and confidence, emboldened not a little by the factual detail which the great plant geographical works of Hooker and Gray were simultaneously supplying. In relation to the distributional problems of species, as apart from those of higher systematic units, his mind dwelt particularly upon the significance of physiographical and climatic barriers in checking, diverting or canalizing migration, and so in determining the form and extent of distributional areas. As we have seen, the idea of climatic change as a force driving migration appealed deeply to him, and he developed the theory of Forbes to an extent well beyond the limits of its author. Thus he inferred a steady southward migration of northern plants and animals under the compulsion of a cooling climate long before the beginning of the Glacial Period, a view again to be fully vindicated by fossil evidence later to be discovered. The bold extension of this idea to explain cases of bipolar distribution—where a species has two widely separate areas in corresponding climatic belts in the northern and southern hemispheres—led him to postulate sufficient cooling to allow transtropical migration of temperate species, notwithstanding the opposition of Hooker, who confessed that "... much as I should like it, I can hardly stomach keeping the tropical genera alive in so very cool a greenhouse!" [9, I, p. 437]. The difficulties of explaining what happened to the tropical flora during the period of maximum cooling caused Darwin to seize upon an hypothesis which appeared to provide a way out—that of Croll, according to which glacial periods alternated in the two hemispheres. In the editions of the *Origin* after the fifth, biogeographical implications of this theory, which would provide a home for the tropical genera in the southern hemisphere whilst the north was glaciated, and vice versa, are developed.

The hypothesis of Croll cannot now be sustained, and the problem of bipolar distribution must be regarded as still outstanding. It falls into the class of distributional problems which broadly admit of two types of explanation: one based upon the acceptance of occasional chance dispersal over very long distances, and the other upon the assumption that suitable migration routes formerly spanned the present physiographic or climatic barriers, and that they have subsequently been destroyed or modified. Darwin, ever ready to believe in the possibility of past changes in climate, was strongly opposed to any suggestion of the impermanence or mobility of the continents, and suspicious to an extreme degree of postulations about land-bridges between them. Whilst following Forbes—and even preceding him—in accepting climatic change as a force motivating migration, he regretted that author's " astounding boldness " in postulating land-links to provide the routes along which migrations took place.

The outcome of rejecting explanations of disjunct distributions based upon slow migration over former land connections is, of course, to throw emphasis upon the means available to the organisms concerned for long-distance dispersal. This aspect, particularly in relation to plants, is one to which Darwin gave much thought, and in a well-known passage in the *Origin* [3, p. 323–330] he enumerates the several methods whereby he believed land plants could achieve occasional trans-oceanic dispersal, by flotation, transport in and on birds, on logs, icebergs and the like. The experiments which he recorded in the *Origin* on the viability of seeds following submergence in salt water were the first of a systematic nature to be directed upon this problem. From them he concluded that on the average the seeds of perhaps 14 per cent of the flora of a country might survive long enough to float a thousand miles in sea currents.

Like the concept of polytopic creation, long-range chance dispersal is readily enough amplified to provide a blanket explanation for all distributional phenomena; but in his many writings on the matter Darwin was rigorous in using the hypothesis to explain just as much as was necessary, and no more. The need not only for dispersal but for establishment in the new home did not escape his attention, nor did the fact that establishment in a closed community would involve far greater difficulties than establishment on an untenanted land surface. In effect, he adopted the position that long-range dispersal might act as an occasional supplement to the more normal processes of migration, and that to assume the possibility of its occurrence is always preferable to invoking major geographical changes without geological backing. In this position he would undoubtedly find himself joined by the bulk of contemporary phytogeographers.

FLOWER FORM AND FUNCTION

Darwin's interest in biogeography was to a large extent an incidental one, arising from his desire to use its evidence in support of the hypothesis of organic evolution. He regarded the gathering of distributional data as a means to an end for, as he wrote to an acquaintance, "... all observation must be for or against some view if it is to be of any service" [7, I, p. 185]. This precept is illustrated again in his work on floral biology and breeding systems, which engaged his attention to a greater or lesser extent from 1839 to the end of his life, and provided material for three major works, *Fertilization of Orchids*, *Cross- and Self-fertilization of Plants* and *Forms of Flowers*, as well as numerous papers.

The arguments which the mass of detailed observation on flowers and flower function was intended to support were essentially teleological, and we see in them not a little of that special Darwinian form of progressive deductive reasoning which is so nearly, but not quite, circular. Commenting to Hooker in 1845 on the principle which he attributed to Knight, that every plant must be occasionally crossed, he wrote, "... I find ... plenty of difficulty in showing even a vague probability of this, especially in Leguminosæ, though their (structure?) is inimitably adapted to favour crossing" [9, I,

p. 52]. Eleven years later, however, he was sufficiently convinced of the view that "Nature abhors perpetual self-fertilization" as to refer to it in a letter to the same correspondent [9, II, p. 250] as "*my* doctrine". The entire research on the pollination mechanisms of the orchids was, in fact, motivated by a desire to demonstrate that its principal

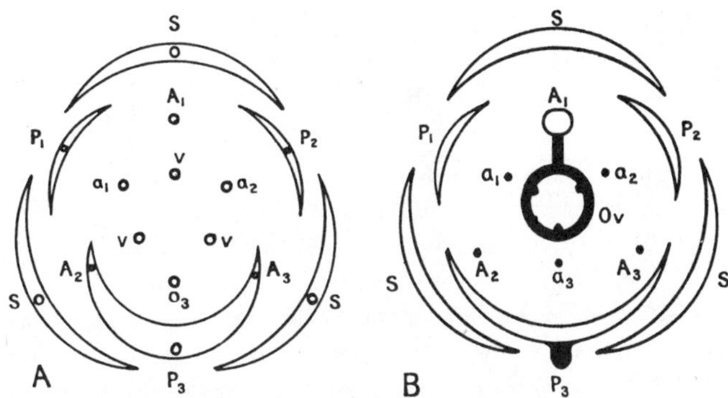

52.—A, Diagram of an orchid flower sent by Darwin to Sir Joseph Hooker. "Look at my diagram (which please return, for I am transported with admiration at it), which shows the vessels I have traced, one bundle to each of fifteen theoretical organs, and no more. . . . In all orchids yet looked at . . . the vessels on the two sides of the labellum are derived from the bundle which goes to the lower sepal, as in the diagram. This leads me to conclude that the labellum is always a compound organ." (*More Letters of Charles Darwin.*)

B, Recent floral diagram of *Orchis*, by P. Vermeulen. The labellum is here regarded as a single modified petal. Vermeulen's diagram includes an interpretation of the structure of the ovary and the single anther, linked to one style to form the column; Darwin's diagram is concerned only with the petals and sepals, giving no more than an indication of the lay-out of the vessels serving the ovary. S, sepals; P^1 and P^2, lateral petals, P^3, the modified lower petal, the labellum; A^1, the single fertile anther; A^2 and A^3, positions of the undeveloped anthers (according to Darwin, fused with P^3 to form the labellum); a_1, a_2, a_3, undeveloped anthers of the inner whorl; Ov, ovary; V, vessels entering the ovary.

purpose was "the fertilization of the flowers with pollen brought by insects from a distinct plant" [1, p. 1], and the same conviction lay behind the investigations into heterostyly and allied phenomena. Having proved the justification for this conviction from the researches on floral mechanisms, we see then a turning of the tables, for the researches reported in *Cross- and Self-fertilization of Plants* were directed to determine the nature of the benefit which he felt must accrue from a process which had now been shown to be so universally favoured.

In the experimental work recorded in *Fertilization of Orchids*, Darwin followed up with great persistence the hypothesis that the many structural complexities of the orchid flower must have functional significance, solving the floral mechanism of genus after genus as material was fed into his hands by his botanical friends at Kew and elsewhere. Among his more spectacular successes was the investigation of *Catasetum*, where he discovered that individuals which were so unlike each other that they had been placed in different taxonomic genera were in fact male, female and hermaphrodite members of the same species! Rarely was Darwin's interpretation of floral structure at fault, and where error did arise, it was usually from the advent of some attendant qualifying circumstance, unknown to him at the time. Thus he met defeat in the genus *Acropera*, in which he concluded from the undeveloped state of the ovules in flowers which he examined that the sexes were separate. John Scott of the Edinburgh Royal Botanic Garden, writing to him shortly after the publication of the first edition of *Fertilization of Orchids*, pointed out that at least some of the flowers of the plants Darwin had considered to be male were capable of setting seed upon artificial pollination. It was not until the following year that Hildebrand showed that in many *Orchidaceæ* the stimulus of pollination is an essential prerequisite for ovule formation, and it is now known that the growth of the ovary, the development of the ovules and even the processes of reduction division and embryo-sac formation are dependent upon the generation of growth-hormone (auxin) in the ovary following the advent of the pollen tubes.

Nothing delighted Darwin more, nor proved to him more convincingly the justification of his approach to the floral structures of the orchids, than to receive verification from correspondents that particular mechanisms functioned in the manner he had deduced, and in conjunction with the types of insects which he had suggested. To him, the orchid work—which on more than one occasion he referred to as the most interesting of his life—provided one of the finest of all test cases for the theory of adaptive evolution. He saw in it a clear demonstration that when an end as significant as he believed cross-fertilization to be was to be attained, the simple basic ground plan of the monocotyledonous flower could become modified in the course of evolution to produce mechanisms transcending " . . . in an incomparable manner the contrivances and adaptations which the most fertile imagination of man could invent " [1, p. 286]. The motivation for these modifications was, of course, natural selection; they were " . . . . slowly acquired through each part occasionally varying in a slight degree but in many ways, with the preservation of those variations which were beneficial to the organism under the complex and ever-varying conditions of life " [1, p. 285].

The desire to establish this viewpoint led him to seek for and analyse structural intergradations, for a demonstration of a series of intermediate forms between two modal types is the surest answer to those who would assert that one could not have been attained from the other by a process involving the slow accumulation of minor changes. The application of the theory of natural selection to the orchid flowers was, in fact, strongly attacked by his critic, Mivart, who adduced the familiar argument that as it is only the final product which can be regarded as being of utility, the incipient beginnings of specialized structures cannot be regarded as being of selective value. Darwin's reply, given in later editions of the *Origin*, was to quote the intergradations especially evident in the pollinium and its associated structures, an organ of the orchid flower which in its most highly developed form shows perhaps the most exquisitely refined example of adaptation of all floral mechanisms—even to that power of spontaneous movement to a functional position after removal from the anther of which Darwin wrote, " . . . . I never saw anything so beautiful " [7, III, p. 263]. To those present-day followers of Mivart who would still argue that the very perfection of adaptation shown by floral structures in such families as *Orchidaceæ* is evidence against their being evolved through a course of slow change, Darwin's catalogue of intermediate states should continue to be instructive!

In developing the argument about the orchids, Darwin attempted to answer another doubt which still assails even those sympathetic to the general conception of natural selection ; namely, whether a force of this nature operating over uniform parent material could have brought into existence the *diversity* of types recognizable even within so natural an alliance as the *Orchidaceæ*. The explanation given in *Fertilization of Orchids* and later editions of the *Origin* is hardly a satisfying one, for it amounts to saying that the exact nature of the end-product of a selective process is decided fortuitously according to the types of variations which turn up, and that because of this several different end-products may be attained, all equally efficient in performing the same function. The factor missing from this argument is the prior condition which must prevail to allow the very first onset of divergence, namely, some form of isolation between the diverging populations which would prevent the erosion by continuous intercrossing of such differences as did arise. The significance of isolation as a factor favouring differentiation did not, of course, escape Darwin. Whilst he found himself unable to agree with Wagner that *geographical* isolation was a necessary factor for the formation of new types, his recognition that intercrossing tended to eliminate differences and so to prevent diversification led him to suppose that some circumstance must always intervene to limit its range or reduce its intensity before such diversification could take place within an initially homogeneous population. Thus he held that in hermaphrodite organisms in which outcrossing is only occasional, variety formation might be a more local matter, and take place more rapidly, than in widely ranging, obligately outbreeding organisms, such as birds, in which race differentiation will normally demand some degree of geographical isolation. The insight which this discrimination indicates is indeed remarkable. Today, those who follow Wagner in insisting that speciation is mostly a result of geographical isolation are mainly ornithologists and others dealing with widely ranging, outbreeding organisms, whilst those that suppose race differentiation can take place within local populations are mostly botanists concerned with sessile organisms capable of close inbreeding.

Nowhere did Darwin advert to the matter of isolation specially in relation to the *Orchidaceæ*, but it is clear from the discussion of the behaviour of pollen carriers in *Cross- and*

Self-Fertilization of Plants that he recognized that floral structure could itself act as an isolating mechanism. This can happen when insects show a preference for one particular type of flower, and, by specializing upon it during their foraging excursions, bring about a degree of assortative cross-fertilization —a process which, as Darwin noted in *Variation of Plants and Animals under Domestication*, favoured the establishment and maintenance of distinct races. Research on insect behaviour has now established the capacity for flower-constancy in a wide range of pollen carriers. Not only may assortative pollination be brought about by the specialization of flower structure for particular types of visitors, but also by the elaboration of particular forms of pattern, colouring and scent in the floral envelopes. Julian Huxley has called such characteristics allæsthetic—exerting their biological effect through the sense organs of another organism. That the corolla, and sometimes other parts of the flower, are allæsthetic organs in insect-pollinated plants was recognized by Conrad Sprengel in *Das entdeckte Geheimniss der Natur* of 1793—a book which, placed in Darwin's hands by Robert Brown in 1841, served greatly to enhance his interest in flower biology. Almost any variation in the allæsthetic organs of a flower might affect the reactions of its more sentient visitors, and it seems entirely conceivable that variation of this type, originally " random ", should initiate processes of assortative pollination which could lead ultimately to race formation and thence even to the establishment of species.

This process of diversification is not necessarily one of biological " improvement " in the sense that some definable function, like securing more effective cross-pollination, is attained. Darwin averred in the *Origin* that a demonstration that structural diversification had taken place " for the sake of mere variety " would be absolutely fatal to his theory; yet in the structure of flowers there is no scarcity of examples of diversification apparently for the sake of diversification, and selection theory is not incapable of giving an adequate explanation of its origin. A glorious example from the *Orchidaceæ* is that of the genus *Ophrys*. In the Mediterranean region numerous species grow together, showing little or no difference in preferred habitat. Many are vegetatively almost indistinguishable, yet they show astonishing diversity in the colours, patterning and sculpturing of the allæsthetic organs of the flower. No rational criterion can be applied to establish whether one form is " better " or " more advanced " than any other; all that can be reasonably stated is that they are all different. Yet when the agency of pollination is examined, the situation is immediately illuminated, for it is found that each flower type simulates the female of a different bee species, and that male bees of each species concentrate upon the appropriate flower type and transfer the pollen from one plant to another as, deceived by the resemblance, they alight again and again to attempt copulation. Evidently, during the course of evolution of the genus, various modal types in an originally randomly varying population have been perpetuated selectively by assortative pollination because they have happened to resemble to some slight degree female insects; generations of this process have then refined the different types to the degree we observe today. This interpretation is not only probable, but indeed would, we may suppose, be demonstrable, were it practicable to breed *Ophrys* in cultivation; for a process of artificial assortative pollination for a few generations of some of the aberrant forms still present in natural populations would readily establish a race bearing flowers resembling a non-existing bee!

The case of *Ophrys*, which to a human observer has something of the attributes of a grotesque practical joke, is an instructive one, since it casts light upon some central problems of flowering plant evolution. More than one modern writer, examining the unbelievable variety of the flowering plants, has concluded that the whole array represents simply a jumble of diverse trends without any particular theme and certainly not capable of any explanation on the basis of natural selection. A much-quoted phrase of Darwin's own about the origin of the group appears in a letter to Hooker of 1879 [9, II, p. 20]: " . . . the rapid development as far as we can judge of all of the higher plants within recent geological times is an abominable mystery." But in this same letter he refers to Saporta's view that the rapid evolution took place as soon as flower-frequenting insects were developed and favoured inter-crossing, and in a letter written two years earlier to Saporta himself the following significant passage appeared: "Your idea that dicotyledonous plants were not developed in force until sucking insects had been evolved seems to me a splendid one. I am surprised that the idea never occurred to me, but this is always the case when one first hears a new and simple explanation of some mysterious phenomenon. . . ." [7, III, p. 284]. Darwin himself never came to trace out the implications of this fertile theory of evolution by the mutual interaction of two major groups—flowering plants and insects—which came to him so late in life, and it is only in recent years that its full import has come to be recognized.

One of the most interesting incidental aspects of Darwin's work on flower function was the attention it caused him to pay to the basis of flower structure, and thus to the fundamental problems of derivation and homology. The concept of homology had for Darwin a perfectly clear meaning, well conveyed in the statement in *Fertilization of Orchids*: ". . . all homologous parts or organs, however much they may be diversified, are modifications of the same ancestral organ" [1, p. 233]. Homology interpreted in this sense he saw as clearing away ". . . the mist from such terms as the scheme of nature, ideal types, archetypal patterns or ideas, etc.; for these terms come to express real facts". As a highly derived group with a striking complexity and diversity of floral structure, the *Orchidaceæ* formed a fine test-case for this evolutionary concept of homology. Darwin attacked it with enthusiasm. Beginning from Robert Brown's dictum that the orchid flower was based upon a standard monocotyledonous pattern of five alternating whorls of three elements, he followed a suggestion of Hooker, and turned to anatomy as a guide, tracing the longitudinal course of the vessels of the vascular system from the floral axis into the tepals and other appendages. Except for one major error, the resultant interpretation of homologies has in most respects stood the test of time, although the conflict about details is by no means yet resolved.

The use of anatomical evidence in tracing homologies, whilst not unknown before Darwin, was carried by him to a new level of precision, and to this extent he must be regarded as the originator of this particular tool of interpretative morphology. Notwithstanding his successful use of the course of vessels as a guide in interpreting homology in the orchids, he by no means fell into the error of assuming that vascular systems are conservative in evolution to a degree which makes their use a sure guide to ancestry. Following a study of the vasculation of the sweet-pea legume, he decided that " . . . the midrib vessel alone gives homologies, and that the vessels on the edge of the carpel leaf often run into the wrong bundle, just like those of the sides of the sepals " [9, II, p. 289], and concerning the structure of the flower of the butterfly orchis, he concluded despairingly that " . . . not the least reliance can be placed upon the course of ducts " [9, II, p. 276]. Not all modern floral morphologists are as wary of the pitfalls attending the use of vascular evidence!

BREEDING BEHAVIOUR AND INHERITANCE

Complementary to the work on the orchids was that on flower polymorphism recorded in *Forms of Flowers*. The discovery that in heterostyled plants like the primrose and loosestrife the different forms " . . . although all are hermaphrodites, are related to one another almost like the males and females of ordinary unisexual animals " [5, p. 2], so that " legitimate " pollinations between plants of different types produce a normal crop of seed while " illegitimate " unions result in impaired fertility, ranks as perhaps the most important single result of Darwin's own experimental work. As is revealed in a letter to Hooker of 1860 [7, III, p. 297], he began with the assumption that the existence of two flower forms (dimorphy) in plants like the primrose was a step on the road to the condition of two separate sexes (dioecism), and dubbed the " pin " plants as female, and the " thrum " plants as male; the finding that each would set seed freely when pollinated from the other disabused him of the idea.

The significance which he finally attached to heterostyly was, of course, that it formed yet another device to promote cross-pollination, comparable therefore with the specialization of flower structure in relation to insect visitation which he had observed in the orchids, and to the phenomena of dichogamy (the maturation of the male and female organs of a given flower at different periods) and diœcism. On self-sterility as a means whereby the same end could be attained, Darwin maintained a certain ambivalence; thus while listing it as a means of promoting outbreeding in *Forms of Flowers*, in *Cross- and Self-fertilization of Plants* he stated that it seemed to be an almost superfluous acquirement for this purpose when so many other devices were available. Nevertheless, he fully accepted the significance as an outbreeding mechanism of what he termed pollen pre-potency, the so frequently manifest ability

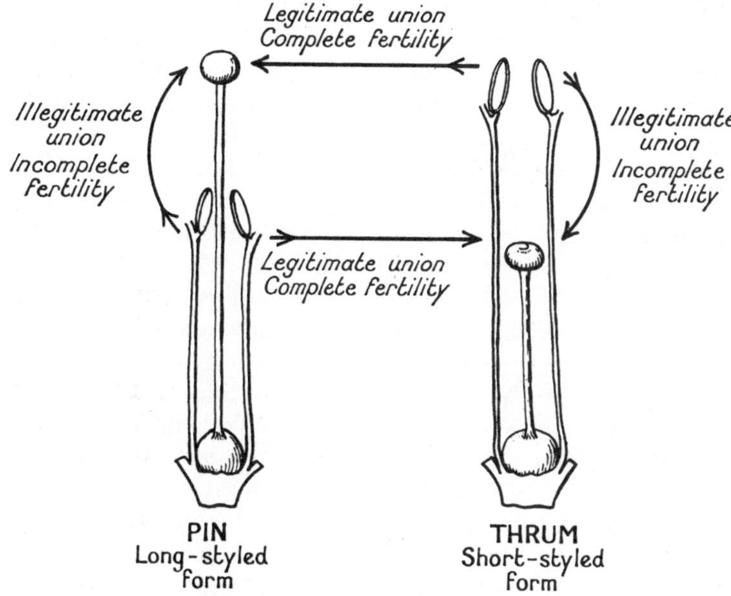

53.—Darwin's diagram illustrating heterostyly in the primrose (from *Forms of Flowers*). In the long-styled form, the anthers are attached halfway down the corolla tube, and in the short-styled, at the mouth. In the illegitimate unions, the pollen tube grows very slowly through the style, so that fertility is impaired.

of foreign pollen to bring about fertilization to the exclusion of own-pollen when both are present upon the stigma together, an incompatibility phenomenon of which total self-sterility is an extreme development.

All of the outbreeding devices which we see investigated in *Fertilization of Orchids* and *Forms of Flowers* added up, then, to a clear indication that " Nature abhors perpetual self-fertilization ". Since natural selection could not have produced such devices lest the function achieved was of substantial significance for survival, it remained to establish what this significance was. The results recorded in *Cross- and Self-Fertilization of Plants* appeared to supply the answer, for Darwin here obtained the first statistically acceptable evidence of inbreeding depression. The first results came to him as a surprise, notwithstanding his conviction of the importance of cross-fertilization, for even in the first generation of inbreeding in *Linaria vulgaris* vigour and size were less than in plants raised from the normally outcrossed seed. " I ought to have reflected," he wrote " that such elaborate provisions favouring cross-fertilization . . . would not have been acquired for the sake of gaining a distant and slight advantage, or of avoiding a distant and slight evil " [4, p. 8].

One problem led to another, however, and the proof that degeneration may result from self-fertilization merely raised the question of what was the special merit of cross-fertilization which prevented such degeneration. The answer which Darwin gave to this was that the " advantages of a cross depend altogether on the differentiation of the sexual elements " [4, p. 443]. The differentiation envisaged was physiological, and in Darwin's view, could as well arise from a difference in the conditions in which the two parents were grown as from any difference in their ancestry.

In the work on cross- and self-fertilization and in the associated passages in *Variation of Animals and Plants,* we see Darwin grappling with some of the most refractory, and yet most significant, of all the issues which he took up in his lifetime. The outcome, if far from being unproductive failure, was yet not success; the reason, as we may now conclude, was that he lacked a securely founded theory of heredity. Much has been written about Darwin's conceptions of inheritance; some of it less than accurate in its basis. It is, for example, often asserted that he accepted a principle of blending inheritance, believing that, in the course of sexual reproduction, the parental contributions blended like ink and water in the offspring, never again to separate.

This is a severe over-simplification of his viewpoint, as study particularly of *Variation of Animals and Plants* serves quickly to reveal. Segregation in the mendelian sense was well enough known to him and many cases from the work of others are quoted in *Variation of Animals and Plants*: even an instance of one of Mendel's own pea characteristics—round and wrinkled seeds in the same pod! [1] From his own botanical breeding work he quotes an instance of segregation in the second hybrid generation from a cross between peloric and normal flowered snapdragons: the first hybrid generation completely resembled the normal plant, and of the second hybrid generation of 127 seedlings, 88 proved normal, 37 perfectly peloric, and 2 imperfectly so [2, p. 46]. This is an excellent example of a mendelian experiment in which there is a single factor difference between the parental plants, with dominance in the first hybrid generation, and a second generation segregation ratio approaching 3 : 1.

From results such as these Darwin arrived at a fundamental fact of inheritance, that a characteristic may be transmitted without being expressed, and from this the conception that the appearance of an organism is not necessarily a guide to its hereditary constitution—or, as we would say today, that genotype cannot always be deduced from phenotype—followed naturally. Darwin was fully conversant with the views of Naudin, who stated clearly the principle of segregation and recombination of character differences in hybrid progeny; the idea that variability was not in fact destroyed irrevocably by the process of crossing was therefore certainly known to him.

Where, then, did " blending inheritance " come into the matter? Again numerous examples in *Variation of Animals and Plants* provide the answer. The cases where Darwin encountered qualitative differences showing mendelian segregation in inheritance were considerably fewer in number than those in which quantitatively varying characters were involved, and in the latter " blending " is the rule following crossing. It is noteworthy that the apparent contradiction between these two modes of inheritance caused the clash of opinion between the mendelians and biometricians in the early days of the present century. It took the better part of two decades for a reconciliation, which came about as a result of the recognition that the control of quantitatively varying characteristics in inheritance is not non-mendelian, but the outcome of segre-

[1] [See also Chapter 3.—Ed.]

gation and recombination of numerous contributory genes all individually with minor effects, but all following mendelian laws.

The racial differences frequently discussed by Darwin, being mostly of the " multiple factor " type, necessarily showed a blending type of inheritance, but it did not escape his attention that even when the first hybrid generation was uniform and intermediate between the parents, the second and subsequent generations could be very variable. Nevertheless, it appeared to him plainly enough that the maintenance of racial distinctions necessitated that crossing should *not* take place, and from this he was led to believe, as he wrote to Gray, that " crossing was one means of eliminating variation " [9, II, p. 254]. This in turn led to what has appeared to some subsequent writers as a paradoxical view of sex itself, namely, that it is a device for restricting variation.

Since Darwin regarded the cases of non-blending or particulate inheritance with which he was familiar rather as exceptions to the general rule of blending, he concluded that variation was, in fact, liable to be continuously eliminated in the course of descent. As is well known, he sought for a potential source from which variability could be replenished—since continuous minor variation was an essential pre-requisite for his evolutionary hypothesis—and he ultimately came to accept that it could arise as a direct result of environmental influences; the outcome of his reasoning along these lines was the theory of pangenesis, that curious, ingenious, but unsubstantiated hypothesis with which he concluded *Variation of Animals and Plants*.

Whilst the pangenesis hypothesis contained Lamarckian elements, in that it allowed for the inheritance of acquired characteristics and so for evolution by direct adaptation, it was in no sense regarded by Darwin as a dispossessor of natural selection. Perhaps the Lamarckian aspect of the hypothesis has been overemphasized by subsequent commentators, for the antecedent chapters of *Variation of Animals and Plants* make clear that the important effect of environmental influence (specifically, " change of conditions ") is conceived to be the enhancement of variation in general. This random variation then became the material over which selective forces acted.

The pangenesis hypothesis did, in fact, allow some sort of logical interpretation of the multitudinous and often divergent facts, rumours and speculation which confronted Darwin in the field of inheritance and variability, but whilst he regarded it with some affection, it is doubtful whether he ever took it with any very great seriousness, if one may judge from the mildness of his reaction to its criticism. As we see it today, the major error of pangenesis lay in giving weight to a mechanism which would allow direct registration of somatic effects on the germ-plasm; but it was only "some such" theory which Darwin hoped to see established [7, I, p. 93], and in the conception of "gemmules" as the carriers of inheritance he was not so distant from the twentieth-century conception of the gene.

THE MOVEMENTS OF PLANTS

In retrospect, one can well sympathize with Darwin in his colossal struggle to apprehend and assess the overwhelming body of data which came under his notice concerning variation and inheritance, and to deduce from it a unifying hypothesis which would supply the essential link necessary to complete his evolutionary argument. In the effort he appears to have felt himself approaching the limits of his intellectual powers; Francis Darwin records that in later life his father frequently referred to the physiological researches as being undertaken in place of work of this trying type for which he felt himself too old. The work on climbing plants, like that on the orchids, was, in fact, taken up as a form of relaxation during the actual writing of *Variation of Animals and Plants*; the results proved of such interest that Darwin felt impelled to divert his attention temporarily from the larger work to prepare them for publication.

The book *Climbing Plants* itself is principally a descriptive catalogue of the structure and behaviour of twiners, tendril- and leaf-climbers of a wide range of flowering-plant families, illuminated by a refined observation which makes it still a standard work on the topic. The central hypothesis was that the climbing habit is an adaptation to allow leaves to reach the light with maximum economy of stem material, and that it has been attained in different families by the development and specialization under the influence of natural selection of powers of movement and sensitivity common to all flowering plants. Darwin found particular satisfaction in tracing intermediate states between different types of climbers, since here again he argued that the presence of intergrading forms gave excellent evidence that the various adaptations had been achieved by the cumulative selection of minor variants. And, indeed, there are few better examples of the adaptation of an organ originally specialized for one function to the performance of another of totally different character than may be seen

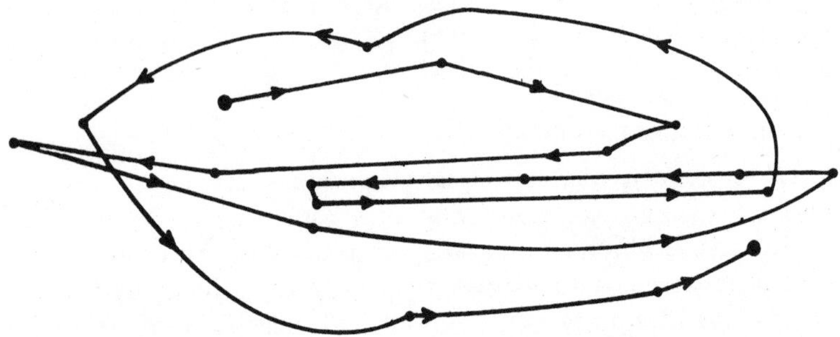

54.—Darwin's diagram of the circumnutation of a seedling of cabbage. The movement was followed by attaching a glass filament to one seed-leaf and tracing the path of a glass bead fixed to its tip upon a glass plate. The bead moved seven times from side to side, and thus described three-and-a-half ellipses in ten-and-threequarter hours, each being completed on an average in three hours four minutes. (From *The Movements of Plants*.)

among plants like the vetches (*Vicia*), where all intergradations between leaflets and tendrils appear.

The conception that specialized activities like climbing had been evolved through the amplification of basic properties of sensitivity and movement which were widespread among plants led Darwin to examine the spiralling movement of the stem tip in a wide variety of species. This movement, known as circumnutation, is due to unequal growth on different sides of the stem, the zone of maximum growth migrating completely around it in the course of a few hours, or sometimes even more rapidly. The result is that the growing apex describes a spiral in space with a diameter varying from a few millimetres in a cactus to as much as a metre and a half in one of the climbers studied by Darwin. Darwin's view of these movements was that they are autonomic, that is to say, due to some internal rhythmical impulse which is not generated by an environmental stimulus. What the nature of this rhythmical impulse was he could form no conception. In the active period of plant physiological study during the closing decades of the nineteenth century, this view of circumnutation came under criticism particularly in Germany, since the movement can be arrested when a plant is rotated slowly in a horizontal plane, indicating a relationship with the stimulus of gravity. We now know Darwin to have been justified, in so far as the existence of autonomic rhythms of widely different periods involving growth, movement and many more closely definable aspects of metabolism is now well established. The basis of such rhythms is in many respects as obscure as it was to Darwin, but there are several indications of what may be involved. In particular, the view that certain autonomic rhythms are comparable with the " hunting " which man-made self-regulating mechanisms undergo is attractive. In such " servo "-mechanisms, an essential part of the operation is a feed-back, whereby the future course of action is in part determined by the result already attained. Dependent upon the amount and delay of the feed-back, a mechanism of this type—an automatic laying radar, or a guided missile—will settle on to its target by a succession of approximations, or will continue to oscillate rhythmically about it. In growing vertically, the stem apex of a plant is, in fact, rather like a beam-riding missile, the guiding field in this case being that of gravity. It is probable that the vertical path is found by successive departures from it in different directions, each departure being corrected by a compensatory movement, the whole adding up to the process of circumnutation.

A connection between autonomic movements and the regulatory movements brought about by external stimuli forms the theme of Darwin's principal physiological work, *The Movement of Plants.* He argued that phototropism and geotropism, the processes whereby a plant becomes oriented during growth to light and gravity, were to be understood as movements arising from the amplification and orientation of that circumnutatory movement which he believed to be present in all growing plant parts. They were thus to be viewed as responses of survival value, acquired in the course of evolution by the natural selection of variants in which the basic tendency to movement took a beneficial form under environmental stimulus. Of the attempt to connect nutational movement with tropic responses, Thiselton-Dyer wrote in 1882: " . . . whether this masterly conception of unity of what has hitherto seemed a chaos of unrelated phenomena will be sustained, time alone will show " [10]. In the sense that the mechanism of each type of movement is the same, namely a curvature brought about by differential growth of the two sides of the organ, there is certainly a connection; but this was known to Darwin, and his hypothesis was intended not simply to describe the fact, but to link the causes behind it.

In recent years it has been suggested that the circumnutation of organs such as the coleoptile of wheat is due to a slow movement around the apex of the zone where maximum production of growth hormone proceeds; were this true, then its proximate cause is the same as that of tropic movements, which are also due to the establishment of a transverse gradient of growth hormone. If the possibility outlined above is correct, in circumnutational movement gravity is the displacing agent. Whereas Darwin thought circumnutation was an autonomic process from which the geotropic response was derived, we now see that it is more likely to be a consequence of geotropic sensitivity. The connection with the phototropic reaction would be of a different order, for here the principal effect appears to be on the formation and destruction of growth hormone rather than upon its transverse displacement.

So, three-quarters of a century later, while the judgment predicted by Thiselton-Dyer can still not be given unequivocally, it is already plain that Darwin's hypothesis can be correct only to a limited extent, and then only so far as it presumes a connection between circumnutation and sensitivity to gravity. With this limitation, the further element of his argument, namely, that tropisms have evolved from nutations under the pressure of selection, cannot be considered valid. For it we must substitute the assumption that the evolution of tropic reactions in plants has been a corollary of the development of hormonally controlled growth; or, to put the matter in Darwinian terms, and rather more remotely, that during the evolution of multicellular plants selection has favoured the establishment of mechanisms of intercellular communication whereby correlation of activity is attained and efficient responses to environmental stimuli in terms of orientation and posture made possible.

Darwin's other deductions from the work on plant movements have proved more significant than the unifying hypothesis which he himself regarded as its principal outcome. Outstanding among these were the linked ideas that the environmental factors bringing about the orientation of plants, ". . . . act not in a direct manner upon growth, but as stimuli" [7, III, p. 337], and that, in the response of organs like the radicle and stem, " it is the tip alone which is acted on, and that this part transmits some influence to the adjoining parts " [6, p. 545]. Darwin and his son Francis, in introducing the grass coleoptile, the first shoot of the seedling plant, into plant physiology, performed a service comparable with that of Morgan in introducing the fruit-fly *Drosophila* into genetics. Perhaps the best known of their experiments with coleoptiles was the demonstration that when the extreme tip is darkened with a metal cap, curvature towards lateral light is much reduced. From this they deduced that maximum sensitivity to light resides in the apex, from which a stimulus is transmitted downwards to the region of maximum growth where curvature results. Research on this reaction still occupies space annually in botanical journals, from which it may be concluded that much remains to be found out about its mechanism.

The reactions of the tip of the radicle Darwin found to be no less remarkable than that of the coleoptile apex. The fact that it could perceive gravitational, moisture and contact stimuli—even integrate all simultaneously—and then transmit an influence upwards which led to an appropriate growth curvature, caused him to compare it with the brain of a lower animal, so perfectly did it seem adapted to the function of penetrating the soil and tapping its moisture. The comparison may be a fanciful one, but as we come to fuller understanding of the functions of stem and root apices as perceptors of stimuli, as centres of co-ordination and correlation and as initiators of the processes of differentiation and morphogenesis which so largely determine the structure and behaviour of the higher plant, it emerges yet more clearly that Darwin was at least in no error in placing the emphasis where he did.

The work of Darwin's later years has given him a position in botany which he never attained in zoology, that of a progenitor of a line of research of a purely physiological nature. Indeed, amongst zoologists, an effect of the Darwinian revolution of thought in the late nineteenth century was to direct attention away from physiology towards systematics and comparative anatomy and morphology. Amongst botanists, Darwin's later work on plant movements had a special impact upon plant physiologists whose concern with his evolutionary arguments was slight indeed. Through a distinguished line of personalities Darwin's work on plant movements may be linked with present-day activity in the field of tropic responses, one outcome of which has been the discovery of the plant growth hormones.

CONCLUDING REMARKS

Whilst historical connections like that just mentioned may be distinguished linking Darwin's experimental botanical research with that of today, it cannot be said that the Darwinian philosophy or mode of reasoning is particularly evident in modern plant physiology. In his technique of experiment, Darwin was irreproachable within the limits of his facilities, and in powers of detailed observation it is doubtful whether any biologist has ever surpassed him. But the motivating impulse was of a different type from that which lies behind much modern physiology. " Observation must be for or against some view ", he wrote; and whilst he could achieve a passionate enthusiasm for investigation of the minutiæ of a structure or process, he never forgot to apply his findings as a critical test of the validity of his current " view ", nor did he shrink from accepting their implications when so applied. His hypotheses led into and overlapped one another, and were often, indeed, hierarchical. Thus the conception that all growth movements of plants represented modifications of a common reaction formed a basic hypothesis for much of the experimental work in this field; at a higher level stood the opinion that such modifications as these were comparable with the universally present structural adaptations of organisms, and above this lay the ever-present belief of steady evolutionary change under the impulse of natural selection operating over minor variation. The preconceptions which these evolutionary views gave Darwin were, of course, responsible for the teleological reasoning which marked all of the botanical work; and it cannot be said that the approach was unfruitful. Much modern botanical research, at least as recorded, has attained an ateleological attitude which verges on sterility, and indeed might signify such, were it not that teleological reasoning is substantially more common in the laboratory and field than in the research paper.

Arising also from the evolutionary view was Darwin's constant attempt to weigh all his observations of structure and function in terms of phylogenetic relationships. This procedure, carried to an illogical extreme in some biological fields, has lapsed, or been consciously rejected, in others. It is possible, on the one hand, for a systematist or morphologist who is blind entirely to the implications of evolution when dealing with the lower levels of organic variation to be dogmatic about supposed evolutionary arrangements of higher systematic categories, and, on the other, for an analytically minded physiologist so to scorn phylogenetical speculation, which he regards as unfounded and incapable of rigorous test, that he rejects a mode of thought not necessarily without value in application to his own material. There is a lesson for both in the spirit—cautious yet enterprising, speculative yet reasoned—in which Darwin used the evolutionary hypothesis, the most important generalization of biology, in his own botanical researches.

13

Darwinism and the Social Sciences

By

DONALD G. MACRAE

DARWINISM is one of the rare cases of a theory in natural science being derived very largely from work done earlier in the social sciences; and it is easily the most important. The genesis of the Darwinian hypothesis has been described by Darwin himself in a famous passage: in October 1838 he wrote, " I happened to read for amusement ' Malthus on Population ', and being well prepared to appreciate the struggle for existence which everywhere goes on from long-continued observation of the habits of animals and plants, it at once struck me that under these circumstances favourable variations would tend to be preserved, and unfavourable ones to be destroyed. The result of this would be the formation of new species " [4].

The decisive importance of Malthus is further underlined by the testimony of Alfred Russel Wallace. Wallace, it will be remembered, arrived quite independently at the same views on the origin of animal species as did Darwin, and it was this independent reaching of the same conclusions which precipitated the famous paper of 1858 to the Linnean Society in which these views were first put before the scientific public. According to Wallace it was again Malthus who had given him the essential clue " to the effective agent in the evolution of organic species " [22]. The coincidence is too striking to be merely accidental, and this first link between Darwinism and the social sciences remains probably their most important connection to this day.

It is probably correct to regard Darwinism as consisting of three main elements. The first of these is pre-Darwinian and is based on the Linnæan classification of the various forms of life, while the second is a consequence of Darwin's personal experience of the distribution, variety and fossil remains of plant and animal species observed by him during his voyage round the world in the *Beagle* from 1831 to 1836. The third element is the explanation of these two in terms of long-term changes taking place under selective influences of a kind similar to those to be found in the theory of population dynamics put forward by T. R. Malthus (1766–1834). These theories of Malthus have, in turn, been thought of as representing a specific response to the problems and distresses of the early period of industrial capitalist society. In consequence, the creation of the Darwinian theory of evolution has been placed in the history of thought as a concomitant of a particular stage of social development. However this may be, it is clearly essential that in any consideration of Darwin and the social sciences we begin by asking ourselves what in fact was the content of Malthus' theory of population.

In each edition of his main book Malthus added historical and statistical data until the sixth edition of *An Essay on the Principle of Population*, published in 1826 [17], was enormously larger than the first. Yet his argument is essentially analytical and constant: it is that the economic resources on which life depends cannot be multiplied indefinitely, nor at any very rapid rate; while potential fecundity is such that populations will, unless somehow checked, grow without limit and with ever greater velocity. There are only three checks on this growth: vice, misery, and self-restraint. By misery Malthus meant those conditions which early nineteenth-century England abundantly manifested—poverty, malnutrition, exposure, disease, " wars, infanticide, plague, and famine ". By vice he meant prostitution, corruption, " unnatural passions and improper arts . . ." [17]. For considerable numbers of people over quite long periods of time his theories have not held true. Elsewhere—usually in the " under-developed " countries—they have seemed plausible enough. Birth control, which was vice to Malthus, alters the picture today even in those countries where religion most vehemently opposes its practice.

Two questions remained unasked. Who, dowered with what qualities, survived the checks on population? How was this survival to be valued morally or biologically? We will return to the second of these questions when we examine the eugenic ideology which rose out of Darwinism. To the first question only one answer was possible. The survivors were, as Herbert Spencer put it in 1852, " the select of their generation " [21], or, as the phrase went, the fittest. In this way the human species might be assumed to progress. Darwinism applied the same reasoning to animal species to explain not just their progress, but their origin and differentiation.

The demonstration of the high probability of this reasoning, the analysis of the mechanisms involved, and the tracing of the main lines of organic evolution required—and still require—enormous labours that belong uniquely to the biological sciences. One may feel that it is here that Darwin's true greatness lies. The clue taken from a doubtful theorem in social science has proved enormously valuable. Almost at once the social sciences, for long enough already developmental in their formulation, tried to reclaim the debt and apply Darwinism to their problems. The consequences have not been altogether happy. The tendency of social scientists to whore after theories drawn from natural science—physical or biological—has a long history. Something has been gained, but the mass of consequent error suggests that the price may well have been too high. It is certainly the case that the place of both specifically Darwinian and more broadly evolutionary ideas is smaller in modern social science than has been true at any time in the past century. To some extent, indeed, this comparative eclipse of Darwinism is a measure of the degree of maturity and self-reliance to which the social sciences have attained.

The volume which celebrated the jubilee of *On the Origin of Species* contained either seven or eight chapters—it is a matter of definition—on or derived from the social sciences, out of a total of twenty-nine [19]. No comparable proportion is imaginable fifty years later. In 1909 sociology, anthropology and the comparative study of religions were predominantly evolutionary sciences, political science was under strong Darwinian influence, some economists were asking in a worried way why their subject was not " evolutionary ", and the new subject of eugenics was at its most fashionable. Today the eugenics of that period, profoundly biased in terms of class and race, has ceased to count. With one exception there has been an anti-evolutionary revulsion in the other disciplines, and only linguistics—which was in a sense Darwinian before Darwin—remains faithful. The explanation is both pragmatic —more fruitful approaches have been found—and ideological. It is also probable that the reaction has gone too far and included too much. Certainly the historical role of Darwinism was considerable.

In the first place Darwinism helped to liberate the social sciences and even to create the possibility of their development. Of the sciences of man only economics had an assured status in the 1850s. A great deal of good work had been done in what we today would call sociology and social anthropology. Scandinavian archæologists had established the technological sequence of stone, bronze and iron tools. History was becoming, by different paths in different countries, more scientific —though what "scientific history" was, and whether its essence was to be found in its method only or in some general conclusions, remained in dispute. Nevertheless a prime difficulty remained. Man was not a part of nature according to the teaching of most religion, and the history, problems and destiny of so ambiguous a creature therefore lay outside the province of science. This non-naturalistic view of man had often enough been attacked, but the attack was given a new weight and authority by *On the Origin of Species* which Darwin's *The Descent of Man* (1871) only confirmed. Man might be more than part of nature, but he was certainly part of it, and therefore open to study by the methods of natural science. The liberation was immense.

Human history no longer had to be conceived as a divine drama, the diversity of human morals and customs was no longer merely a consequence of sin, and ethics could be studied anthropologically as well as philosophically and theologically. What was new was the freedom to do these things. Montesquieu and Voltaire in France, Smith, Ferguson and Millar in Scotland, to some extent Vico in Italy and Herder in Germany had paved the way. An enormously popular and now forgotten book, Humboldt's *Kosmos*, had displayed to the early nineteenth century a panorama of nature including man. Darwinism seemed to guarantee the intellectual reputability of all such endeavours.

What was more, the personal experiences of Darwin on the *Beagle*, Huxley on the *Rattlesnake* and Wallace in South America and Malaya all seemed to point the way. In biology the understanding of the past was the key to the understanding of the present. The key to that past had been found in the geographical variety of living things: might not the same be true of human societies, and the wild places of the world reveal early man living in the most simple and instructive patterns of social organization?

Two Darwinian analogies contributed to the understanding of social organization. Social scientists, confronted by their complex data and with very limited scope for experiment in their work, have always been eager to find helpful models in other disciplines. I think that this is in general an unfortunate habit of mind; but the biological and geological models borrowed under Darwinian influence have been by no means uniformly harmful, and in two instances they have been beneficial. The geological record revealed stages of biological development as the archæological record revealed stages of technological change. Could an analogous evolution be found in social life? Certainly Greek and Roman writers, as well as some modern scholars, had thought in terms of such an evolution, but Darwinism enormously reinforced such theories and encouraged their investigation.

Equally venerable was the idea that societies resembled organisms in consisting of the interdependent unity of functionally specialized parts. The organic analogy as such belongs rather to the history of conservative ideology than to that of science, but the very anthropologists who were most ready to reject theories of social evolution wholeheartedly adopted and developed, for specifically sociological purposes, the idea of function and functional interdependence; these they used as essential tools for the analysis of social structure. In this they had been anticipated by Herbert Spencer to a degree which is seldom, even today, fully realized. Functionalism therefore has its roots deep in the evolutionary tradition, though perhaps more in the philosophy and sociology of Spencer than in the biology of Darwin.

It was, however, the influence of Darwin's most direct borrowing from early nineteenth-century *laissez-faire* economics that had the greatest and worst effect, not merely on social science but also on administration, practical politics and the conduct of social work. It is much to our advantage in the ordinary business of living, as well as in the pursuit of scientific knowledge about society, that the phrase " the survival of the fittest " is seldom or never heard in the contemporary discussion of human affairs. Racist theory is no responsibility of Darwin's and can be found long before him, while its bible—Gobineau's *Essai sur l'inégalité des races humaines*—first appeared in 1854–5. Nevertheless the racists claimed in Darwin a scientific confirmation of their views which is still exploited in South Africa and the American South. The theory that public provision and private charity should not be bestowed on individuals or groups unable to take a full part in the social contest can be traced back to at least the sixteenth century, but here again Darwinism was, quite illegitimately, appropriated in order to hinder or to forbid charitable endeavour as being essentially opposed to the best interests of a freely competitive economic order. This took place at the very moment when the purely economic claims for such an order were beginning to seem particularly thin. What was more, charity was thought to be " dysgenic ", that is, opposed to the best interests of the " race ". It is therefore appropriate to say something about eugenics before going on to discuss some of the other effects of Darwinism on the social sciences.

EUGENICS AND SOCIAL SELECTION

It was Darwin's relative, Francis Galton, who invented eugenics as a discipline. (Erasmus Darwin was the grandfather of Charles Darwin by his first wife, of Galton by his second. He anticipated the evolutionary theories of the one and the eugenic theories of the other of his grandsons.) Francis Galton engaged on genealogical studies to prove the heritability of talent and genius, and in so doing advanced in a most genuine way the content of statistics. (Galton, indeed, turned statistics to some odd uses, including a paper on *Statistical Inquiries into the Efficacy of Prayer* [7, *cf.* 8]). He was profoundly impressed by the continued ability of those family connections which have regularly produced in Cambridge some of the best second order talents in British society ; he believed that the component of biological inheritance was the main agent in this process, and that of environment only marginal. Curiously enough the lively mind of J. M. Keynes, himself a first order product of one of these connections, was fascinated by their study.) Galton's eugenic work was continued by Karl Pearson, philosopher of science, biologist and outstanding statistician, at University College, London. For a time it must have seemed that through eugenics a fruitful union had been achieved between biology, psychology and sociology—indeed, that the master-key to the latter disciplines had been found in statistical biology.

Alas, the facts were otherwise. Much was achieved, some harm was done, but so far as the social sciences are concerned only the gain of some useful statistical tools remains as of real importance. Partly this is due to a fundamental and elementary error in logic which we can find at the foundations of eugenic theory. In what is still the best short critical account of eugenics [9] Ginsberg has summarized the primary propositions of the creed:

> (i) That individual differences in mental and bodily character are determined mainly by heredity and only to a very small extent by environment. . . . (ii) That progress depends upon natural selection. (iii) That modern conditions are tending to suspend the selective death-rate, while . . . " they have allowed prepotent birth-rate to be associated with a tabid and wilted stock." (iv) That consequently degeneration has set in and must continue unless measures are taken to counteract the evil. . . .

Now nothing is to be gained by calling those who—partly because they breed more—are surviving in greatest number " a tabid and wilted stock ". It was quite reasonable to object to the fittest proving their fitness by survival, and insisting on other standards of human excellence, but it is futile simultaneously to praise the survival of the fittest and to damn the survivors. Nor can one follow the error of J. S. Huxley and try to deduce the validity of values from a biological theorem.[1] Nor in fact do we know now—let alone knowing fifty years ago —enough about the mechanisms of inheritance and the conse-

[1] [See Chapter 15.—Ed.]

quences of differing environments to put mankind to stud, even if one's values were to be guaranteed and agreed by whole benches of bishops and laboratories of biologists. Happily the demands that one should do so are becoming less insistent and are far less often heard. This does not mean that an adequate human biology would not be of great value to both sociology and criminology, or that recent work on the biology of mental defect—to take but one example—has not been of value in the concrete tasks of social administration.

In one area, that of intelligence testing and its application to social and educational selection, there is still a certain Galtonian hangover. It has been demonstrated [10] that in modern England whether a boy or girl is to be socially mobile upwards depends centrally on access to the grammar school or some similar form of education. There are, of course, other channels of movement between classes, but this is statistically the decisive one. Access to these schools is by various forms of examination which usually has an intelligence test as one of its major components—or as the major one. It is argued that such tests are largely free of environmental bias and cultural advantage. But is this so?

On the whole the poor are more fertile than the rich (though this difference is becoming less marked) and, as measured by psychological tests, less " intelligent ". There should therefore be a measurable decline of national intelligence with time. The most adequate evidence, however, suggests that during the fifteen-year period 1932–47 the " intelligence " of eleven-year-olds increased. This applied to Scotland. I am not nationalist enough to suggest that this is due to some special endowment of my fellow-countrymen; in any case, similar results have been obtained from smaller-scale enquiries in England. Nor am I convinced by attempts to explain the result away [3, 20]. The vigour of these attempts does, however, suggest a continuation of the strong class bias of the heroic days of eugenics earlier in this century—a class bias to be detected in even a socialist such as Pearson. In fact, cultural and environmental factors are being more and more admitted by psychologists to play a role in testing. It may well be that such testing is *de facto* the best available method of separating children for differing educational purposes—given the validity of these purposes; but the tests do not reflect in any single way the genetical component of intelligence.

(It is perhaps worth remarking that Darwin himself when young would probably not have qualified for entry to a grammar school, since he was considered slow and of poor ntellectual calibre. In his day, though, it was phrenology that was popular and occasionally used for selecting people for jobs. On sound phrenological grounds the captain of the *Beagle* came within an ace of refusing Darwin for the voyage.)

THE STRUGGLE FOR LIFE

A peculiarity of Darwinism, both in biology and in other fields, is that it explains too much. It is very hard to imagine a condition of things which could not be explained in terms of natural selection.[1] If the state of various elements at a given moment is such and such then these elements have displayed their survival value under the existing circumstances, and that is that. Natural selection explains why things are as they are: it does not enable us, in general, to say how they will change and vary. It is in a sense rather a historical than a predictive principle and, as is well known, it is rather a necessary than a sufficient principle for modern biology. In consequence its results when applied to social affairs were often rather odd.

How far should a mid-twentieth-century account of Darwinism and the social sciences concern itself with this oddity? C. Bouglé, a French sociologist, wrote an admirable, modest and sceptical account of all the eager Darwinian sociological speculators of the late nineteenth century [1], but I do not think that ground needs be traversed today. Nobody now reads Gumplowicz, Ammon, Novicov or others of the same kind. Nobody—and this is more important—is influenced in research or speculation by their work. Demographers know more than ever about differential fertility, but Lapouge is not a source. We can therefore afford to be summary in what we say.

Human life is a struggle: that is in part the point of Homer, and of the Mesopotamian saga of Gilgamesh, and of the oldest known cosmological myth, the battle of Marduk and Tiamat. What aspects of this struggle are sociologically significant or

[1] [Compare Chapter 1.—Ed.]

can be shown to parallel the Darwinian contest? Are the contestants races, biologically separate with different physical and psychological characteristics? Did the contest result in social stratification with the conquerors forming the ruling class? Such views have been widely held. Sometimes ruling classes have been recruited accordingly and, as the history of English colonization often shows, have tried to restrict marriage to within their own ranks. Yet race struggle and the assumption of significant innate differentia are both probably nonsense. At best only a Scots verdict of " not proven " can be returned about it; at worst it results in genocide and the total evil of the Nazis.

But perhaps the clue lay in the struggle between individuals battling for subsistence, or for mates,[1] or for power, prestige or other values such as money. According to Hobbes man is a wolf to man, and this is a fact to be at once accepted and deplored. But need it be deplored—was not this nature's way to perfect the species and to produce its finest flower, the successful capitalist who had triumphed in the jungles of industry and commerce and proved his right to survive? It might be an affront to Christianity that the weakest should go to the wall, but it was " nature's law ". It was difficult enough to believe so improbable a proposition, but it was impossible to find in it an explanation of any genuine problem of social anthropology or sociology concerning social structure and social change. More plausible was the countering view of the Russian aristocrat, geographer and anarchist that mutual aid between individuals was a factor in natural and social selection. Yet even Kropotkin's argument that the meek did, in some sense, inherit the earth lacked any specific explanatory power in particular instances.

This left class struggle. Marxists claim that the *Communist Manifesto*, in which all previous history was held to be the history of class struggles, at once anticipated and completed the perspective on the total world of nature given by the *Origin*. Perhaps for this reason, Marx wished to dedicate the first volume of *Capital* to Darwin. Paradoxically, in fact, both conservatives and socialists turned to Darwinism for a justification of their views. The philosophy and sociology of neither,

[1] [Compare Chapter 10.—Ed.]

however, owed anything intrinsically to Darwinism, and the sociology of Marxism is therefore excluded from this survey [15].

There were others. A constant theme on the lunatic fringe of what might be called the nineteenth-century intellectual underworld was the idea that struggle and development were preparing the way for a new and better race as superior to man as was man to the gorilla. It was the task of social science by eugenic or other reforms to prepare the way for this new arrival. This was not specifically Darwinian in inspiration. For example, as early as 1847, in Disraeli's *Tancred* [5], we read of Lady Constance telling the hero:

> " You must read the *Revelations*. . . . You know, all is development. . . . First there was nothing, then there was something; then I forget the next, I think there were shells, then fishes; then we came: let me see, did we come next? . . . And the next change there will be something very superior to us, something with wings. Ah! that's it: we were fishes and I believe we shall be crows. . . ."
>
> " I do not believe I ever was a fish," said Tancred.

The upshot of all this was Nietzsche and Shaw and the gospel of the superman.

DEVELOPMENTAL ANTHROPOLOGY AND SOCIOLOGY

More important is the programme put forward by Huxley: " We must have History; treated not as a succession of battles and dynasties; not as a series of biographies; not as evidence that Providence was always on the side of either Whigs or Tories; but as the development of man in times past, and in other conditions than our own " [14]. Darwinism had given a new impetus to such attempts, and we must turn to them for they made a great contribution to the social science of their day, are still to be found among the half-forgotten foundations of social anthropology and sociology, and are, I think, too much neglected.

Four names in Britain may be taken as typical: Spencer, Tylor, Frazer and Hobhouse. A list of equal distinction could be offered for France, and one only slightly less imposing for the United States [2, 13]. As we know, Spencer's biological theories and his view of cosmic evolution were independent of and prior to Darwin, but the parallel is so close that it is reasonable to consider them under a Darwinian title. Unfortunately Spencer is remembered for the least important and influential portions of his enormous output. His theory of social evolution from military to industrial society and his account of the rôle of fear in the development of religion are repeated in a score of textbooks, but these are not so much wrong as lacking in contemporary relevance and explanatory power. His extreme *laissez faire* position in politics and economics might find a Darwinian sanction but is irretrievably dated. His strength is to be found in what can be discovered only by the comparison of his writings with those of his sociological and anthropological contemporaries: Spencer invented and established the working concepts and vocabulary of the modern sociologist or social anthropologist, whether engaged on field or on theoretical work.

It was Spencer who first used the terms " social structure " and " social function " in their modern senses, that is, to refer to the essential framework of institutions without which no continuing association of human beings in society is possible. The term institution had been employed in a fashion still contemporary as early as the eighteenth century, but Spencer re-established its usage as a key to the problems of comparative social study; he also put the classification of institutions on a firm basis.

His more general formula, which he believed could describe all evolution—physical, biological and social—has also proved of unexpected value. Evolutionary change ran, he asserted, " not simply from homogeneity to heterogeneity, but from an indefinite homogeneity to a definite heterogeneity; and this trait of increasing definiteness, which accompanies the trait of increasing heterogeneity, is, like it, exhibited in the totality of things . . ." [21]. It was this formula that, I believe, underlay Durkheim's classic study of the division of labour, his account of the segmentary principle of organization in primitive societies, and the development of functional specialization and interdependence in advanced industrial communities [6]. Durkheim's work remains fundamental and not merely for the understanding of industrial society: it is the essential foundation for any understanding of social change in the " underdeveloped " countries of the modern world.

Tylor's reputation has never faded as much as Spencer's, but for all that I think his importance is today much less. In saying this I do not refer to his ethnographic researches, but to his more theoretical interests. Of these, his concept of culture, as an anthropological category including both material artefacts and customs and institutions, is most remembered. Despite its employment in the United States as a definition of what anthropologists study, the whole idea of culture is too inclusive, large and awkward to be of much use for either analysis or comparison, and it is not surprising that the chapter on culture in most worthwhile monographs follows an unconsciously Spencerian account of institutions and is concerned to gather up loose ends rather than to synthesize the material. And, quite frequently, such a chapter is often no longer to be found in ethnographic writing at all.

Specifically Darwinian is Tylor's concern with primitive societies as " survivals " into the present of earlier stages of social development, and his concept of certain customs and beliefs in advanced societies as representing " vestiges " of earlier ways of social behaviour, stuck like fossils in the soil of modern living [18]. Much more important was his insistence on the use of the comparative method in social science and his formulation of the conditions—including the statistical conditions—on which it might proceed [23]. It is here that the greatest promise of the social sciences—with the possible exception of economics—still lies.

The case for Frazer—who like Spencer is rather under a cloud today—is too complex and technical to be argued briefly here. His use of the comparative method on an enormous scale can be faulted, though the fascinating detail it reveals and the charm of his Augustan style ensure that he is still read. His industry was truly Darwinian, and I believe that his success in subsuming vast masses of data under a few leading ideas was considerable. Unfortunately the anti-evolutionary reaction, largely led by Malinowski, has resulted in neglect of Frazer's achievement. Such a reaction was not surprising, for hypothetical yet untestable evolutionary theories had multiplied endlessly in the early years of the present century. In rejecting these a new freedom was gained, but, alas, much that was solid in the work of a Frazer of a Westermarck was forgotten.

Westermarck's colleague at the London School of Economics and Political Science, L. T. Hobhouse, established the first continuing school of sociology in a British university and applied Tylor's methods to a great variety of data. At the same time he continued the task, which Spencer had begun, of clarifying the concepts and vocabulary of social science. His greatest weakness was to neglect industrial society in his sociological—as distinct from his political—writings. His strength was in his breadth of learning combined with meticulous scholarship. His evolutionary scheme gives what is still the most adequate and probable account of social evolution. Even if, as he saw, such a scheme is ultimately impracticable in the complex, imitative social behaviour of men, yet his attempt to establish it provides a basis for the classification of societies superior—so far as non-industrial societies are concerned—to any other, and the most adequate basis so far available for comparative studies.

With Hobhouse the ghost of the " survival of the fittest " at last disappears. Material culture is an adequate basis for the classification of simple societies, but a number of criteria are required to deal adequately with more complex forms of organization, and these Hobhouse provided. In doing so he returned to an older tradition and took the social sciences away from a too imitative Darwinism to the more general study, not of social evolution, but as the title of his book has it, to the investigation of *Social Development* [12].

This is not to say that no part of the content of social life displays a genuinely evolutionary form of development. The evolution of language, says Ginsberg, " in the sense of descent with modifications is in many respects more clearly and securely established than is descent in biology " [9, p. 197]. Such cases, though important themselves, are however of little help for the explanatory tasks of sociology, social anthropology and political science. Unfortunately the realization of this produced a reaction which has hampered the development of these subjects during the last generation or so.

THE MID-TWENTIETH CENTURY

The reaction was essentially one against the element of history in the social sciences. One social science, perhaps the most advanced, had repudiated history during a prolonged and essentially Central European controversy at the end of the nineteenth century. If economics could gain from such a repudiation might not the same be true of the other disciplines? Certainly hypothetical history, guaranteed only by pseudo-Darwinian sociology, was dangerous and misleading. Certainly the attempt to demonstrate that all societies had passed and must pass through a sequence of invariant stages, unilinearly arranged, was both an affront to human moral dignity—though such an affront of itself could prove nothing—and extremely implausible in the face of the growing data of archæology, the history of Asian societies, and ethnography. It did not follow that the past was irrelevant to the understanding of the present or that relationships of temporal dependence might not be found to be as real as those of simultaneous functional interdependence. Yet in anthropology it was often, if as a rule tacitly, assumed that this did follow, and something of the same attitude influenced sociology; though it did not affect the least developed of the social subjects, political science.

This period of reaction is, in its turn, now coming to an end. Perhaps even economics is changing. Until recently, interest in economic dynamics was confined almost entirely to the trade cycle, the sequence of boom and slump, which had presented the major practical economic problem of industrial societies in the present century, and which was blamed for many of the political disasters of our time. It is now, however, fashionable and respectable—as well as realistic—to be concerned with problems of economic growth, development and accumulation. This is partly a consequence of the political circumstances of our time—the needs, economic and ideological, of the " under-developed " countries, the industrialization of socialist states, and the struggle for world power between the U.S.S.R. and the United States. It is also partly a consequence of the comparative exhaustion or trivialization of traditional economic problems, of technical advances in analysis, and the flood of information available from economies about which there was neither much known nor much to know only a short time ago.

I do not mean to say that economics, however much interested in growth and forced to have recourse to the comparative method, is going to become in any nineteenth-century sense an evolutionary science; but it would be surprising if it did not become more developmental, comparative and institutional. Similarly the dimension of time is returning to anthropology and sociology. Two contemporary trends have helped in this. First, we have social change and the exhaustion of really primitive primitives; second, there are the problems of urbanism as a new way of life for people born into comparatively static tribal societies. Both of these raise developmental problems for anthropologists and sociologists [16]. At the same time many social anthropologists have begun to rebel against a deletion of historical data from their material; indeed, some of them had never accepted it [11], and experience and theory alike were proving it impractical. Again we are not witnessing a rebirth of social Darwinism, but a recovery from a too violent rejection of some of the necessary factors of social explanation which had been recognized only in their Darwinian guise.

This recovery was probably least necessary in sociology as a theoretical discipline. The classical period at the beginning of this century had been at once acutely conscious of problems of social change and development and very incompletely Darwinian in its standpoint: consider for example its greatest names, Durkheim, Hobhouse, Pareto and Weber, none of whom neglected these problems and none of whom is in any strict sense a social Darwinist. It would undoubtedly be of advantage if today sociologists were again to interest themselves more in questions about social change and development; but the problem is one of degree, of appropriate focus of attention, rather than one of radical revision and a fresh start. What, in all probability, sociology most needs at the moment is not either a Newton or a Darwin, but a Linnæus to elaborate a really workable classification of social structures and of the range and variety of institutional patterns and sequences. Only after theoretical sociology has achieved this will it be possible to return with adequate equipment to the problems which an earlier generation believed could quickly and simply be solved by the application of a few Darwinian analogies.

As for political science—other than administrative and descriptive studies—I suspect that much the same holds good.

The situation in this subject is, however, now so complex that any such judgment is inevitably tentative and partial. Certainly here it is hard to detect much of a past, any present, or any future for Darwinian influence.

CONCLUSION

Discoveries in human genetics, physiology or psychology may at any time alter the whole relation of the social to the biological sciences. Without such discoveries it is probably fair to say that at no time in the last century has the influence of these sciences in general and Darwinism in particular been less in the scientific study of society, or their future influence seemed more problematical. Their past influence has been by no means wholly bad, but the present situation is not to be regretted if it is taken as a measure of the maturity and autonomy of the social sciences.

It used to be said that the last word of biology was the first of sociology. This is in no sense true now. Logically it ought to be the case but, for it to become so, biology will have to offer the social sciences something other than Darwinism—or at least something additional.

14

Natural Selection and Biological Progress

By

J. M. Thoday

On the last page of *On the Origin of Species* Darwin wrote: " as Natural Selection works solely by and for the good of each being, all corporeal and mental endowments will tend to progress towards perfection ". Thus the *Origin* put forward not only the view that species as we now find them result from processes of evolutionary change and that this change was directed by " Natural Selection ", but also the view that such evolution must necessarily be progressive. It appeared to Darwin inevitable that natural selection must promote progress towards ever higher states.

There have of course been difficulties with this view. Not only is it logically assailable, but to many it is and has been emotionally objectionable, because it has been tied up with the concept of individuals struggling for survival: " nature red in tooth and claw " does not seem too nice a basis for progress towards perfection. Such difficulties were felt by T. H. Huxley who found it impossible to reconcile his concepts of human progress with this concept of biological progress, and found himself forced to deny the applicability of biological progress to man [6]. Progress for man required that he turn his back on nature's " Cosmic Code ", and actively oppose it with the " Ethical Code " that man has made or part-made for himself.[1]

In effect, Huxley was forced to deny that knowledge of the mechanism of evolution can have any ethical message for us, which of course he was entitled to do because any argument of ethics from evolution must be based on the *a priori* premise that what evolution does is desirable. We may either deny this premise or accept it. We cannot justify it. Huxley denied the premise, but others accepted it and some still do, and it is

[1] [Huxley's ethical views are more fully discussed in chapter 15.—Ed.]

therefore important that, at the biological level, we should try to be clear what such acceptance might imply in the light of contemporary knowledge, for this knowledge may lead to conclusions differing from those that Huxley reached. To be so clear requires that we have some carefully considered concept of biological progress and of the mechanisms that may bring such progress about.

This is an important, though negative, incentive for thought about biological progress. The biologist must do his best to ensure that those who are prepared to base ethical conclusions on evolutionary concepts do not base them on erroneous biological concepts. Some feel, however, that there are positive reasons too. As Dampier has written:

> Regarded as a whole, as in natural history, any organism shows a synthetic unity as its characteristic expression of life, and man, carrying further what is seen in other animals, displays a higher unity in his mind and consciousness—a new aspect of life. The theory of evolution carries this synthetic process a step onward, and discloses an underlying unity in the whole organic creation. Life from a single cell of protoplasm to that infinitely complex structure, fearfully and wonderfully made, which we call man, is linked in all its parts by evolutionary ties. It forms one problem . . . the solution of which, could we reach it, would give us also the solution of subordinate problems, and give us a firm basis for ethics, æsthetics and metaphysics, the inner meaning of the Good, the Beautiful and the True. And one clue to the solution is the theory of evolution elucidated by Darwin's principle of natural selection. [1, p. 344]

Defining biological progress is essential to following that clue.

T. H. Huxley's grandson, Julian Huxley, with such considerations in mind, has done much to develop discussion of biological progress, taking into account our modern knowledge of the course and of the mechanism of evolution, and our knowledge that man is placed in a special position by his ability to learn from the experience of previous generations [4, 5].

Able to weigh the great body of evidence that has accumulated since Darwin's times concerning the course that evolution has actually taken, he has abstracted certain general trends which he calls progressive. These are broadly three: first, increasing adaptation to the more general aspects of the environment; second, increasing independence of the environment or increasing ability to maintain internal conditions despite external variation; and third, increasing rate of evolutionary change. Huxley derives this definition of progress from a consideration of the differences that have at various times distinguished the dominant groups of animals from their less successful contemporaries, and that have distinguished particular successful groups from those that they replaced; he has explicitly departed from Darwin's original attitude that progress must be inevitable and hence universal.

While Huxley's has been a notable contribution to the study of our problem, it has an inherent weakness which leaves many dissatisfied. The weakness is that in essence it defines the progressive as that which has generally (though not always) occurred. Huxley, in effect, assumes that progress has occurred, and then in masterly fashion shows us what we must on this assumption label as progressive. But it needs to be demonstrated that this basic assumption is sound.[1]

Some feel all attempts to define progress must be open to some such objection. Haldane has said " When we speak of progress in evolution we are already leaving the relatively firm ground of scientific objectivity for the shifting morass of human values " and these values are themselves the product of evolution [3]. However, it seems to me, as it seemed to Julian Huxley, that there is no need to involve human values in defining progress *biologically* and that it is possible to stick to the firm ground of objectivity, though of course there is need to involve human logic which itself involves value judgments at a different level.

I have accordingly defined biological progress by basing the definition on objective consideration of the general conditions of living [10]. This approach does not directly involve the " morass of human values " for it leads to a definition of biological progress that does not depend on our preferences. At the same time the approach does not involve the assumption that progress has actually occurred in evolution.

THE DEFINITION OF BIOLOGICAL PROGRESS

Since Darwin's time we have obtained a great deal of knowledge, based on critical experiment, concerning the nature of

[1] [See also chapter 15.—Ed.]

the inheritable individual differences upon which natural selection must operate if it is to produce evolutionary change. In the process, the belief that the inheritance of acquired characters can supply any important part of such variation has receded.[1] The raw material on which natural selection operates is mutation, largely undirected change in the heritable properties of the individual.

This process of mutation provides the basis from which an array of genetically different individuals are produced. In any particular set of environmental conditions some of the kinds of individual will succeed in producing more offspring than others. Their kinds will therefore multiply more than the others and, since the numbers that can survive are limited, will gradually replace the others. This is the basis of evolution by natural selection.

In the classic phrase, this process of natural selection promotes the survival of those relatively more fit to survive. The only form of progress, therefore, that it can promote is progress in fitness to survive, an apparent dilemma that has often been pointed out. For example, Dampier [1] writes: " Herbert Spencer's phrase for natural selection, the survival of the fittest, standing alone begs the question. What is the fittest? The answer is: The fittest is that which best fits the existing environment . . . that which is fit survives, and that which survives is fit." The dilemma, however, is not real and does not need evading. Progress by natural selection can only be increase of fitness for survival, but fitness itself requires elucidation. We can begin to understand the nature of biological progress only if we can satisfactorily answer the question " What does fitness for survival mean? " In particular, is it true to say " The fittest is that which best fits the existing environment "? And if it is true, is this the whole truth?

Defining biological progress as increase in fitness for survival requires some quantitative definition of fitness. The quantity involved will be the probability that, after a given lapse of time, the organism or organisms under consideration will be alive or have left descendants, that is to say, will have been naturally selected. Biological progress involves increase of this probability and any evolutionary change that leaves this

[1] [See also chapter 1, pp. 6 *et seq.*, and chapter 3.—ED.]

probability unaffected will be neutral in regard to biological progress, no matter how large it may be in other respects. In seeking to understand biological progress, therefore, we must look into the factors that may favourably or adversely affect the probability of survival.

The actual value of this probability will clearly depend upon the time we choose to consider. The probability that the author will be alive or have live descendants tomorrow is high. The probability that he will be represented by live descendants 1,000 years hence is lower. The probability that he will be represented by live descendants after a million years is lower still. Our choice of time scale is, however, limited, for we are considering progress in evolution and must consider fitness in terms of a time-scale that can permit significant evolution. We are therefore involved in long periods of time such as a million or ten million years.

This being so, we see at once that " The fittest is that which best fits the existing environment " is probably untrue, and is certainly not the whole truth. Environments change, and that which best fits the existing environment will not necessarily fit the future environment at all. That which best fits the existing environment will succeed temporarily, but unless it can also adapt to future environments it will in time fail completely. That which fits the present environment merely adequately may be well equipped to survive the change that is going to occur. *The fit are those who fit their existing environments and whose descendants will fit future environments.*

Environmental change and ability to accommodate to that change are therefore the keys to understanding the nature of fitness and progress. Relative adaptation to the contemporary environment is comparatively unimportant. Here lies the weakness of much thought on our subject and of arguments, frequently met, which claim to show natural selection cannot be the source of biological progress. Such argument was well expressed by Joad, who felt forced to reject natural selection as the source and directive agent of evolutionary change, because he believed natural selection must have an end point when all organisms fit the environment perfectly and no further natural selection is possible [7]. With natural selection Joad threw out all materialist philosophy as well.

> Why, then, it may be asked, does life still develop? [he wrote]. Why does evolution go on, and go on to complicate our structure so unnecessarily that, instead of becoming more fitted to our physical environment than we used to be, we are less? A degree of adaptation which, from the purely physical point of view, would put the average human being to shame has been achieved by living organisms thousands of years ago.
>
> The inference is irresistible, that the achievement by life of mere adaptation is not enough, but that living beings are evolved at more complicated and therefore more dangerous levels, in the endeavour to attain *higher* forms of life. The amœba, in short, is superseded by the man, not because the man is better-adapted life, but because he is better-quality life. In making this inference, however, we are admitting the suggestion that evolution is not a haphazard but a purposive process—an admission which is incompatible with Materialism.

Here is implied an idea of progress independent of fitness and resulting from some force, mystical in kind, opposed to natural selection. But the argument on which it is based is invalid, for environmental change is in the nature of life and there is no evidence that a stable equilibrium is yet established. The origin of life was itself a change, and since life involves the removal of raw materials from the environment and their return in changed form, it must involve perpetual change of the environment. Furthermore, every evolutionary change involves change in the way the environment is exploited with consequent change of the environment. Since living things are important components of each other's environment, every evolutionary change is a change of the environment to which all organisms must be able to adapt if they are to survive. Evolution must therefore continue if life is to continue ; for, quite apart from inorganically caused change, organisms must always be catching up with the change caused by their own and each other's evolution. If organisms do not continue to evolve, their probability of survival must necessarily fall as environmental changes leave life behind. Failure to evolve must be retrogressive.

Clearly, then, evolution is inevitable if life is to go on. But equally clearly evolution does not inevitably involve biological progress, for much of evolutionary change must be discounted as merely maintaining adaptation as the environment changes.

Progressive evolution can occur only if evolutionary changes do more than this, that is, if they actually increase the probability of survival. There has, on our definition, been biological progress in the last million years only if the probability of survival for a million years from now is greater than the probability of survival for a million years was a million years ago.

One question that has caused difficulty must be made clear at this point. It is the probability of survival, not the fact of survival, that is relevant to our discussion. Our primitive ancestor in whom life on this planet originated has living descendants now. This does not mean that its probability of survival was high. A low probability of survival means a high probability of extinction, but it does not mean inevitable extinction. Likewise a high probability of survival does not imply inevitable survival.

GENERAL FACTORS INFLUENCING THE PROBABILITY OF SURVIVAL

Let us now consider what is required of a solitary individual organism if it is to survive in any way for a million years. First, it must be adapted to the existing environment. Since it is alive we must suppose that it is. Second, it must reproduce, or sooner or later, purely by chance, it will be extinguished by a flash of lightning or some other "random" environmental fluctuation. The first progressive step in evolution, which some would define as the origin of life, is the establishment by reproduction of a *population* of individuals.

Now it seems most unlikely that such reproduction at its origin could be perfect: there would be bound to be mistakes (mutations). The population produced would therefore contain a variety of related types, and natural selection would automatically come into force.

The population must now maintain itself in a slowly changing environment, change partly inorganic in cause, partly determined by the activities of the population itself. Natural selection will first operate so that the members of the population become more and more adapted to contemporary conditions. The more perfectly adapted they become, the greater will be the advantage of their producing offspring adapted like themselves and natural selection will consequently tend to reduce the frequency and the magnitude of reproductive errors (mutations). The probability of survival, therefore, will depend upon the ability of individuals to produce offspring exactly or very like themselves, which we may refer to as the *genetical stability* of the population. However, during any long period considerable environmental change will occur and, unless it too can change, the population will find itself in an environment to which it is not adapted. It can change and, therefore, can survive, only if its individuals are capable of producing offspring unlike themselves, that is to say, if there is *variability*. The probability of survival therefore depends both on the ability of organisms to produce offspring *like* their parents and upon the clearly opposite ability to produce offspring *unlike* their parents. Progress must depend upon the establishment of the best compromise between these antagonistic needs, and upon any resolution of the antagonism that may be possible. The probability of survival will also depend upon the amount of change that will occur in environmental conditions during the period under consideration.

The probability of survival of the population, therefore, will depend upon adaptation, the genetical stability of the population (its capacity to remain adapted), the variability of the population (its capacity to change) and the stability of the population's environment (which determines its need to change). Increase of any of these, provided it does not involve corresponding decrease of another, will be biological progress as defined.

Clearly, genetical stability must depend upon the mechanisms controlling the similarity of parent and offspring—the " genetic system " as Darlington called it [2]. But variability too will be affected by this genetical system, that is to say, by the *genetical versatility* of the population...(Closely similar is the *genetical diversity* of the population. The environment is heterogeneous, that is, there are many environments. If a population includes different kinds of individuals these may be adapted to different adjacent environments. The probability that an environment in which the population can survive will be available in the future will then be relatively large.)

Resolution of the antagonism between stability and variability must therefore be, at least, in part a function of evolving genetical systems, and we must look at these systems for evidence of progress. Variability, however, is not solely to be achieved by genetical variation. It also depends upon the *versatility of individuals*, that is, on the capacity of individuals to accommodate themselves to environmental differences. If they can do so they can survive corresponding environmental change, and their probability of survival is correspondingly large. Our definition of biological progress requires us to call the evolution of such versatility progressive.

Such individual versatility must depend upon a measure of physiological complexity, a complexity that permits the organism to survive unaffected by certain environmental changes; in effect this enables the organism, by becoming independent of some environmental variables, to reduce the effective rate of environmental change. In addition, rather similar complexity may permit the individual organism, or more often a community of organisms, so to *control the environment* that the rate of environmental change is actually reduced. This too will increase the probability of survival, as also will control of the environment directed towards adapting the environment to the organism and making adaptation of the organism itself less necessary.

Biological progress then may be in principle brought about by any evolutionary change that so improves the genetical system that the antagonism between genetical stability and genetical versatility is reduced, any change that increases individual versatility, any change that increases the diversity of types and hence the number of environments to which there are adapted forms, and any change that permits the organism to control its environment in such a way that the environment remains suitable or becomes more suitable for the organism.

GENETICAL STABILITY AND GENETICAL VERSATILITY

Studies of the mechanism of heredity in a wide variety of organisms are beginning to show us that organisms have in the past developed genetical systems (systems controlling heredity and variation) that go a long way to resolving the antagonism between the conditions of survival they have had to meet

Mutation is the source of the variation that a population must exploit if it is to be genetically adaptable, but mutation, a sudden change of hereditary properties, is necessarily a breakdown of genetical stability. Organisms have, however, evolved mechanisms that permit them to carry potential variation without showing corresponding actual variation, and thus to retain genetical versatility while increasing stability.

There are several ways of doing this [9], all dependent upon the organism having sexual reproduction (or some such means of combining genes from different individuals), but only one will be described here. It depends on the organism possessing a number of pairs of alternative genes affecting the same character. (The difference between the members of such a pair of genes will have been produced by mutation.) Stability is achieved as a result of these genes being tied together in the same mechanical system, the chromosomes.

We will illustrate the principle in greatly simplified form. Suppose the character is size (though less evident physiological characters are likely to be far more important). Then we have what we may call " plus " (+) genes that increase size, and " minus " (−) genes alternative to them which decrease size. We will suppose the system to comprise only two pairs of alternative genes, both equally effective, and that the optimum size of the organism is given by the combination of two + and two − genes. There are four kinds of chromosome $\underline{++}$, $\underline{+-}$, $\underline{-+}$ and $\underline{--}$, and, since each individual has a pair of the relevant chromosomes, these can be combined in four different ways to give adaptive sizes, thus :

$$\frac{++}{--},\ \frac{+-}{+-},\ \frac{-+}{-+} \text{ and } \frac{+-}{-+}.$$

The first of these is a variable arrangement. When an organism reproduces sexually, each of the gametes (eggs or sperm) it forms receives one member of each of the pairs of chromosomes it contains. Thus a $\frac{++}{--}$ individual will form $\underline{++}$ and $\underline{--}$ gametes. Two such individuals mating will therefore produce a proportion (one-quarter) of offspring by combination of a $\underline{++}$ egg and a $\underline{++}$ sperm. These will be $\frac{++}{++}$ and will be too large. Likewise a proportion (one-quarter) of the offspring will be $\frac{--}{--}$ and will be too small. Only half the offspring will be $\frac{++}{--}$. This arrangement of the genes therefore produces a wide variety of offspring. It is versatile but unstable.

By contrast the $\frac{+-}{-+}$ arrangement of genes is one which is both stable and versatile. Individuals of this type will form two kinds of gamete, $\underline{+-}$ and $\underline{-+}$, and these can be combined to produce offspring of three types, $\frac{+-}{+-}$, $\frac{-+}{-+}$ and $\frac{-+}{+-}$. Each of these types has two + and two − genes, and all of them will therefore have the same genic size properties as their parents. The size is stable in heredity. However, in gamete production there are exchanges of material (crossovers) between members of a pair of chromosomes, such that a $\frac{-+}{+-}$ individual will produce a few gametes carrying $\underline{++}$ chromosomes and a corresponding number containing $\underline{--}$ chromosomes. These provide a source of variability that can be exploited to produce $\frac{++}{++}$ or $\frac{--}{--}$ individuals if and when a change of size is necessitated by a change in environmental conditions. The system is therefore in principle stable but, at the price of regularly producing a few ill-adapted variants (the crossovers), permits the maintenance of adaptability that can be exploited rapidly if the need arises. This model is grossly over-simplified, but it illustrates how the principle works, and how the tying together of genes in chromosomes permits some resolution of the antagonism between stability and variability to which we have referred.

It is in fact quite astonishing how stable and at the same time how versatile their genetical systems permit some natural species to be. An example from the author's own work will suffice to illustrate this. The figure on page 325 shows the results of experiments on some flies, all of which were the descendants of one fertilized female captured in the wild. The character studied, chosen purely because it is experimentally convenient, was the number of hairs on a certain part of each fly. Graph A shows the average number of hairs per fly in a population in which the parents of each generation (four pairs of flies) were selected at random. The number of hairs remains fairly steady, showing the stability of the character. Graph B shows the average number of hairs per fly in a population in which the 5 per cent of flies with most bristles were deliberately selected in each generation to be parents of the next generation. Graph C shows a population artificially selected for low bristle-number in the same way. It is clear from these graphs that the population responded readily to artificial selection in either direction. Further, the artificial selection produces something new. The fly with the most hairs found in generation 0 had 23 hairs; that with the least had 13. By generation 10 artificial selection for high hair-number had produced a fly with 32 hairs, and artificial selection for low hair-number had produced a fly with only 9 hairs. The selection has not only changed average hair-number but has produced new types of fly.

It is known that mutations occurring during the experiment cannot account for these responses. The genetical make-up of the original wild female and its mate, from whom all these populations were descended, was therefore such as to determine a high stability of hair-number but yet to preserve great potential variation; so that hair-number could be altered quickly if new conditions (here the artificial selection) required it. The system is both stable and versatile. The antagonism between stability and variability is partially resolved.

This stability and versatility depend upon what we call "balanced heterozygosity". $\frac{+}{+}$ and $\frac{-}{-}$ individuals are said to be homozygous, and can each produce only one kind of gamete. $\frac{+}{-}$ individuals are said to be heterozygous and can produce two kinds of gamete. Genetical versatility is promoted by heterozygosity, but balanced heterozygous systems permit stability as well. Homozygous systems also permit stable populations but cannot preserve versatility, which we consider to be the reason why populations with heterozygous systems are the more usual. Natural selection may and does establish species with homozygous systems (for instance, self-fertilizing species), but in *the long run* it must eliminate these and preserve only the versatile heterozygous species. These depend for their heterozygosity on *out-breeding* (cross-breeding) systems which limit the amount of mating between near relatives: we must regard the origin of such genetical and breeding systems, and of sexual reproduction and the chromosome mechanism that are part of them, as having involved evolutionary progress.

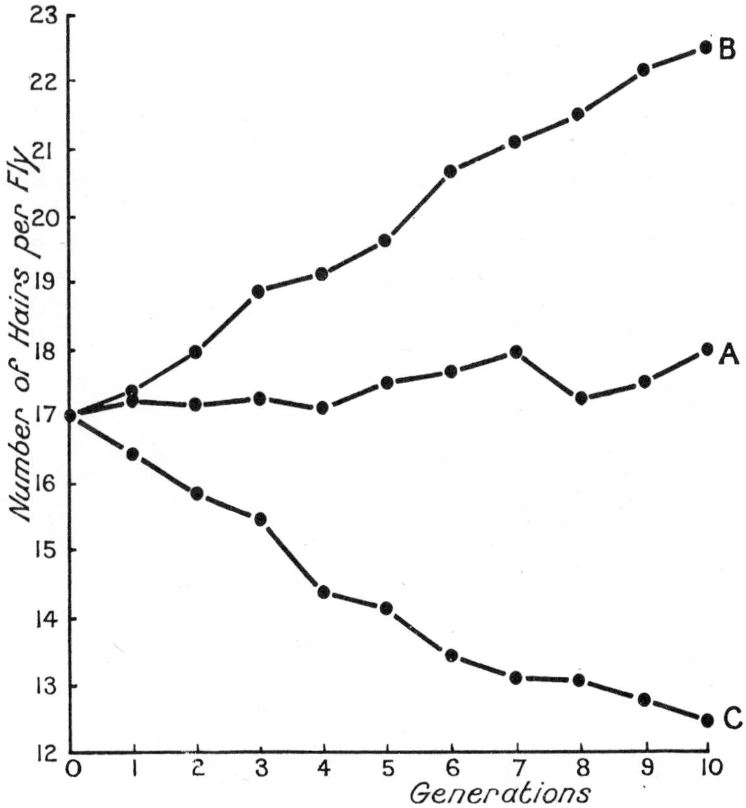

55.—Changes in average hair-number resulting from artificial selection in populations of flies. For explanation see text.

A—average hair-numbers in a population not selected (the "control").

B—average hair-numbers in a population selected for high hair-number.

C—average hair-numbers in a population selected for low hair-number.

Each point is the average obtained from 160 flies.

GENETICAL DIVERSITY

A population can also partially resolve the antagonism between stability and variability by splitting into two populations. Let us take the same simplified example as before, two pairs of genes affecting size. These, it will be recalled, can exist together in a number of combinations giving intermediate size, namely $\frac{+-}{+-}$, $\frac{-+}{-+}$, $\frac{+-}{-+}$ and $\frac{++}{--}$. Now a population can achieve a stable intermediate size, and yet preserve all these genes by splitting up into two separate populations one of which is $\frac{+-}{+-}$ and the other $\frac{-+}{-+}$. Each of these is homozygous and will be genetically stable apart from mutation, but each has lost its genetical versatility. However, if individuals from the two new populations occasionally cross, they will reconstitute the versatile $\frac{+-}{-+}$ type and thus, *provided that the two subpopulations do not separate completely*, genetical versatility may be preserved. The separation of a population into two populations that can never cross with one another is the sort of thing that is involved in the formation of new species. We therefore see that the origin of species is not necessarily evolutionary progress by our definition. To be progress it must produce something that compensates for the loss of genetical versatility that in some measure it must involve.

A further principle may be illustrated to strengthen this point, though it is necessary to consider more pairs of genes for this purposes. Four pairs will do. $\frac{+-+-}{-+-+}$ may be regarded as our starting point. Let us suppose a population contains these two types of chromosome, and lives in an environment where 4 + and 4 − genes together give adaptive size. Now let us suppose that this environment changes so as to produce two different environments in one of which 6 + and 2 − genes would give optimal size, and in the other 2 + and 6 − genes. Our population is versatile, since it contains heterozygous individuals, and these can produce new kinds of chromosome by exchange of parts between the old kinds, for example, $\underline{++-+}$ and $\underline{--+-}$. Natural selection will pick these out and may produce two sub-populations of different sizes, occupying the two new environments, namely, $\frac{++-+}{++-+}$ and $\frac{--+-}{--+-}$. The old types will be eliminated. Complete separation of these two new populations would form two species. However, they would retain greater probability of survival if they did not separate completely. For suppose the change in the two environments were to continue until 8 + genes were needed in the one, and 8 − genes in the other. Apart from mutation, only a cross between the two populations could produce the type, namely $\frac{++-+}{--+-}$, that could produce these. The two populations would therefore have a higher probability of survival if they retained the tendency to cross.

Such environmental situations are common in nature. In fact, most species occupy heterogeneous environments. Hence we would expect natural selection to ensure corresponding genetical diversity. Such diversity provides a source of genetical versatility. But it also ensures that a species shall consist of a number of types each adapted to different environments. The probability that all these environments will disappear is smaller than the probability that one of them will. Such genetical diversity therefore increases the probability of survival in two ways, first by providing individuals adapted to a number of environments and second by preserving genetical versatility in a stable system.

Natural selection, therefore, is not to be expected to give rise to uniformity, a best, ideal type. Fitness, which natural selection must promote, depends upon diversity. Further, species formation, though it involves increased diversity, is not necessarily progressive, for it involves some loss of genetical versatility. Progress is more likely if diversity can be achieved without speciation, that is, by the formation of separate races which to some extent interbreed. This is a conclusion strikingly different from that of Darwin's followers to whom speciation was of the essence of evolutionary progress, a view that has been held by many thinkers on the subject ever since. Sir Arthur Keith, for instance, wrote: " If a tribe loses its integrity . . . by free interbreeding with neighbours and thus scattering its genes, then that tribe as an evolutionary venture has come to an untimely end. For evolutionary purposes it has proved a failure " [8]. There is no warrant for a statement of this generality. A tribe is at least as likely to increase as to decrease its fitness by mixing genetically (and culturally) with others. Even if purity of race were achievable, it would not be desirable.

INDIVIDUAL VERSATILITY

As indicated above, the probability, that a population of organisms will survive a given amount of environmental change, is not only dependent upon the versatility of its genetical system, but may also be influenced by the versatility of the individuals comprising the population. If the individuals can cope only with a restricted range of environmental conditions, the amount of environmental change the population can survive without genetical versatility will be correspondingly restricted. If, on the other hand, the individuals comprising the population can each cope with a wider range of environmental conditions, not only will the population be able at any one time to occupy a wider range of environments (and hence to exist in larger numbers), but the probability that all the environments the individuals can occupy will disappear will be correspondingly smaller.

There can therefore be no doubt that, unless it be achieved at the expense of a greater amount of genetical versatility, any increase in individual versatility is progressive. It is this aspect of progress that has most attracted the attention of biologists. Reptiles are capable of withstanding a greater range of humidity than are amphibians. Mammals and birds are capable of functioning in a greater range of temperatures than are reptiles. The reader is referred to the writings of Julian Huxley [4] for detailed discussions of such biological progress.

Such individual versatility must depend upon a certain measure of physiological complexity. Complexity therefore may be essential to this aspect of biological progress, and some authors have in fact regarded increasing complexity as itself evidence of progress, though there seems to be no warrant for this. Increased complexity of a particular kind (such as to result in increased individual versatility) is biological progress, but not increased complexity of any kind.

ENVIRONMENTAL CONTROL

Progressive increase of individual versatility, an increase itself dependent on genetical versatility, has in time brought some organisms to a degree of complexity permitting them not only to become independent of some aspects of environmental change, but also to control the environment so as to render it less liable to change.

All organisms affect their environment, but this is not to say that all organisms affect their environment in such a way as to increase their fitness or probability of survival by reducing the rate at which the environment becomes unsuitable for them. The fungus attacking an apple rots the apple and destroys it. The parasite may kill the host on whom it depends, and may even extinguish the host species and hence its own. On the other hand beavers build and maintain dams and birds build nests, thus altering the environment to make it more suitable for themselves. Likewise, as we have learnt at some cost, the stability of most, if not all, soils depend upon the vegetation, especially trees and grasses, whose roots bind the soil together and prevent it from being blown or washed away. Such control of the environment depends to a great degree on co-operation between individuals (and between species), for a single grass plant can seldom hold any soil on a slope. It requires a sward. Thus, once such capacity to control the environment is evolved, natural selection will tend to promote co-operation.

Supreme in ability to control his environment is man, because his ability to communicate information renders him able to co-operate in the control of nature in an entirely new way. It is this that determines man's dominance in contemporary nature.

BIOLOGICAL AND HUMAN PROGRESS

As we have seen, the organisms that have the greatest probability of survival are those which are not only adapted to their contemporary environment but which also to the greatest degree possess genetical stability and versatility, racial and individual diversity, individual versatility and the capacity to control their environment so as to maintain its suitability for them. There can be no doubt that man is one of the species that possesses these attributes in most marked degree, and it seems unlikely that any other species can match man in fitness as defined here. Increasing, though very limited, knowledge of human genetics leaves little doubt that man's genetical system provides considerable versatility and diversity within most human populations, and there is also considerable racial diversity so that potential variation may be exploited through racial hybridization provided that those with wrong-headed ideas of the biological function of races do not prevail.

In addition, man possesses a degree of individual versatility and environmental control that no other species can equal. He is capable of functioning successfully in a wider variety of environments than any other species. He can function in the arctic and the tropics, he can cross the sea and penetrate its depths, he can climb mountains and burrow into the earth, and he can fly in the air. A lot of nonsense is written about man's unfitness by comparison with other animals. No other species can be warm in the arctic and cool in the tropics: stoves and air conditioners are adaptations just as much as are hair and sweat glands. In fact, man can travel along the ground faster than any animal, on water faster than any fish, and fly the air faster than any bird. It is nonsense to deny these facts just because they depend upon machines not limbs, and because these attributes of man have been evolved by social, not biological, means. If all the factors involved in fitness as here defined had to be measured by one approximate empirical measure, that measure would be the range of environmental conditions in which the species can function, for this is closely related to the range of environmental change the species could survive and hence to its probability of survival. Man with his aids is at least as fit as and probably far fitter than any other species by this measure.

Man has reached this stage because, with his evolution by natural selection, a new form of evolution became effective. Man evolves not only by natural selection of genetical constitution (biological evolution) but by the non-biological inheritance of acquired environment (social evolution). What one man learns can be taught to others without the necessity of biological relationship, and what each generation achieves can be handed to the next by the path of social inheritance. Man has thus acquired a degree of control of the environment that could not otherwise have been achieved.

In this new phase of evolution man has not escaped the biological necessities. Adaptation, genetical stability and versatility are still required for maximum fitness. But stability and versatility are now also to be considered in a new context, social evolution, in which stability is maintained by tradition, and versatility by the continuous production of new ideas. Once again we have the old antagonism at a new level, and progress will be much dependent upon resolution of the antagonism within political systems in which new ideas and old traditions can react upon one another with minimum discord so that change may be harmonious. Thus man may come into increasing harmony with his environment, living and non-living.

However, fitness, or the probability of survival, is not necessarily increased and may be decreased by environmental control. Control of the environment not only permits stabilization of the environment and the improvement of the environment in the directions that will increase the probability of survival, but also permits the exploitation of the environment for short-term ends that make long-term survival less probable. For fitness to be increased the control must be directed to the right end. In the main the end must be increasing harmony between the controlling organism and the environment, which of course includes all other organisms. This requires not only the intention that the end shall be attained, but also increasing knowledge of the environment such that the results of attempts to control it shall become more predictable.

Once knowledge of the environment has reached this predictive level man may be ready to embark upon a third phase of evolution, in which not only the environment, but also genetical constitution is controlled, so that, by genetical and environmental control, both the genetical constitution of the species and the environment are brought into ever more perfect harmony. Once again such control will be biologically progressive only if directed to the right end, increasing harmony with an increasing range of environments. And since a most important component of the environment for each man is other men, this includes harmony amongst ourselves. Such control of evolution would seem to be the natural outcome of natural selection.

PROGRESS OF LIFE

So far we have only considered biological progress with respect to the probability of survival of a population or species. We have not considered the general progress of life itself. By the same argument that we applied to a population we must here be concerned with the probability that any life at all will survive after the lapse of a long period of time. This probability is clearly related to the probabilities of survival of all the separate species of which life is composed. Life therefore progresses as each of its component species progresses in adaptation, in stability, in versatility, and in their control of the environment towards its improvement for their own long-term ends. It will also tend to progress as the diversity of species increases.

It does not, however, follow that mere increase of number of species is necessarily biological progress for life, for species interact on one another because they are components of each other's environment. They therefore determine one another's probability of survival. The origin of a new species, therefore, may or may not be progressive. It will be progressive in so far as the new species does not reduce the probability of survival of old species by as much as the probability of survival of the new species itself.

This will be so if the new species occupies a new environment to which hitherto no species had been adapted, as occurred when aquatic forms gave rise to others capable of living in land marshes, and when the marsh forms gave rise to those living on dry land. (In the same way the conquest of space may be progress for life.) This will also be so if the new species occupies an old environment, but exploits that environment in a way that harmonizes with its use by old species, and even, as do the various species involved in the nitrogen cycle, makes that environment more suitable for the other species that occupy it. On the other hand, the origin of new species will not involve biological progress for life as a whole if the new species merely competes with the old species for limited environmental resources.

Biological progress, therefore, not only involves increasing versatility of individual species, but also increasing diversity of species *harmoniously adapted to one another.* It would seem therefore that nature's code is far nearer to Huxley's "ethical code" than to the "cosmic code" of struggle which he rejected. As was pointed out at the beginning of this article, we cannot justify any argument of ethical principles from evolutionary principles. If, however, we were to accept the premise that what evolution does is desirable, the conclusions we would reach today would be the opposite of those that T. H. Huxley reached.

15

Darwinism and Ethics

By

D. Daiches Raphael

The theory of evolution has been connected with ethics in three different ways. The first suggestion is that ethics is a product of evolution. Secondly, some people have thought that evolution should guide the future course of ethical ideas. A third, more recent, suggestion is that ethical ideas can affect the future course of evolution.

These three notions may or may not be tied up with each other. There is little or no dispute about the general truth of the first proposition, though details of the process of the evolution of ethics are still speculative. The filling in of these details is a matter for further scientific investigation. The second proposition, that the facts of evolution show the correct standard to be used in ethical judgment, is very much a matter of dispute. The dispute is of a logical nature. Those who believe that evolution provides a guide for ethics think this is a consequence of the truth of the first proposition that ethics is a product of evolution. Those who deny the second proposition may (and usually do) accept the first, but they deny that the evolution of ethics implies anything about the standards that ought to be used in ethical judgment. The third proposition, that ethics can determine the future course of evolution, has been put forward by Julian Huxley, who also believes both the first and the second propositions. He evidently thinks that all three propositions are connected with each other. But it is not necessary to accept the second proposition in order to agree with Huxley about the third.

THE EVOLUTION OF ETHICS

The first of the three notions I have distinguished is that our moral capacities ("conscience") and our moral ideas have evolved by a process which is part of the general process of evolution or at least analogous to the general process of evolution. By " conscience " is meant: (*a*) the capacity of human beings to make moral *judgments*, to think of actions as right and wrong, and of motives, ends, and characters as good and bad; (*b*) the specific *motive* of the sense of duty, this being often contrasted with motivation by desires; (*c*) the painful *feelings* of remorse that are apt to follow action done deliberately against the sense of duty.

One of these functions of conscience, the making of moral *judgments*, shows itself in the use of certain words that are said to express ethical ideas. Some of these ideas are highly general, like good and bad, right and wrong, virtue and vice. Others are more specific, and are often called particular virtues and vices, like justice and injustice, courage and cowardice, benevolence and malice, and so forth.

What does it mean to say that our moral capacities and ideas have *evolved*? Sometimes " evolved " is just a synonym for " developed ". In that event the term " evolution " has little or no explanatory value. The Darwinian theory of evolution had explanatory force because it tried to show how all the different species of living things could be correlated together in a *single* process of development, in which an enormous diversity of results could all be put down to a single complex of causes, namely, natural selection of varieties in " the struggle for existence " between forms of life all of which reproduced themselves in far greater numbers than could survive. The theory had great explanatory force because it substituted the simple hypothesis of a single process for the alternative hypothesis of a special creation for *each* of the many different species of plants and animals. Darwin wrote, in the *Descent of Man*, that in the *Origin* he " had two distinct objects in view; firstly, to shew that species had not been separately created, and secondly, that natural selection had been the chief agent of change ". And while he was ready to acknowledge that his positive hypothesis might have exaggerated the power of natural selection, he justly claimed to have " done good service in aiding to overthrow the dogma of separate creations " [2, ch. 2]. Whether or not the theory was correct in its details, it substituted the notion of an empirically verifiable, uniform process for a purely metaphysical concept of causation invoked *ad hoc* for every species. While the theory of natural selection still leaves room for (though it does not necessarily require) a metaphysical foundation in *an* original creation, it gives an intelligible causal explanation of the differences in the forms of living things.

A theory of the evolution of ethics will have similar explanatory force if it can show that the development of conscience and of ethical ideas is a part of the same process of evolution by natural selection which explains the origin of biological species. It will have less, but still some, explanatory force if it takes the more limited form of showing that conscience, instead of being a "separate creation", has developed from *other* human endowments responding and adapting themselves to the stimuli of natural and social environment. A theory of the first type was put forward by Darwin, in the *Descent of Man*, where he tried to show how the intellectual and moral capacities of man could have owed their development at least partly to the process of natural selection. Modern psychologists do not try to use natural selection as the medium whereby the moral faculties have developed. They have instead constructed theories of the second type, showing how conscience is built up from other mental endowments, such as love and fear, in response to the stimuli of family and wider social environment. Sociologists and anthropologists have provided similar evidence of changing environmental stimuli to account for changes in moral ideas, i.e. for changes in *what* is approved and disapproved by conscience as right and wrong, just and unjust, and so on.

The evolution of ethics as portrayed by modern psychologists and social scientists, therefore, has little in common with the specifically Darwinian concept of evolution through natural selection. The type of evolutionary explanation they give follows lines laid down before Darwin. This is particularly true of accounts of the development of the moral capacities. Darwin himself knew something of the views of contemporary and earlier philosophers about the development of conscience from social impulses and especially from sympathy. He drew largely on the work of Alexander Bain, but he knew that the idea originated with David Hume and Adam Smith in the eighteenth century. Adam Smith's highly interesting theory of conscience has much in common with the Freudian theory of the super-ego. This is not to say that the researches of recent psychology have added nothing to the theories of eighteenth-century philosophers, but simply that the general line of investigation was laid down before Darwin and has little to do with the Darwinian concept of evolution.

Like every genius, Darwin built upon the work of his predecessors. The idea of the struggle for existence was taken from the views of Malthus on population. " It is ", wrote Darwin, " the doctrine of Malthus applied with manifold force to the whole animal and vegetable kingdoms; for in this case there can be no artificial increase of food, and no prudential restraint from marriage " [1, ch. 3]. But to note Darwin's debt to Malthus is not to disparage his own achievement. The originality of genius lies not so much in thinking of an idea as in seeing how to make the best use of it. Malthus had held that population tends to increase faster than the means of subsistence, but is restrained by the " positive checks " of famine, war, and disease, and by the " preventive checks " of " moral restraint " and " vice ". His theory thus allows a distinction between natural tendencies that cause a lot of people to be born and then killed off, and premeditated measures to prevent them being born. Darwin followed out the implications of the natural causes in the absence of preventive measures, and used this situation as a hypothesis to account not merely for the numbers of animals and plants but also for the extinction or survival of their different varieties. His ability to use the theory in this way depended in part on his intellectual powers of analysis, but more on the stimulus to those powers provided by his astonishingly wide and careful observation of similarities and differences in the data of biology. Although Darwin's theory was an " application " of the doctrine of Malthus, it was none the less an epoch-making discovery.

It is quite otherwise with Darwin's borrowings from Adam Smith and Bain on the development of the moral capacities, and Darwin would not have claimed any special originality here. When he came to discuss this topic in the *Descent of Man*, his purpose was not to provide a new theory, but simply to show that the peculiar endowments of man were no objection to the evolution of man from another form of animal life. He paid special attention to the moral capacities because, he said, he agreed with "those writers who maintain that of all the differences between man and the lower animals, the moral sense or conscience is by far the most important" [2, ch. 4].

He did, however, add something to previous theories of the development of conscience from sympathy and other social impulses. The previous theories of philosophers and psychologists had confined themselves to man. They showed how conscience develops from feelings and desires with which all normal human beings are naturally endowed, modified by reaction to the attitudes of their fellows in social intercourse. To this Darwin adds three things, all due to his biological interests: (*a*) In the first place, he suggests that the initial social impulses that form the basis on which conscience is built are to be found in animals also. In particular, he cites some interesting examples of animal behaviour which he thinks can only be motivated by sympathy. Unfortunately his evidence is mainly anecdotal and would not, so I am advised, satisfy modern standards of scientific reliability. (*b*) Next, he suggests that natural selection is likely to have been an important causal factor in the development of "social instincts". An individual of strong social instincts would be less likely to survive than a selfish individual intent on saving his skin, and therefore less likely to leave offspring inheriting his instincts. But because his type of behaviour is more conducive to the welfare of the group, he would, as Hume had said, be praised by his fellows, and the praise of his qualities would cause others to emulate it and to counsel emulation in the teaching of their own young. To this Darwin adds that an increase of such qualities in a group by the force of example would give the group a greater survival-value than would be possessed by groups whose members acted only for their individual welfare. (*c*) Finally, Darwin was inclined to believe that the qualities of mind and behaviour acquired by way of example and education were transmitted by heredity to offspring.

The first of these three points is not altogether trite. Few people would deny, especially in view of later researches in animal psychology, that many animals exercise processes of intelligence that are in principle the same as the lower reaches of intelligence exercised by man. But relatively little attention has been paid to the parallel analogy in moral capacities. If it were clearly shown that the behaviour of animals in helping others is at times an expression of sympathy and not merely an instinctive reaction, such instances could quite properly be regarded as a rudimentary form of moral behaviour. Strictly speaking, moral behaviour occurs only within the context of moral concepts, that is, of the understanding and use of moral language. All the same, it would be perverse to deny that sympathetic behaviour in animals contains the beginnings of moral behaviour, just as it is perverse to say that animals do not think if they do not use language.

Darwin's second point would hardly be disputed when applied, as he applied it, only to the early stages of ethical evolution. It would, however, be doubtful if applied to later stages. As T. H. Huxley argued in the Prolegomena to his famous Romanes Lecture on "Evolution and Ethics", a policy of action based on sympathy, on doing as you would be done by, leads in the end to the suppression of the qualities best fitted for success in the struggle for existence:

> If I put myself in the place of the man who has robbed me, I find that I am possessed by an exceeding desire not to be fined or imprisoned; if in that of the man who has smitten me on one cheek, I contemplate with satisfaction the absence of any worse result than the turning of the other cheek for like treatment. Strictly observed, the "golden rule" involves the negation of law by the refusal to put it in motion against law-breakers; and, as regards the external relations of a polity, it is the refusal to continue the struggle for existence. It can be obeyed, even partially, only under the protection of a society which repudiates it. Without such shelter, the followers of the "golden rule" may indulge in hopes of heaven, but they must reckon with the certainty that other people will be masters of the earth. [3, p. 52]

However, Huxley agrees that in the early stages of social development, "every increase in the duration of the family ties, with the resulting co-operation of a larger and larger number of descendants for protection and defence, would give the families in which such modification took place a distinct advantage over the others " [3, p. 48]; and this is all that Darwin claimed for the principle of natural selection in the evolution of ethics.

In his third point, that the acquired virtues of parents would be inherited by their children, Darwin was mistaken. It is an example of his retention of Lamarckian ideas on inheritance. At the same time Darwin noted a flaw in the Lamarckian thesis. If acquired dispositions can be inherited, he observed, this should not apply only to those that are socially useful. " My chief source of doubt with respect to any such inheritance, is that senseless customs, superstitions, and tastes, such as the horror of a Hindoo for unclean food, ought on the same principle to be transmitted. I have not met with any evidence in support of the transmission of superstitious customs or senseless habits " [2, ch. 4].

In his account of the development of the moral capacities, Darwin found it a problem to explain why the promptings of conscience should be given *precedence* over self-regarding motives. This cannot be because they are stronger, for everyone knows that in general this is untrue.

> Although some instincts are more powerful than others, and thus lead to corresponding actions, yet it is untenable, that in man the social instincts . . . possess greater strength, or have, through long habit, acquired greater strength than the instincts of self-preservation, hunger, lust, vengeance, &c. Why then does man regret . . . that he has followed the one natural impulse rather than the other; and why does he further feel that he ought to regret his conduct?

Darwin's answer to his question is that " the social instincts are ever present and persistent ". This is due to " reflection: past impressions and images are incessantly and clearly passing through [the] mind ".

> Even when we are quite alone, how often do we think with pleasure or pain of what others think of us,—of their imagined approbation or disapprobation; and all this follows from sympathy, a fundamental element of the social instincts. A man who possessed no trace of such instincts would be an unnatural monster. On the other hand, the desire to satisfy hunger, or any

passion such as vengeance, is in its nature temporary, and can for a time be fully satisfied. . . . The wish for another man's property is perhaps as persistent a desire as any that can be named; but even in this case the satisfaction of actual possession is generally a weaker feeling than the desire. [2, ch. 4]

Now plainly this answer, that the social " instincts " are more enduring than the self-regarding, will not do. It is easy enough to say that the desire to satisfy hunger is temporary, but, as Darwin observes in the last sentence I have quoted, it is not so easy to say this of all self-regarding motives. Reflection, with its persistent imagination of what pleases and what displeases, is not confined to social feelings. Bishop Butler, in the eighteenth century, drew a distinction between two types of motive, which he labelled reflective " principles " and " particular passions ". Hunger, and likewise a surge of pity, would be " particular passions ". Butler does not seem to have made up his mind clearly whether there is a reflective " principle " of sympathy—he uses the word " benevolence " —or whether benevolence is just the name of a group of particular passions. But he was quite clear that there is in all men a reflective principle of self-regard—his word is " self-love "—which *makes use* of particular passions like hunger. And it cannot be denied that we reflect upon the effect of our actions on our own welfare, as well as reflecting upon the way they affect other people. Our use of language confirms the drawing of such a parallel. Darwin's problem here is to explain the " ought " of conscience, and he holds, as Hume and Adam Smith had held, that it comes from reflection upon the approval and disapproval shown by our fellows. Now we use the word " ought " of self-regarding actions also. To say that one ought to act in a certain way may mean that one is morally obliged, but it may equally mean that one is prudentially obliged. " I ought not to have done that " may express remorse for morally wrong action, or it may express regret for imprudent action. It can be used to imply " I was a knave to have acted so ", or it can equally well be used to imply " I was a fool to have acted so ".

Suppose someone were trying to explain why we use the word " ought " of prudential action. His explanation could be a parody of Darwin's explanation why we use the word

" ought " of moral action. He would note first, as Butler noted, that we cannot say the " principle " of " cool self-love " is stronger than " particular passions ". A starving man faced with a banquet feels an almost uncontrollable desire to eat and eat, more than is good for him, while at the same time he wants to avoid making himself ill. If he does eat more than is good for him, why does he regret " that he has followed the one natural impulse rather than the other; and why does he further feel that he ought to regret his conduct "? He feels this, not only when he has the pain of indigestion, but afterwards also. We might answer that he feels it because the desire for one's interest on the whole, what Butler called cool self-love, is " ever present and persistent ", owing to " reflection ". " Past impressions and images " of the effects of giving way to particular impulses, and of the effects of restraining such impulses, " are incessantly and clearly passing through his mind ". As Butler puts it, it is natural for an animal that does not reflect to give way to each particular passion as it comes; but man, who has also the capacity of reflective self-love, tells himself that often he ought not to do so. While a man without sympathy would be, as Darwin says, " an unnatural monster ", so would a man without prudence.

If we are to give a genetic account of conscience, including its judgments that we ought to perform some actions and refrain from others, we must recognize that self-regard, with *its* judgments of " ought " and " ought not ", is similarly a reflective and late-developed capacity. If a highly-developed form of unselfishness is peculiar to man owing to his capacity for reflection, so is a highly-developed form of selfishness. Darwin's omission to observe this does not of course weaken in the least his argument that conscience has evolved from capacities which man shares with other animals. But it is highly relevant to any contention that the direction of evolution is the direction of morality. While selfishness is aimed at the good of the person concerned, and does not itself aim at harm, it can, in many situations, have the collateral effect of harm to others. In its aim it is no more moral or immoral than its rudimentary counterpart in " the struggle for existence ". It can be immoral at the reflective stage just because it is accompanied at that stage by the capacity for moral reflection, for thinking about the interests of others as well as of ourselves. As T. H. Huxley said, while there is little doubt that the moral sentiments have evolved, " the immoral sentiments have no less been evolved " [3, p. 80].

EVOLUTIONARY ETHICS

The thesis that a study of evolution can teach us what to think good finds no place at all in the works of Darwin. But ideas which have produced striking advances in any one field of inquiry are always (and rightly) tried out in others. Often the first step comes in philosophical speculation, and the theory of evolution soon gave birth to a crop of metaphysical and moral philosophies. Nineteenth-century attempts to base ethics on evolution were forcibly criticized in T. H. Huxley's Romanes Lecture of 1893 and the Prolegomena added in 1894. These two essays are by far the clearest and the most important discussion of the topic.

> The propounders of what are called the " ethics of evolution ", when the " evolution of ethics " would better express the object of their speculations, adduce a number of more or less interesting facts and more or less sound arguments, in favour of the origin of the moral sentiments, in the same way as other natural phenomena, by a process of evolution. I have little doubt, for my own part, that they are on the right track; but as the immoral sentiments have no less been evolved, there is, so far, as much natural sanction for the one as the other. . . . Cosmic evolution may teach us how the good and the evil tendencies of man may have come about; but, in itself, it is incompetent to furnish any better reason why what we call good is preferable to what we call evil than we had before. [3, p. 80]

Not only our capacity to do evil, but likewise our capacity to suffer it, evolves along with the things we call good.

> Where the cosmopoietic energy works through sentient beings, there arises, among its other manifestations, that which we call pain or suffering. This baleful product of evolution increases in quantity, and in intensity, with advancing grades of animal organization, until it attains its highest level in man. Further, the consummation is not reached in man, the mere animal; nor in man, the whole or half savage; but only in man, the member of an organized polity. [3, p. 63]

Nobody thinks that things become better just because they are later in time, that the hangover is preferable to the party because it comes on the morning after. Why, then, should it be supposed that the direction of evolution can be a guide for ethics? One reason is that " more evolved ", unlike " later ", includes in its meaning the idea of being superior, higher on a scale of value. But now we need to ask whether the scale of value implied is the same as that used in ethics.

What criterion of evaluation is in fact used when one organism is said to be " higher " in the evolutionary scale than another? At first sight, it might seem that the natural results of the process of evolution supply the criterion. For natural selection results in " the survival of the fittest ", and it is easy to suppose that " fittest " means " best ". So indeed it does if we are thinking of what is best fitted to survive. But in that case the survival of the fittest means no more than the survival of those *most able* to survive. Perhaps that is a slight exaggeration, at least so far as ordinary language is concerned. Normally, to call something " good " or " best " is to express commendation of it. So to say of one thing that it is better fitted than another to do something, is to express commendation of its greater ability as well as to say that it has the greater ability. Nevertheless, this does not necessarily imply commendation of that which it is able to do. Bill Sykes may be best fitted to crack a safe; in the *art* of safe-cracking, we take off our hats to him. But this does not mean that we take off our hats to safe-cracking in preference to leaving other people's property alone. To be sure, we do not feel the same about survival as we do about safe-cracking. Most of us do not want to crack safes but do want to survive for as long as we can; and we also sympathize with the desire of others to survive, while having no sympathy with Bill Sykes's desire to crack safes. Still, it is not survival as such that suffices to induce our sympathy. We have no sympathy with the efforts of an influenza virus to survive and multiply. A late-evolved strain of virus is fitter than earlier strains which have been eliminated by natural selection; yet no doctor feels morally obliged to let it go on surviving so successfully, any more than a policeman thinks himself morally obliged to let Bill Sykes go on with the safe-cracking which he does so successfully.

Different standards of commendation are used for different purposes. So far as safe-cracking goes, Bill Sykes is at the top. So far as ethics goes, he is near the bottom. High ability at performing any function, whether this be safe-cracking, survival, or anything else, does not necessarily go along with high ethical value. The standards of ethical evaluation do not depend upon the performance of a specific function. To speak of a good man is not to commend by the same standards as those we use when we speak of a good runner or a good burglar or even a good survivor.

Furthermore, superior ability at surviving does not necessarily go along with superiority in the evolutionary scale. It is not at all clear that there is a single standard of evaluation for the purpose of grading organisms as higher or lower on the scale of evolution. Darwin pointed out [1, ch. 4] that different standards are used for different biological kingdoms, for animals, for fishes, and for plants; and even within any one biological kingdom different biologists may use different standards of evaluation in determining evolutionary superiority. On the whole Darwin favoured using, as a standard, increasing differentiation and specialization of function among the parts of the organism. But he noted that continuing advance in this direction does not necessarily accompany the continuing ability of a species to survive.

Superior fitness to survive is, then, no guarantee of biological "superiority". Nor is biological superiority any guarantee of superior ethical value. Herbert Spencer, who was the most prominent exponent of evolutionary ethics in Darwin's day, felt it necessary to argue that biological superiority implied ethical superiority *only if* it brought with it "a surplus of agreeable feeling". According to Spencer, an increase of life in length and breadth (where "breadth" refers to complexity of organization) can be called good in the ethical sense if and only if human beings find life more pleasant than painful. Plainly, therefore, his standard of ethical goodness is pleasure, not length and breadth of life.

Pleasure is, of course, used very commonly as a standard of evaluation. (Whether it is the sole standard of ethical evaluation is another question.) But the use of this standard is not confined to ethical matters. It is extended to situations of choice which we should not commonly call ethical at all. Human choice covers a wider range of action than the ethical, and it is worth noting that in non-ethical as in ethical choices the standards of value we employ are independent of anything that might be derived from the course of natural evolution.

T. H. Huxley, in the Prolegomena to his Romanes Lecture, brings out sharply the distinction between natural process and artificial process. The gardener who cultivates a piece of earth works in opposition to the state of nature, eliminating weeds that would otherwise win out in the natural struggle for existence between the different plants. Huxley of course agrees that man, and what man does, are part of nature in the broadest sense. Why, then, should the struggle between the gardener and the weeds not be called a part of the general evolutionary process of natural selection ? Just because the process of natural selection is *not* a process of *selection*, properly speaking. Darwin himself insisted [1, ch. 4] that his use of the term " selection " was metaphorical, since it did not imply conscious choice. Like Huxley, Darwin contrasted nature with art. The point of the expression " natural selection " lay not so much in drawing, with the word " selection ", an analogy with human choice. It lay rather in the contrast between the " immeasurably superior " power of " the works of Nature " over " those of Art " [1, ch. 3]. Huxley's gardener, unlike the weeds and the other plants in his garden, has a purpose. It is to ensure the survival of the plants that he likes or needs, not of those that will survive in unrestricted competition. Consequently his art works against the process of natural " selection ". If he relaxes his efforts, the natural process will resume its course, and the weeds will swamp the garden. The gardener's art protects his favoured plants from the natural course of events. As Huxley says, the gardener selects certain varieties of plants for survival by reference to his own ideas of what is useful or beautiful. Natural selection, by contrast, leaves surviving those varieties that are stronger than their competitors. Among the standards of value used in horticulture, utility for human needs and the satisfaction of human ideas of beauty take precedence over strength.

T. H. Huxley was not implying that the standards of value used in horticulture are the same as those used in ethics. On the contrary, he went on to point out that the ethical guidance of human affairs can be opposed not only, like horticulture, to the course of natural evolution but to the principles of horticulture as well. Although the gardener goes against the principle of natural selection in pulling up the dandelions, he goes with it in eliminating the weaker plants among the varieties he wants to grow. A ruler of a human society who thought his job was strictly analogous to horticulture, would not only protect his social garden from the ravages of nature, but would also eliminate the weaker members of his society. And while there have been societies in which such a policy was followed, we do not praise them on that account for the superiority of their ethical principles. T. H. Huxley concluded, therefore, that the standards of ethics, unlike those of horticulture, are *completely* opposed to the principle of natural selection.

> The practice of that which is ethically best—what we call goodness or virtue—involves a course of conduct which, in all respects, is opposed to that which leads to success in the cosmic struggle for existence. . . . Its influence is directed, not so much to the survival of the fittest, as to the fitting of as many as possible to survive. . . .
>
> Let us understand, once for all, that the ethical progress of society depends, not on imitating the cosmic process, still less in running away from it, but in combating it. [3, pp. 81–2]

*

This conclusion of T. H. Huxley has been challenged by some present-day scientists, who try to present arguments for deriving our standards of ethical evaluation from biological and other scientific knowledge. The most notable exponents of the thesis in Great Britain have been C. H. Waddington and Julian Huxley.

I take first the views of Waddington. In an essay on " The Relations between Science and Ethics ", he first refers to psychological and sociological evidence for the evolution of conscience and of ethical ideas. This, of course, is not a matter of dispute. Waddington then asks why the direction of evolution should be accepted as good. He quotes, and disagrees with, T. H. Huxley's verdict that ethical improvement lies in going against the direction of evolution. He then says this :

> To return to our question, we must accept the direction of evolution as good simply because it *is* good according to any realist definition of that concept. We defined ethical principles as actual psychological compulsions derived from the experience of the nature of society; we stated that the nature of society is such that, in general, it develops in a certain direction; then the ethical principles which mediate the motion in that direction are in fact those adopted by that society. [4, p. 18]

The first sentence of this quotation tells us what Waddington thinks his argument is : " we must accept the direction of evolution as good simply because it *is* good according to any realist definition of that concept ". But what follows the first sentence does not give us any definition of the concept of good. What follows tells us that people's " ethical principles " are derived from the experience of their society and express the direction in which that society is developing. That is to say, people *think good* the way their society is going, and judge other things to be good by reference to that standard. This, whether true or false, is beside the point. To ask " Is the direction of this process good? " is not to ask " Do people think the direction of this process good? " If we raise the question whether we are to accept the direction of a process as good, we presuppose that we have it in our power to accept it as good or not to accept it as good. So it is of no use to answer : " That is what you *do* think good, being formed by society as you are ". After all, we can raise the question whether the direction in which our society is moving is good, as well as the question whether the direction of evolution in general is good. And some people, having raised the question, answer that their society is going to the dogs.

Waddington is giving us an answer to the question : " When people commend things, do they use as their standard of commendation the direction of evolution, and if so, why? " He answers the first part of the question in the affirmative, and he answers the second part by giving us a causal account of the evolution of ethical judgment. Now this question which he has answered is quite different from the question: " Am I to use the direction of evolution as my standard of commendation, and if so, why?" It is no answer to this question to say, "Yes, you should, and the reason why you should is that you always have done in the past, and so has everyone else". Consider an analogous situation outside the field of ethics. Suppose a child says to me: "I have always believed the earth is flat, but my friend says it is spherical. What am I to believe?" It would be absurd to answer: "You should believe it is flat because you have done so in the past". Nor would it be much better to answer: "You should believe it is spherical because other people nowadays believe that".

Waddington has failed to distinguish two different kinds of question and the two different kinds of answer they seek. Plainly, the question "Am I to . . .?" or "Should I . . .?" is different from the question "Do I . . .?" The first question seeks advice, the second information. In addition, the question "Why should I . . .?" uses the word "why" in a different way from that employed in the question "Why do I . . .?" The question "Why should I . . .?" seeks justifying reasons, the question "Why do I . . .?" seeks explaining causes. To ask for facts and the causes of those facts is one thing. To ask for advice and reasons for that advice is another thing.

In publishing his essay, together with the comments of other people, in book form, Waddington included a discussion between himself and H. Dingle. During the course of it, Waddington says: "You suggest there are two different problems involved in our discussion; 'how did I make my ethical choice' and 'how shall I make it now'. I cannot see that there is any real distinction here" [4, p. 101]. The distinction might perhaps have been clearer to him if the second question had been phrased "How *am I to* make it?" instead of "How shall I make it?" The word "shall" is ambiguous. A question put in the form "Shall I . . .?" may be asking for advice, or it may be inviting a prediction. There is no logical difference (relevant for our purpose) between a report of a past action, "You did *X*", and a prediction of a future action, "You will do *X*". Both are used to make statements of fact. But there *is* a logical difference between these two forms of statement on the one hand and, on the other, a sentence of the form, "You are to (or "You should" or

"You ought to") do *X*". A sentence of this form is not used to make a statement of fact, but to give a piece of advice or an instruction.

If we ask "What do we think good?" we are seeking an answer that states matters of fact. If we ask "What are we to think good?" (or "What is good?") we are not seeking information, but advice. The answer, "You should think *X* good" (or "*X* is good") is not a statement of fact; it is a prescription for choice. The sentence "*X* is good" looks like a statement of fact, because its grammatical structure is similar to that of a statement of fact like "*X* is spherical". But here grammatical structure is misleading. "*X* is good" is more or less equivalent to "*X* is something that ought to be", and this is more or less equivalent to "*X* is something that anyone able to do so ought to produce"; and a sentence telling us what we ought to do is a prescription.

Again, if we ask "Why do we think *X* good?" (e.g. "Why do we think monogamy good, although other societies have thought polygamy good?"), we are seeking a causal explanation of the fact that our opinions are of one sort, while those of other groups are of a different sort. But if we ask "Why should we think monogamy good?" (or "Why is monogamy good?"), we are seeking justifying reasons, and not causal explanations.

Let us return for a moment to the non-ethical example of the child who asks whether he should believe that the earth is spherical. We have seen that the request for advice, "*Should I approve of* the practice of monogamy?" can be framed in the form, "*Is it good* to practise monogamy?"—a form that has the appearance of a question seeking a fact-stating answer. Similarly, the request for advice, "*Should I believe* that the earth is spherical?" can be put in the form, "*Is it true* that the earth is spherical?"—again a form of question that has the appearance of seeking a fact-stating answer. The answer, "*It is true* that the earth is spherical", appears to state a fact additional to the fact stated by the sentence, "The earth is spherical". But the sentence, "It is true that the earth is spherical", is logically equivalent to the sentence, "You (and everyone else) should believe that the earth is spherical", which is a prescription, not a statement of fact (though of course it implies that the sentence, "The earth is spherical", states a fact).

Now suppose the child asks me *why* he should believe that the earth is spherical. This is equivalent to asking for *evidence justifying* the acceptance of the proposition that the earth is spherical as a statement of fact. He is not asking for the *causes* of the fact that the earth is spherical, nor even (and here we come to the crux of Waddington's argument) for the *causes of* my and other people's *acceptance* of the proposition. To see that justifying evidence for a proposition is not the same as the cause of acceptance of that proposition, consider the situation when I have given the child some evidence for the proposition that the earth is spherical. I tell him, for instance, that an aeroplane flying in a straight line comes back to its starting point after 25,000 miles. In the light of this evidence, he decides to accept the proposition that the earth is spherical. Can we now say, as I think Waddington wishes to say, that the cause of the child's acceptance is the same as the evidence justifying the proposition he now accepts? No, we cannot. The cause of his acceptance of the proposition is *my having told him* a moment ago what some of the evidence is, and *his seeing*, also a moment ago, that the evidence justifies the proposition. The evidence for the truth of the proposition he now accepts is not the noises I uttered a moment ago or the inference he made a moment ago. The evidence is what my utterance and his thought referred to (namely, what happens to aeroplanes flying in a straight line for 25,000 miles), and *this* is not something that took place a moment ago.

Waddington thinks the answer to the question, "What am I to do, and for what reasons?" is the same as the answer to the question, "What will I do, and from what causes?" This is why he thinks that a causal account of how ethical judgments have come to be what they are can supply a criterion, or rational ground, for the ethical judgments we should make. But since the two questions, and the kinds of answer they seek, are of different logical types, Waddington's argument for using the direction of evolution as the criterion for ethical judgment rests on a logical confusion.

*

T. H. Huxley's Romanes Lecture at Oxford University was delivered in 1893. Fifty years later the University invited his grandson, Julian Huxley, to give the Romanes Lecture and suggested that he might deal with the same topic. Julian Huxley has added some further remarks to the theme of his lecture in a book, *Evolution and Ethics 1893–1943*, in which he prints the two lectures side by side.

Like Waddington, Julian Huxley disagrees with T. H. Huxley's contrast between the direction of ethical progress and the course of natural evolution.

> For T. H. Huxley, fifty years ago, there was a fundamental contradiction between the ethical process and the cosmic process. . . .
>
> To-day, that contradiction can, I believe, be resolved—on the one hand by extending the concept of evolution both backward into the inorganic and forward into the human domain, and on the other by considering ethics not as a body of fixed principles, but as a product of evolution, and itself evolving. [3, p. 105]

The opposition which T. H. Huxley found between ethics and evolution was between " the ethical process " and the process of *natural selection*, and Julian Huxley agrees that human progress, ethical or non-ethical, is quite independent of that. But he extends the term " evolution " to cover three phases of development, first of the inorganic world, second of biological species, and third of human societies. Only in the second of these phases is the agency of evolution the Darwinian process of natural selection.

Giving the term " evolution " this extended sense has the advantage of enabling us to fit together the whole history of the earth and its inhabitants in a single process of development. On the other hand, the extended concept does not have the same explanatory force as Darwin's theory of evolution had, just because the entire process of evolution in Julian Huxley's sense cannot be explained by a single set of causes. We still need three different sets of causes for the three different phases. The first stage of evolution was succeeded by the second when matter turned into self-reproductive organism, and the second was succeeded by the third, Huxley says, when *society* became self-reproducing. It became so, he tells us, through thought and language, which enable the product of experience to be handed on in tradition and education. We have, then, a picture of three phases of evolution, in which the really vital causal agencies are those that permit a leap from one phase to the next, from matter to living organism, and from biological organisms to social organization. What is common to the two leaps is that each involves the emergence of a type of self-reproduction. The causes of the two leaps are of course different. The causes that enabled matter to form self-reproducing organism, however obscure they may be, are plainly altogether different from the causes that enabled one species of biological organism, man, to form self-reproducing societies, the causes of the latter possibility being thought and language and the consequent transmission of experience by way of tradition and education. In the generalized picture, the causal process of natural selection within the biological phase takes a relatively subordinate place. Huxley's connection between evolution and standards of ethics, therefore, has nothing to do with the Darwinian process of natural selection.

How does he see the connection? In the first place, he notes, as Waddington does, that men's ethical ideas tend to be determined by psychological and sociological factors. Nevertheless, he thinks that we can learn to discount our prejudices and to find " external standards . . . for the validity of our moral sense " [3, p. 111]. These standards are found by two different methods.

The first method is simply the familiar intuitive approach used by T. H. Huxley when he wrote of our ideas of beauty and utility providing standards for evaluation in horticulture, and by Herbert Spencer when he took it as obvious that our enjoyment of pleasure is the ultimate standard for ethical evaluation. We all see at once, Julian Huxley says, that certain human experiences are " good in themselves "—such experiences as " enjoyment of natural beauty, æsthetic experience, the acquisition of knowledge and understanding, personal love, devotion to an active cause, and so on " [3, p. 214]. To these experiences we assign " higher and lower degrees of value, the higher values being those which are intrinsically or more permanently satisfying, or involve a greater degree of perfection " [3, p. 125]. Using this criterion, we shall be able to judge other things good if they help to produce these " satisfying " or " perfect " experiences.

In his second method, however, Huxley hopes to derive a standard of value from examination of the evolutionary process. If we consider the process of evolution as a whole, in all its three phases, he argues, we may observe two things that enable us to make this derivation. (*a*) The general direction of evolution is from the less to the more complex, and in consequence it is continually producing new forms of existence. (*b*) Since this process of evolution includes, in its latest phase, the things that we value highly, namely certain human experiences, we may say that the direction of the process is from what has little or no value to what has greater value. Huxley does not think that everything which appears later in the process of evolution is better than what appeared earlier. If he did, we should have to recall T. H. Huxley's observation that the immoral sentiments have evolved as much as the moral, and that pain is no less a late-developed product of evolution than is pleasure. Julian Huxley in fact agrees " that there are more and less desirable or valuable directions in evolution " [3, p. 32]. His argument is that at any rate one of these directions can be called valuable just because it ends up with things that we all think are valuable.

Now if his argument told us nothing more than this, it would be useless. It would say: " A change may be called good if it ends up with what we know to be good ". This would not give us a criterion for making value judgments, since it presupposes that we exercise beforehand the non-evolutionary criterion in order to make the essential value judgment about the end-product. The point of Huxley's argument is that, once we have, by use of the non-evolutionary criterion, judged the process of change to be good because it leads to good end-products, we can then use a *further* property of the process (namely, increasing complexity with continuing scope for future novelty) as a second criterion for valuing *other* things than those experiences which we already know to be good and the processes of change which lead to such experiences. By the use of the second criterion we may judge to be good other changes which have the property of increasing complexity with scope for continuing novelty.

But if the argument is to be a conclusive one, we need to know that the property of complexity with scope for novelty is *not* a property of any process of change that does not have good end-products. Would Huxley claim to know that a direction of evolution which leads to undesirable end-products —pain, for instance, or cruelty—cannot have the property of increasing complexity with scope for novelty? I imagine not. But perhaps he might say this. Since the process of evolution as a whole has led to results in which we think the good far outweighs the evil, we may infer that a majority of the changes within the process lead to good rather than bad results. Consequently, application of the criterion of complexity plus novelty will yield a *probability*, but not a certain guarantee, that any change exhibiting the property will lead to good rather than bad results. (The inference in the last sentence but one is in fact dubious, since the preponderance of good results may not be due to a preponderance in the *number* of beneficial changes. One powerful change may compete successfully with many weaker ones.)

Huxley has given us two kinds of criterion for evaluation. The one is the kind of criterion we all use, depending on our judgments that certain kinds of experience are good for their own sake. The other criterion, complexity with scope for novelty, is derived from an examination of evolution together with the assumption of the first criterion.

Unfortunately, Huxley writes as if he thought that the first criterion can also be confirmed from a knowledge of evolution. He says, for instance, that " the facts of nature, as demonstrated in evolution, give us assurance that knowledge, love, beauty, selfless morality, and firm purpose are ethically good " [3, p. 214]. He gives the impression of thinking (though he does not explicitly say) that the goodness of these states of mind can be derived from the fact that they are late products of a progressive process. But this would be to argue in a circle. He judges the process to be progressive because he has *already* judged good some of the experiences that it produces. He cannot therefore now use the progressiveness, which *depends* on the goodness of the end-products, to confirm his evaluation of the end-products.

Huxley next proceeds to apply his criteria of value, and it is worth considering which of his conclusions depend on the non-evolutionary criterion and which on the evolutionary.

> In the broadest possible terms evolutionary ethics must be based on a combination of a few main principles: that it is right to realize ever new possibilities in evolution, notably those which are valued for their own sake; that it is right both to respect human individuality and to encourage its fullest development; that it is right to construct a mechanism for further social evolution which shall satisfy these prior conditions as fully, efficiently, and as rapidly as possible. [3, p. 124]

We have here three principles of right action: (*a*) to realize new possibilities, " notably "—but not only—those that we value for their own sake; (*b*) to respect and develop human individuality; and (*c*) to take the means necessary for the most effective realization of the first two principles. Let us see how they are justified.

(*a*) Why is it right " to realize ever new possibilities in evolution "? Of the new possibilities afforded by the continuing process of evolution, some will be found good, some bad, and some indifferent. Huxley says that it is " notably " right to realize the first, but this implies that it is also right, though in a lesser degree, to realize others as well. We shall all agree that it is right to realize new possibilities that are " valued for their own sake ". We shall agree to this because it is virtually a tautology to say that it is right to realize what is good. But is it also right to realize new possibilities that are *disvalued*? The realizing of new possibilities that are valued is right because they are valued, not because they are new. The realizing of new possibilities that are disvalued is wrong because they are disvalued, and it is not made any less wrong because they are new. It would be equally right to realize old possibilities that are good, and equally wrong to realize old possibilities that are bad. Newness or oldness has nothing to do with the case.

If we take Huxley's principle strictly as counselling us to realize new possibilities even though we may *know* that the new possibilities in question will turn out to be bad or to have no value, it is absurd. For instance, it would be absurd to say that we ought to increase mutations in human beings by adding to the amount of radiation in the atmosphere, when biologists know that such mutations are far more likely to have baneful than beneficial results. But principle (*a*) can be framed more sensibly as follows. "If we can foresee that a new possibility is likely to be good, obviously we should try to realize it, because it will be good, not because it will be new. If we can foresee that a new possibility is likely to be bad, obviously we should try to prevent its realization, because it will be bad. But if we cannot foresee of new possibilities whether they will be good or bad, we should not sit tight but should enable them to be realized, judging it to be more probable that they will lead to good results than that they will lead to bad ones, because the realization of new possibilities in evolution has led on the whole to far more good than evil." The last part of this principle is not absurd (though, as we saw, the reasoning about probabilities on which it rests is pretty shaky), and it *is* a maxim that depends on scientific knowledge of the course of evolution.

(*b*) Why is it right to respect and develop human individuality? Huxley answers:

> It is clear on evolutionary grounds that the individual is in a real sense higher than the State or the social organism. The possibilities which are of value for their own sake, and whose realization must be one of our primary aims, are not experienced by society as a unit, but by some or all of the human beings which compose it. [3, p. 126]

What is clear is that Huxley's evaluation of the individual above the social organism is *not* based on evolutionary grounds. The evolutionary criterion requires us to say that *A* is higher on the scale than *B* if *A* is more complex while still leaving room for new possibilities to be developed. Huxley has also said that the new and more complex possibilities available in the third phase of evolution are due to "a new type of *organization* . . . that of self-reproducing *society*" [3, p. 123; my italics]. All this quite certainly implies that, by the standards of evolutionary development, the social organism is "higher" than the individual. The ground that Huxley in fact gives for valuing the individual above the social organism, does not depend on the evolutionary criterion of complexity plus novelty; it depends on the criterion of finding certain experiences to be valuable "for their *own sake*" (and not because they are complex or likely to lead to further novelties).

(*c*) The third principle, that it is right to take effective means to fulfil principles (*a*) and (*b*), does not receive any special justification from Huxley, and of course it needs none. This third principle simply says that if we are going to do something we had better do it efficiently. The maxim holds good whether our principles go along with evolution or against evolution or have nothing to do with evolution either way. What is more, it holds good whether we are talking about ethical choice or non-ethical choice. Bill Sykes does not need evolutionary ethics, or anything else, to advise him that, if he chooses to go in for burgling, he had better be efficient.

Principle (*c*), therefore, can be ignored. Principle (*b*) depends entirely on the criterion of experience and goes against the criterion of evolution. Principle (*a*) tells us two things. First, it counsels us to realize what we know to be good for its own sake. This maxim, of course, does not in the least depend on the evolutionary criterion. Secondly, however, principle (*a*) counsels us to realize what is *new* where we cannot foresee whether the novelty will turn out to be good or bad. This one part of principle (*a*) does depend on the evolutionary criterion. It is the only part of Huxley's set of ethical principles that can be called " evolutionary ethics ".

ETHICAL EVOLUTION

In his Romanes Lecture, Julian Huxley not only subscribes to the two propositions that ethics is a product of evolution and can ground its standards on the character of evolution. He adds the further proposition that ethics can help to determine the course of evolution in its present, third phase.

> It is only through social evolution that the world-stuff can now realize radically new possibilities. Mechanical interaction and natural selection still operate, but have become of secondary importance. For good or evil, the mechanism of evolution has in the main been transferred onto the social or conscious level . . . . The slow methods of variation and heredity are outstripped by the speedier processes of acquiring and transmitting experience. . . .
>
> And in so far as the mechanism of evolution ceases to be blind and automatic and becomes conscious, ethics can be injected into the evolutionary process. Before man that process was merely amoral. After his emergence onto life's stage it became possible to introduce faith, courage, love of truth, goodness—in a word moral purpose—into evolution. It became possible, but the possibility has been and is too often unrealized. [3, p. 123]

This needs only the briefest of comments. We should note two phrases in the quotation: "*for good or evil*", the future of evolution depends largely on what men do; and "it became *possible* to introduce . . . moral purpose into evolution". It also became possible to introduce *immoral* purpose into the process of evolution. "Before man that process was merely amoral." Now that conscious choice is an effective factor in the world, evolution is no longer merely amoral. What Huxley's point comes to, then, is not so much that "*ethics* is seen not only as a product but also as an agency in the later history of the process" [3, p. 32; my italics]. It is rather that human action, whether moral, immoral, or amoral, has become such an agency. Just as the evolution of ethics is accompanied by the evolution of the unethical, so the future course of evolution can be, and is likely to be, affected by ethics and the unethical alike. Scientists, and perhaps especially biologists, are in a position to predict, to some extent, the effects of human action on the future course of evolution. Consequently the work of present-day biologists carries, both for them and for the layman who can learn from them, an ethical responsibility that was not carried by the work of Darwin and his contemporaries. A hundred years ago the advance of biological knowledge made a difference to human beliefs. Today it makes a difference to human life.

REFERENCES

PREFACE

[1] BONGER, W. A. (1943). *Race and Crime.* Oxford.

[2] HALDANE, J. B. S. (1954). "The statics of evolution." In *Evolution as a Process.* London.

[3] KETTLEWELL, H. B. D. (1955). "Selection experiments on industrial melanism in the *Lepidoptera.*" *Heredity,* **9,** 323–342.

[4] LACK, D. (1957). *Evolutionary Theory and Christian Belief.* London.

[5] MONTAGU, M. F. A. (1952). *Darwin: Competition and Coöperation.* New York.

[6] Waddington, C. H. (1939). *An Introduction to Modern Genetics.* London.

CHAPTER 1

[1] SIMPSON, G. G. (1949). *The Meaning of Evolution.* New Haven.

[2] WADDINGTON, C. H. (1940). *An Introduction to Modern Genetics.* London.

[3] WADDINGTON, C. H. (1958). *The Strategy of the Genes.* London.

CHAPTER 2

[1] DOBZHANSKY, TH. (1941). *Genetics and the Origin of Species.* New York.

[2] JEPSEN, G. L., MAYR, E., and SIMPSON, G. G. (1949). *Genetics, Paleontology and Evolution.* Princeton, N.J.

[3] MAYR, E. (1942). *Systematics and the Origin of Species.* New York.

ACKNOWLEDGMENTS TO CHAPTER 2

The author is greatly indebted to several colleagues, particularly to Dr. J. A. Beardmore, Mrs. Lee Ehrman, Professors E. Mayr, J. A. Moore, G. L. Stebbins and Mr. L. Van Valen for critical readings of the manuscript of this essay. Professor R. C. Stebbins has kindly supplied the drawing for figure 5, prepared under his direction by Mr. Gene Christman. Mr. L. Van Valen has drawn figure 4. The essay has been written during the author's stay at the South-western Research Station of the American Museum of Natural History, in the Chiricahua Mountains, Arizona. Thanks are due to Dr. Mont Cazier, the Director of the Station for his hospitality and for many useful discussions.

CHAPTER 3

[1] Beale, G. H. (1954). *The Genetics of* Paramecium aurelia. Cambridge.

[2] Billingham, R. E., Brent, L., and Medawar, P. B. (1956). "Quantitative studies on tissue transplantation immunity. III. Actively acquired tolerance". *Phil. Trans.* B, **239**, 357–414.

[3] Billingham, R. E., and Medawar, P. B. (1948). "Pigment spread and cell heredity in guinea-pig's skin." *Heredity*, **2**, 29–47.

[4] Briggs, R., and King, T. J. (1957). "Changes in the nuclei of differentiating endoderm cells as revealed by nuclear transplantation." *J. Morph.*, **100**, 269–311.

[5] Danielli, J. F., Lorch, I. J., Ord, M. J., and Wilson, E. G. (1955). "Nucleus and cytoplasm in cellular inheritance." *Nature*, **176**, 1114–1115.

[6] Darlington, C. D. (1953). *The Facts of Life.* London.

[7] Darlington, C. D., and Mather, K. (1949). *The Elements of Genetics.* London.

[8] Darwin, C. (1876–1878). *The Effects of Cross and Self Fertilization in the Vegetable Kingdom.* London.

[9] Darwin, C. (1909). *The Foundations of the Origin of Species.* Cambridge.

[10] Glavinic, R. (1955). "L'hybridation vegetative comme methode dans la selection des tomates." *Report of the XIVth Int. Hort. Congress.* Netherlands. See also Glavinic, R. (1957). (In Russian) *Studies in Michurinist Genetics.* Moscow.

[11] Glushchenko, I. E. (1948). (In Russian) *Vegetative Hybridization in Plants.* Moscow.

[12] Haldane, J. B. S. (1955). "Some alternatives to sex." *New Biol.*, **19**, 7–26.

[13] Hall, O. L. (1954). "Hybridization of wheat and rye after embryo transplantation." *Hereditas*, **40**, 453–458.

[14] Hudson, P. S., and Richens, R. H. (1946). *The New Genetics in the Soviet Union.* Cambridge.

[15] Lansing, A. I. (1952). *Cowdry's Problems of Ageing*, 3rd edition. Baltimore.

[16] Lederberg, E. M., and Lederberg, J. (1953). "Genetic studies of lysogenicity in *Escherichia coli.*" *Genetics*, **38**, 51–64.

[17] L'Heritier, Ph. (1948). "Sensitivity to CO_2 in *Drosophila*—a review." *Heredity*, **2**, 325–348.

[18] Lorch, I. J., and Danielli, J. F. (1950). "Transplantation of nuclei from cell to cell." *Nature*, **166**, 329.

[19] Morse, M. L., Lederberg, E. M., and Lederberg, J. (1956). "Transductional heterogenotes in *Escherichia coli.*" *Genetics*, **42**, 758–779.

[20] Sonneborn, T. M. (1955). Lectures delivered at University College, London, summarized in an editorial article in *The Lancet*, **268**, 656–657.

[21] Stockard, C. R. (1941). "The genetic and endocrine basis of differences in form and behaviour." *Amer. Anat. Memoirs*, **19**.

[22] WEISMANN, A. (1892). *Das Keimplasma.* Jena. Trans. *The Germ-Plasm.* 1893. London.

[23] WOLLMAN, E. L. (1953). "Sur le determinisme genetique de la lysogenie." *Ann. Inst. Pasteur,* **84,** 281–293.

CHAPTER 4

[1] BONSMA, J. C. (1949). "Breeding cattle for increased adaptability to tropical and subtropical environments." *J. agric. Sci.,* **39,** 204–221.

[2] ELY, F., and PETERSEN, W. E. (1941). "Factors involved in the ejection of milk." *J. Dairy Sci.,* **24,** 211–223.

[3] HAMMOND, J. (1932). *Growth and Development of Mutton Qualities in the Sheep.* Edinburgh.

[4] HAMMOND, J. (1947). "Animal breeding in relation to nutrition and environmental conditions." *Biol. Rev.,* **22,** 195–232.

[5] HAMMOND, J. (1952). *Farm Animals.* London.

[6] HUNTER, G. L. (1956). "The maternal influence on size in sheep." *J. agric. Sci.,* **48,** 36–60.

[7] LOOMIS, F. B. (1926). *The Evolution of the Horse.* Boston.

[8] MCMEEKAN, C. P. (1940). "Growth and development in the pig, with special reference to carcass quality characters." *J. agric. Sci.,* **30,** 276–343.

[9] PÅLSSON, H. (1955). "Conformation and body composition", in J. Hammond (ed.), *Progress in the Physiology of Farm Animals,* pp. 430–542. London.

[10] PÅLSSON, H., and VERGES, J. B. (1952). "Effects of the plane of nutrition on growth and the development of carcass quality in lambs." *J. agric. Sci.,* **42,** 1–92 and 93–149.

[11] WALLACE, L. R. (1948). "The growth of lambs before and after birth in relation to the level of nutrition." *J. agric. Sci.,* **38,** 93–153, 243–302 and 367–401.

[12] WHETHAM, E. C. (1933). "Factors modifying egg production with special reference to seasonal changes." *J. Agric. Sci.,* **23,** 383–395.

[13] YEATES, N. T. M. (1949). "Breeding season of the sheep with particular reference to its modification." *J. agric. Sci.,* **39,** 1–43.

CHAPTER 5

[1] DARWIN, C. (1872). *The Origin of Species.* 6th edition. London.

[2] DARWIN, C. *A Monograph of the Cirripedia.* Ray Society (Vol. I, 1852, Vol. II, 1854).

[3] DE BEER, G. R. (1951). *Embryos and Ancestors.* Revised edition. Oxford.

[4] SEWARD, A. C., *et al.* (1909). *Darwin and Modern Science.* Symposium. Cambridge.

[5] YOUNG, J. Z. (1950). *The Life of Vertebrates.* Oxford.

CHAPTER 7

[1] DARWIN, C. (1958). "Sketch" (1842), "Essay" (1844), in *Evolution by Natural Selection.* Cambridge.

[2] DARWIN, C. (1859). *On the Origin of Species.* 1st ed., London; 6th ed., London, 1872 (reprinted in World's Classics Edition, London, 1956).
[3] DE BEER, G. (1958). *Embryos and Ancestors.* Oxford.
[4] HUXLEY, T. H. (1887). In *Life and Letters of Charles Darwin*, Vol. 2, p. 190. London.
[5] LEMCHE, H. (1957). " A new living deep sea mollusc of the Cambro-Devonian class Monoplacophora." *Nature*, **179**, 413.

CHAPTER 8

[1] DE BEER, G. R. (1951). *Embryos and Ancestors.* Oxford.
[2] BOYDEN, A. A. (1953). " Fifty Years of Systematic Serology." *Systematic Serology*, **2**, 19–30.
[3] LE GROS CLARK, W. E. (1955). *The Fossil Evidence for Human Evolution.* Chicago.
[4] NUTTALL, G. H. F. (1904). *Blood Immunity and Blood Relationship.* Cambridge.
[5] SCHULTZ, A. H. (1936). " Characters Common to Higher Primates and Characters Specific for Man." *Quart. Rev. Biol.*, **11**, 259–455.
[6] SCHULTZ, A. H. (1957). " Past and present views on man's specializations." *Irish J. med. Sci.*, pp. 341–356.
[7] SIMPSON, G. G. (1950). *The Meaning of Evolution.* Oxford.

CHAPTER 9

[1] BARNETT, S. A. (1955). " ' Displacement ' behaviour and ' psychosomatic ' disorder." *Lancet*, **269**, 1203–8.
[2] BARNETT, S. A. (1958). " Exploratory Behaviour." *Brit. J. Psychol.* **49**, 289–310.
[3] BASTOCK, M., MORRIS, D., and MOYNIHAN, M. (1953). " Some comments on conflict and thwarting in animals." *Behaviour*, **6**, 66–84.
[4] CANNON, W. B. (1915). *Bodily Changes in Pain, Hunger, Fear and Rage.* New York.
[5] DARWIN, C. (1859). *On the Origin of Species.* London.
[6] DARWIN, C. (1872). *The Expression of the Emotions in Man and Animals.* London.
[7] DOLLARD, J., *et al.* (1944). *Frustration and Aggression.* London.
[8] HILGARD, E. R. (1952). In *Psychoanalysis as Science.* Stanford.
[9] LORENZ, K. Z. (1937). " The companion in the bird's world." *Auk*, **54**, 245–73.
[10] MASSERMAN, J. H. (1946). *Principles of Dynamic Psychiatry.* Philadelphia.
[11] MOWRER, O. H. (1940). " An experimental analogue of ' regression ' with incidental observations on ' reaction-formation '." *J. abnorm. (soc.) Psychol.*, **35**, 56–87.

[12] THORPE, W. H. (1951). "The definition of terms used in animal behaviour studies." *Bull. anim. Behav.*, **3**, 34–40.

[13] TINBERGEN, N. (1948). "Social releasers and the experimental method required for their study." *Wilson Bull.*, **60**, 6–51.

[14] TINBERGEN, N. (1952). "'Derived' activities." *Q. Rev. Biol.*, **27**, 1–32.

[15] WHITEHORN, J. C. (1932). "Concerning emotion as impulsion and instinct as orientation." *Amer. J. Psychiat.*, **11**, 1093–1106.

[Chapter 9 is based on an article which originally appeared in *New Biology*, **22**, 1957.]

CHAPTER 10

[1] BASTOCK, M. (1956). "A gene mutation which changes a behaviour pattern." *Evolution*, **10**, 421–439.

[2] DARWIN, C. (1889). *The Descent of Man and Selection in Relation to Sex.* London: John Murray.

[3] DOBZHANSKY, TH. (1951). *Genetics and the Origin of Species.* Columbia University Press.

[4] HOWARD, L. (1952). *Birds as Individuals.* London: Collins.

[5] HUXLEY, J. S. (1914). "The courtship of the Great Crested Grebe (*Podiceps cristatus*); with an addition to the theory of sexual selection." *Proc. zool. Soc. Lond.*, 491–562.

[6] KOOPMAN, K. F. (1950). "Natural selection for reproductive isolation between *Drosophila pseudoobscura* and *Drosophila persimilis.*" *Evolution*, **4**, 135–148.

[7] LACK, D. (1953). *The Life of the Robin.* London: Penguin Books.

[8] MAYNARD SMITH, J. (1956). "Fertility, mating behaviour and sexual selection in *Drosophila subobscura.*" *J. Genet.*, **54**, 261–279.

[9] MILANI, R. (1951). "Osservazione comparitive ed esperimenti sulle modalita del corteggiamento nelle cinque species europee del gruppa '*obscura*'." *R.C. 1st lombardo*, **84**, 48–58.

[10] RENDEL, J. M. (1945). "The genetics and cytology of *Drosophila subobscura.* II. Normal and selective matings in *Drosophila subobscura*". *J. Genet.*, **48**, 287–302.

[11] TINBERGEN, N. (1951). *The Study of Instinct.* Oxford University Press.

[12] TINBERGEN, N., and MOYNIHAN, M. (1952). "Head flagging in the black-headed gull; its function and origin." *Brit. Birds*, **45**, 19–22.

CHAPTER 11

[1] DALY, R. A. (1934). *The Changing World of the Ice Age.* New Haven.

[2] DAVIS, W. M. (1928). *The Coral Reef Problem.* New York.

[3] GARDINER, J. S. (1931). *Coral Reefs and Atolls.* London.

[4] KUENEN, PH. H. (1947). "Two Problems of Marine Geology: Atolls and Canyons." *Verh. Koninklijke Ned. Akad. Wetenschappen*, **43**, 1–69.

[5] LADD, H. S., INGERSON, E., TOWNSEND, R. C., RUSSELL, M., and STEPHENSON, H. K. (1953). "Drilling on Eniwetok Atoll, Marshall Islands." *Bull. Amer. Assoc. Petroleum Geologists*, **37**, 2257–2280.

[6] STEPHENSON, T. A., STEPHENSON, A., TANDY, G., and SPENDER, M. (1931). "The Structure and Ecology of Low Isles and Other Reefs." *Sci. Rpts. G. Barrier Reef Exped.*, 1928–29, *Brit. Mus.* (*Nat. Hist.*), **3**, 17–112.

[7] YONGE, C. M. (1940). "The Biology of Reef-Building Corals." *Ibid.*, **1**, 353–391.

CHAPTER 12

[1] DARWIN, C. (1862). *On the Various Contrivances by which Orchids are Fertilized by Insects.* London.

[2] DARWIN, C. (1868). *The Variation of Animals and Plants under Domestication.* London.

[3] DARWIN, C. (1872). *On the Origin of Species.* 6th Edition, London.

[4] DARWIN, C. (1876). *The Effects of Cross- and Self-fertilization in the Vegetable Kingdom.* London.

[5] DARWIN, C. (1877). *The Different Forms of Flowers on Plants of the Same Species.* London.

[6] DARWIN, C. (1880), assisted by FRANCIS DARWIN. *The Power of Movement in Plants.* London.

[7] DARWIN, F. (1887). *The Life and Letters of Charles Darwin.* 3 vols., London.

[8] DARWIN, F. (1899). "The botanical work of Darwin." *Ann. Bot.*, **13**, ix–xix.

[9] DARWIN, F., and SEWARD, A. C. (1903). *More Letters of Charles Darwin.* 2 vols., London.

[10] THISELTON-DYER, W. (1882). *Charles Darwin.* London.

CHAPTER 13

[1] BOUGLÉ, C. (1910). in Seward, A. C. (ed.) *Darwin and Modern Science.* Cambridge.

[2] BOUGLÉ, C. (1935). *Bilan de la sociologie français contemporaine.* Paris.

[3] BURT, C. (1950). "The Trend of National Intelligence", *Brit. J. Sociol.*, **1**, 154–168.

[4] DARWIN, C. F. (1887). *The Life and Letters of Charles Darwin.* London.

[5] DISRAELI, B. (1847). *Tancred.* London.

[6] DURKHEIM, E. (1893). *De la Division du Travail Social.* Paris.

[7] GALTON, F. (1872). "Statistical Inquiries into the Efficacy of Prayer." *Fortnightly Review.* (Reprinted by the Eugenics Society, 1951.)

[8] GALTON, F. (1873). *Inquiries into Human Faculty and Its Development.* London.

[9] GINSBERG, M. (1956). "The claims of eugenics", in *On the Diversity of Morals.* London.

[10] GLASS, D. V. (ed.) (1954). *Social Mobility in Britain.* London.

[11] GLUCKMAN, M. (1949). "Malinowski's sociological theories." Rhodes-Livingstone Papers, No. 16. Oxford.

[12] HOBHOUSE, L. T. (1924). *Social Development.* London.

[13] HOFSTADTER, R. (1945). *Social Darwinism in American Thought, 1860–1915*. Philadelphia.
[14] HUXLEY, T. H. (1910). "A liberal education", in *Lectures and Lay Sermons*. London.
[15] MACRAE, D. G. (1953–4). "Social stratification", *Current Sociology*, **2**, xxx.
[16] MACRAE, D. G. (1956). "Sociology in transitional societies." *Universitas, Accra*, **2**, 107–109.
[17] MALTHUS, T. R. (1830). *A Summary View of the Principles of Population*, reprinted in D. V. Glass (ed.) *Introduction to Malthus*, London, 1953.
[18] MARETT, R. R. (1936). *Tylor*. London.
[19] SEWARD, A. C. (1910). *Darwin and Modern Science*. Cambridge.
[20] THOMPSON, G. (1949). *The Trend of Scottish Intelligence*. London.
[21] THOMSON, G. A. (1906). *Herbert Spencer*, 193. London.
[22] THOMSON, J. A. (1910). in Seward, A. C. (ed.) *Darwin and Modern Science*. Cambridge.
[23] TYLOR, E. B. (1889). "On a method of investigating the development of institutions." *J. R. anthrop. Inst.*, **18**, 245–272.

CHAPTER 14

[1] DAMPIER, WILLIAM (1929). *History of Science*. Cambridge.
[2] DARLINGTON, C. D. (1939). *The Evolution of Genetic Systems*. Cambridge.
[3] HALDANE, J. B. S. (1932). *The Causes of Evolution*. London.
[4] HUXLEY, J. (1942). *Evolution*. London.
[5] HUXLEY, J. (1943). In *Evolution and Ethics*. London.
[6] HUXLEY, T. H. (1893). In *Evolution and Ethics*. London.
[7] JOAD, C. E. M. (1928). *The Meaning of Life*. London.
[8] KEITH, ARTHUR (1946). *Essays in Human Evolution*. London.
[9] MATHER, K. (1943). "Polygenic inheritance and natural selection." *Biol. Rev.* **18**, 32–64.
[10] THODAY, J. M. (1953). "Components of fitness." *Symp. Soc. exp. Biol.*, **7**, 96–113.

CHAPTER 15

[1] DARWIN, CHARLES. *On the Origin of Species*, 6th ed.
[2] DARWIN, CHARLES. *The Descent of Man*, 2nd ed.
[3] HUXLEY, T. H., and HUXLEY, JULIAN (1947). *Evolution and Ethics 1893–1943*. London: Pilot Press. (Contains the Romanes Lectures of the two authors, together with T. H. Huxley's Prolegomena and further essays by Julian Huxley.)
[4] WADDINGTON, C. H., and others (1942). *Science and Ethics*. London: Allen and Unwin.

INDEX

Acquired character, 8, 15
— inheritance of, 56, 63, 316
Acropera, 278
Adaptation, xiii, 15, 314, 317, 320
" Adaptive peaks", 25, 33
" Adaptive valleys", 25
Agassiz, A., 263
Agassiz, J. L. R., 140, 144, 148, 152, 156, 270
Ageing, 77, 78
Air conditioners as environmental adaptations, 330
Allaesthetic characters, 281
Allele, 13, 22, 23, 29
Allopatric groups, 39, 40
Allopolyploidy, 51
Ammonite
— fossil, 159
— pædomorphosis in, 161
Amphioxus, 165
Amœba, 59, 62, 75
— *proteus*, 75
— *discoides*, 75
Amphibia, 107, 111, 146, 328
Anatomy
— comparative, 139
Andean elevation, 131
Anatomical design, unity of, 176
Anatomical evidence, 283
Androgynoceras, 161
Anecdotal method, 207
Anthropology
— developmental, 306
Anthropomorphism, 210
Antibody, 63
Antigen, 63–65, 70–72
— individual-specific, 71
— organ-specific, 71
— transplantation, 71, 72
Anxiety, 220
Apogamy, 43
Appeasement ceremonies, 235
Apes, 197
— fossil, 194
— modern, 194
Approximation, " vegetative", 76
Archæopteryx, 146
Aristotle, 35
Asexual clones, 43
— organisms, 41 *e.s.*
— reproduction, 23, 42, 55
Atoll, 245 *e.s.*
Atoll, Eniwetok, 264, 265
— Funafuti, 264, 265
— Maldive, 254, 257, 259
Auricularia larva, 165
Australopithecus, 166, 190, 192, 194, 196, 204
Auxin, 278
Aves, 111, 113, 119, 125
Axolotl, 106

Bacteriophage, 79, 82
Bacterium, 41, 42, 59, 62, 65, 78, 80, 84
Baer, K. E. von, 154, 198, 202
— laws of, 154
Bain, A., 336, 337
Balanoglossus, 165
Barnacles, 118, 123, 124, 127
Barnett, S. A., 363
Barrande, J., 152
Barrier reefs, 245, 250, 252, 254, 255, 259, 260, 265
Bastock, M., 226, 239, 363, 364
Beach, raised, 130
Beagle, H.M.S., 3, 130, 134, 245–247, 249, 251, 255, 262, 268
Beale, G. H., 361
Beer, G. R. de, 106, 362, 363
Bees, 282
Behaviour, 207, 208
— abnormal, 218. *e.s.*
— appetitive, 217
— breeding, 231 *e.s.*
— displacement, 224, 226, 229, 238
— exploratory, 217, 218
— instinctive, 211, 216
— psychoneurotic, 226
— psychosomatic, 226
Behaviourism, 228
Betula nana, 272
Billingham, R. E., 361
Biochemistry and taxonomy, 110
"Biogenetic law", 158
Biological progress, Ch. 14 *passim*
Birds, 111, 113, 119, 125, 328
Blackbird, 236
Blair, A. P., 53
Blakeslee, A. F., 33
"Blastæa", 159
Blending inheritance, 6, 9
Blood groups, 40
Bog, peat, 273

Bolk, L., 166
Bonger, W. A., xv, 360
Bonsma, J. C., 362
Bouglé, C., 304, 365
Boyden, A. A., 185, 363
Brain, ape, 186
— human, 186, 192
Breeding systems, 325
Brent, L., 361
Briggs, R., 74, 361
British Museum, 121, 122
Brown, R., 281, 283
Buffon, G. L. L., 19
Bufo, 53
Burt, C., 365
Butler, J., 341, 342
Butter fat colour, 91

Cambrian Age, 148
Candolle, A. de, 270
Cannon, W. B., 206, 212, 213, 220, 363
Carbon dioxide, effect on *Drosophila*, 66, 80
Carlyle, T., 19
Carnivore, 132, 232
Catasetum, 278
Cattle, 95
— beef, 89
— European, 99
— in India, 99
— Jamaican Hope, 89
— Jersey, 86
— Zebu, 89, 99
Causes and reasons, 349, 350, 351
Cell, 59, 60
— germ, 22, 59
— somatic, 59
Challenger Expedition, 263
Chamisso, A., von, 255, 259
Chelonia, 111
Chen, T. T., 44
Chiff-chaff, 119
Chimpanzee, 167, 197
Chin of Java man, 193
— of modern man, 193
Chordates, 164
— pædomorphic origin of, 166
Chromosome, 59–62, 64, 68, 72, 77–80, 322 *e.s.*
— complement, 51
— Morgan's theory, 56
Circumcision, 8
Circumnutation, 290, 292
Cirripedia, 124, 125
Classification
— categories, 36
— natural, 128
— phylogenetic, 103, 105, 107, 109–114
Classification
— of fossils, 109
— of Vertebrata, 110
Climatic optimum, post-glacial, 273
Clone, 23, 41, 42, 59, 62, 65, 67, 75, 76, 78
— ageing, 78
— asexual, 43
Coccyx, 199, 203
Cocos-Keeling, 245, 247, 253, 254, 256
Cœlacanth, 112, 116
Cœnozoic Era, 131
Coleoptile, 292
— grass, 293
Comoro Islands, 148
Communist Manifesto, 305
Communities, mating, 25, 26
Conjugation, 44
Conscience, 334–338, 340–342
Consummatory act, 217
— situation, 217
— state, 218
Convergence, 182, 183
Cook, J. 253
Copulation, 31
— cross-, 44
Co-operation, 329
Corals and Coral Islands, 261
Coral growth subsidence, 258
"Cosmic Code", 313, 333
Courtship, 31, 235, 239–242
Cow
— Aberdeen Angus, 88
— Ayrshire, 86
— Guernsey, 91
— Jersey, 91
Cranial capacity, 187
Crataegus, 43
Creation, xiv, 148, 150
Cretaceous Age, Upper, 148
Criteria of value, 344, 353, 355, 357, 358
Crocodilia, 111
Crofts, D. R., 164
Croll, J., 143, 275
Cross, reciprocal, 86, 87
Cross-breeding, 6, 9, 12, 325
Crossopterygii, 111
Crossover, 323
Crustacea, 124, 125
Cusp patterns of teeth, 178
Cuvier, L. C. F. D., 139, 140
Cypris larva, 124
— stage, 124
Cytoplasm, 59, 60, 62–68, 70, 76, 77, 82

Daly, R. A., 265, 364
Dampier, W., 314, 316, 366
Dana, J. D., 253, 257, 261, 262
Danielli, J. F., 75, 361

Daphnia, 124
Darlington, C. D., 56, 60, 320, 361, 366
Darwin, E., 1, 19, 310
Darwin, F., 251, 267, 269, 289, 293, 365
Darwin-Dana Theory, 262
Datura, 33
Davis, W. M., 262, 364
Daylight
— and breeding, 89
— and growth, 89
Dentition, 177
— cusp patterns, 178
— Pongidæ and Hominidæ, 195
Descartes, R., 19
Descent of Man, 335–337
Desoxyribonucleic acid, 79, 83, 84
Development and gene action, 15
Deviation, 164
Devonian Era, 146, 147
Dichogamy, 284
Dicotyledons, 125
Differentiation, 62, 69, 70, 73
— of cells, 67, 72, 75
— of races, 280
Dimorphy, 284
Dingle, H., 349
Diœcism, 284
Dipnoi, 110, 111, 116
Discontinuity
— and classification, 35
— genetical basis of, 23
— of organic diversity, 37
Dispersal, 275, 276
Displacement, 224
— syndrome, 226
Disraeli, B., 306, 365
Distribution
— bipolar, 274
— geographical, 270
Divergence in evolution, xii
Diversification of species, 281
DNA, 79, 83, 84
Dobzhansky, Th., 235, 360, 364
Dog, hybrids, 62
Dollard, J., 363
Dollo's Law, 110
"Drive", 210, 217
Drosophila, 16, 26, 33 *e.s.*, 59, 61, 66, 80, 119, 123, 235 *e.s.*, 241 *e.s.*, 293
Dualism, 66–68, 70
Dualist philosophy, 62, 69, 77
Ducks and DNA, 83
Durkheim, E., 307, 311, 365

Echinoderm, 165
Economics and Darwinism, 310
Economic biology, 128
Ectoderm, 74, 75
Edentata, 131
Egg production, 90, 91
Egg-tooth, 172
Einstein, A., 5
Elephant, 113, 114, 125
Ellis, J., 251
Ely, F., 362
Embryo, 67, 72, 73, 75, 76, 77
— frog, 72, 73, 74
— "endoderm", 75
— human, 198
Embryology, 102, 198 *e.s.*
— and evolution, 105
— comparative, 106, 198
Embryonic resemblance, 154
Emotion
— definition, 210, 212
— physiology, 212
Endoderm
— nuclei, 74, 75
— embryo, 75
Ensatina eschscholtzi, 47, 48
Environment, 15, 332
— exploitation, 331
— heterogeneous, 327
Environmental change, 317, 318
— control, 329 *e.s.*
— influences, 288
Enzymes, 110
Eocene deposit, 115
Eohippus, 142
Epiceratodus, 111, 116
Essay on the Principle of Population, 297
Ethics, xv
— evolutionary, 356, 358
— evolution, 334, 336, 343, 359
— standards, 347
"Ethical Code", 313, 333
Ethical conclusions, 314
— judgements, criteria, 351
Eugenics, 301
Evolution
— biological, 330–331
— causes, 22
— clandestine, 169
— control, 332
— convergent, 87, 182
— direction, 347, 348, 352, 354, 355
— divergent, 87
— ethics, xv, 334, 336, 346, 359
— fish, 146
— flowering plants, 282
— man, 166
— progressive, 319
— ratio, 109, 110
— selection, 316
— social, 330, 331
— theory, objections to, 140

Evolutionary genetics, 24
Evolutionists, classical, 24

Falconer, D. S., 152
Family, 20, 35, 114, 123
Feed-back, 219
Ferguson, A., 299
Ferns, 125
Ferrets, 89
Fertilization, assortative cross, 281
— cross, 30, 55, 286
— self, 286
Finches, Galapagos Islands, 3, 4
Fish
— evolution, 146
— teleost, 148
Fission, 41
Fitness, x, xi, xii
— definition, 317
Fitz Roy, R., 248, 249, 256
Fixed action-patterns, 211
Flies, populations, 324, 325
Flock books, 87
Flora
— form and function, 276
— mountain, 271
— northern, 271
— tropical, 274
Floral structure, 281
Foot, human, 180
Forbes, E., 152, 273, 275
Fossils
— as evolutionary evidence, 149, *e.s.*
— evidence, 125
— "living", 102, 114, 116
— mammals, 131
— phyletic patterns of, 144
— "prophetic", 160
— record and classification, 109, 114
Fowl, jungle, 91
Frazer, J., 306
Freud, S., 222, 337
Friesian cattle, 88
Frog embryo, 72, 73, 74
Fruit-fly, 123
Fungi, 42

Galapagos Islands, 3, 4, 246
Galathea expedition, 148
Galton, F., 301, 365
Gardiner, J. S., 364
Garstang, W., 162
"Gastræa", 159
Gastropod, 163
Gastrulation, 170
Gates, R. R., 43
Geikie, A., 261
Geological
— age, 114
— time, 142
Gemmules, 8, 69, 289
Gene, 10 *e.s.*, 21, 22, 61, 64, 65, 79, 322 *e.s.*
— chromosomal, 69–71, 80, 82
— combinations, 23–25, 33, 38
— complex, 171
— cytoplasmic, 67
— and development, 15
— dominant, 11
— exchange, 41
— frequencies, 10
— locus, 23
— patterns, 41
— pool, 12, 28
— recessive, 11
— transfer, 42
— "transfusion", 80
— wild-type, 12
Genetics, 9, 56, 59, 60, 122
— and species, 20
Genetical
— constitution, controlled, 331
— basis of discontinuity, 23
— diversity, 326 *e.s.*
— knowledge, 118
— stability, 321 *e.s.*
— variation, xiii
— versatility, 320 *e.s.*
— workers, 123
Genitalia, 44
Genotype, 24, 25, 55, 287
Genus, 20, 35, 111, 113–115, 123, 125
Geographical
— distribution, 270
— separation, 39, 40
Geological Society, 250
Geotropism, 291
Germ cell, 59
Gill arches, 199
Gill-pouches, 155
Ginsberg, M., 302, 365
Glacial period, 271, 273
— late, 273
Glass, D. V., 365
Glavinic, R., 361
Gluckman, M., 365
Glushchenko, I. E., 361
Gluteus maximus muscle, 179
Glyptodonts, 132
— armour, 137
Goat, 31, 91
Gobineau, J. A., 301
Godwin, H., 273
Good, 347, 348, 350, 353, 358
Gorilla, skull, 188, 191, 195, 197
Grafting, 70, 72, 76
— hybrids, 83

Grafting, tomatoes, 80
Graptolites, 116
Gray, A., 268, 273, 274, 288
Great Barrier Reef, 252, 264
Grebe, Great Crested, 235
Growth curve, 94
Growth hormone, 278, 292, 294
Guinea-pig, 66, 67
Gull
— black headed, 235
— heming, 235
Gymnosperms, 125

Hadži, J., 170
Haeckel, E., 158, 173, 180, 202
Haldane, J. B. S., xii, xiii, 55, 316, 360, 361, 366
Hall, O. L., 361
Hammond, J., 362
Hardy, A. C., 169
Heat tolerance
— lack in European cattle, 88
Herbivores, 132
Herd, 87
Herder, J. G. von, 299
Heredity, 6, 8
— chromosome theory, 60–62
— theory, 286
— "one-way theory", 57, 58, 60, 62, 66, 68
— "two-way theory", 57, 63, 69
— unit, 9, 21
— Weismannian diagram, 67
Heterostyly, 277, 284, 285
Heterozygote, 324
— multiple, 23
"Heterogzyosity-balanced", 324
Hilgard, E. R., 223, 363
Hill, J. P., 172
Hinny, 87
Hinshelwood, C., 65
Hip bone, 180, 196
His, W., 159
Hitler, A., 34
Hobhouse, L. T., 306, 311, 365
Hofstadter, R., 365
Hominidæ, 174, 175, 180, 184, 186, 190 *e.s.*
— dentition, 195
Hominoidea, 185, 186
Homo sapiens, 167, 193 *e.s.*
Homology, 283
Hooker, J. D., 267, 274, 276, 282, 283, 284
Horn-shape, 178
Horse, 87, 88, 91
— shire, 86
— Suffolk, 88
Howard, L., 238, 364
Hudson, P. S., 361
Human
— as a colloquial term, 190
— populations, 330
— progress, 329
— teeth, 203
Humboldt, F. H. A., 299
Hume, D., 336, 338, 341
Hunter, G. L., 362
Hurst, C. H., 161
Huxley, J. S., 235, 236, 281, 314, 315, 328, 334, 347, 352–359, 364, 366
Huxley, T. H., xi, xv, 122, 123, 127, 153, 173, 180, 186, 302, 313, 339, 346, 347, 351–354, 363, 366
Hyatt, A., 159
Hybrid, 34, 51, 52, 83, 113, 118
— interracial, 52
— inviability, 34
— sterility, 31
— vigour, 89
Hybridism, 117
Hybridization, 22, 31, 39, 41, 47, 52, 76
Hydra, 62
Hyracotherium, 141
Hysteria, 226

Ichthyosaurs, 144
Ideal type, 327
Inbreeding depression, 286
Incompatibility, 285
Inheritance, 284
— blending, 6, 286, 287, 288
— particulate, 288
— of acquired characters, 316
— social, 331
Insecta, 113–115, 125, 166
— flower-pollinating, 282
Insemination, interspecific, 144
"Instinct", 208, 210, 211
Instinctive behaviour, 211
Intelligence-testing, 303
Interbreeding between species, 28
Intergradation, 48
International Code, 127
International Rules of Zoological Nomenclature, 126
Introgression, 51
Iris, 51
Isolate, 26
Isolating mechanisms, 28, 30, 39, 42
Isolation, 280
— geographical, 280
— reproductive, 32, 33, 40, 41, 49–52
— sexual, 30, 33, 34

Java man, 167, 189
— skull, 193

Jepsen, G. L., 360
Joad, C. E. M., 317, 366
Johannsen, W. L., 56
Journal of Researches, 245, 247, 249
Jurassic era, early, 148

Keith, A., 327, 366
Kettlewell, H. B. D., xiii, 360
Keynes, J. M., 302
King, T. J., 361
Kingsley, C., 186
Knight, J. R., 33, 276
Kimball, R. F., 44
Koopman, K. F., 33, 235, 236, 364
Kropotkin, P., 305
Kuenen, P. L. H., 265, 364

Lacertilia, 110
Lack, D., xv, 233, 360, 364
Ladd, H. S., 364
Lagomorpha, 125
Lagoon, 255, 262, 265
— Channel, 260, 265
Lake sediment, 273
Lamarck, J.-B., 1, 6, 7, 9, 14, 19, 56, 153
— two-way theory, 63
Lamarckian
— behaviour, 63
— elements, 288
Lamarckism, xiii, 6, 208, 340
Lamp-shell, 117
Land bridges, 275
Lansing, A. I., 65, 66, 361
Lanugo, 203
Larva
— auricularia, 165
— cypris, 124
— nauplius, 156
— trochophore, 156
Larval forms, 163
Latimeria, 111, 116, 148
Lederberg, E. M. and J, 42, 361
Le Gros Clark, W. E., 363
Leguminosæ, 276
Lemche, H., 163, 363
L'Heritier, P. L., 66, 361
Life
— progress, 332 *e.s.*
— origin, 319
Limbs
— arterial pattern, 200
— dimensions, 201
Lingula quadrata, 116, 117
Linnæus, C., 19, 35, 37, 46, 121, 311
Liphistius malayanus, 116
Liperoceras cheltiense, 161
Liperoceratidæ, 161
Lizard, 110, 111
Locke, J., 19
Locomotion, erect bipedal, 196
Loomis, F. B., 362
Lorch, I. J., 361
Lorenz, K. Z., 214, 363
Lung fish, 110, 116
Lusitanian element, 274
Lyell, C., 137, 152, 153, 250, 273, 274
"Lysenko controversy", 81
Lysenko, T., 7, 81

MacBride, E. W., 162
McMeekan, C. P., 92, 94, 95, 97, 98, 362
MacRae, D. G., 366
Macrauchenia, 132, 137
Malthus, T. R., 4, 296, 297, 337, 366
Mammalia, 111, 113, 114, 125, 328
— fossil, 131
— hoofed, 132
Man, 26
— as a colloquial term, 187, 190
— dominance, 329
— fitness in, 330
— Java, 167
— modern, 167
— Neanderthal, 167
— skull, 188; 193
Marett, R. R., 366
Marriage, likelihood, 26
Marsupial, 132
Maternal influence, 86
Mather, K., 361, 366
Marx, K., xiv, 305
Masserman, H. J., 219, 220, 363
Mauritius, 249, 253, 257
Maynard Smith, J., 364
Mayr, E., 35, 51, 360
Mechanisms, self-regulating, 291
Medawar, P. B., 67, 70, 71, 72, 361
Megatherium, 132, 134
Melanism in moths, xiii
Mendel, G., 9, 20, 22, 56, 85, 287
Mendelism, 9, 85
— and animal breeding, 85
Merriam, C. H., 43
Mesoderm, 74, 75
Mice, 59, 71, 72
— skin grafting in, 70
Michurin, I. V., 76
Migration, 38, 271, 274, 276
Milani, R., 243, 364
Milk
— "let down," 99
— secretion, 100
— yield, 100
Miscegenation of incipient species, 34, 38
Mobility of the continents, 275

Mollusca
— archaic segmented, 148
— gastropod, 163
Monkey, 131
Monocotyledons, 125
Monogamous species, 232–234, 243
Monograph of the Cirrepedia, 103, 118, 120, 122, 123, 126, 127
Montague, M. F. A., xv, 360
Montesquieu, C., 299
Moore, J. A., 34
Morgan, T. H., 56, 59, 293
— chromosome theory, 56
Morphological pattern, 184
Morris, D., 363
Morse, M. L., 361
Mosquito, 45, 119
Motivation, 216 *e.s.*
— and psychoanalysis, 222
— unconscious, 221
Moths
— dark dagger, 119
— grey dagger, 119
— melanism in, xiii
Mountain building, 131
Mowrer, O. H., 222, 223, 363
Moynihan, M., 235, 363
Mule, 30, 87
Müller, F., 157
Muller, H. J., 13, 34
Murchison, R. I., 152
Murray, J., 263, 264
Muscular system, 179
— muscles, 203
Mutation, 12–14, 22–24, 58, 61, 316, 319, 322, 324

Nanney, D. L., 44
Natural history, 102, 119, 122 *e.s.*
Natural selection, xi, xii, xiii, 4, 9 *e.s.* 22 *e.s.*, 33, 34, 51, 54, 313 *e.s.*, 335 *e.s.*
— consequences, 38
Naturalists, amateur, 128
Naudin, C., 287
Nauplius larva, 156
Neanderthal man, 166, 167, 189, 194
Neo-Darwinism, 10
Neoteny, 166
Nervous system, 213
— autonomic, 213
Newt, 62
— perennibranchiate, 106–108
Nomenclature, biological, 118, 126
— International Rules, 126
Nutrition, 91, 92, 93, 95, 97
Nuttal, G. H. F., 185, 363
Nucleus, Ch. 3 *passim*
— " endoderm ", 74, 75
Nucleus, somatic, 75
— transplantation, 72, 74

Object displacement, 226
Œstrogen, 99
Old Stone Age cultures, 189
Old Testament, 142
On the Origin of Species, 2, 130, 138, 140, 141, 149, 152, 313, 335
Ophidia, 110
Ophrys, 281, 282
Opossum, 131
Orchidaceæ, 278, 279, 280, 283
— butterfly orchis, 284
Orchis latifolia, 271
Ord, M. J., 361
Order, 35, 111, 114, 124
Ordovician Age, 148
Organ
— rudimentary, 102, 110
— vestigial, 172, 203
Origin of life, 319
— polytopic, 270
Outbreeding, 285, 325
Owen, R., 137, 138, 140

Pædomorphosis, 106–108, 163, 165, 167
Palæozoic Age, 116, 143, 144
Pålsson, H., 92, 98, 362
Pampas, 136
Pangenesis, 9, 57, 68, 69, 83, 288, 289
Parallelism, 156, 182, 183
Paramecium, 44, 63, 64, 66, 78, 82
— *aurelia*, 63, 77
Pareto, V. F. D., 311
Parthenogenesis, 41, 55
Particles
— self-replicating, 65–67, 69
— virus-like, 67
Pasteels, J., 170
Patagonia, 136
Patterson, J. T., 33, 36
Pavlov, A., 160
Pavlov, I. P., 219, 228
Pearson, K., 302
Pelvic bone, 196, 197
Peroneus tertius muscle, 179
Phenotype, 287
Phototropism, 291
" Phyllopod ", 125
Phylogenetic
— diagram, 104
— relationships, 295
Phylum, 35
Pig, 92 *e.s.*
Pigeons, 92
Pisces, classification, 111

Pithecanthropus, 167, 189, 192, 194, 196, 204
"Pithecometra thesis", 180
Plants, climbing, 289
Plant communities, arctic-alpine, 273
Plasmagenes, 67, 69
Pleistocene Age, 132, 136, 190
Plesiosaurs, 144
Political system, 331
Pollen
— carriers, 280
— pre-potency, 285
Pollination, 278
— assortative, 281, 282
— cross-pollination, 30, 284
— illegitimate, 284
— legitimate, 284
— mechanisms, 277
Pollinium, 279
Polygamous species, 232, 234, 243
Polymorphism, flower, 284
Polyploidy, 270
Pongidæ, 174, 175, 184, 186, 194, 201
— dental characters, 195
Pontecorvo, G., 42
Pony
— Shetland, 86
— Welsh, 88
Population
— development, 319
— fly, 324, 325
— genetical stability, 320
— human, 330
— Mendelian, 26, 27, 34, 38, 42, 45, 50
— panmictic, 26, 28
— reproduction, 319
Posture, erect, 179
Potato, 62
Pre-Cambrian Age, 149
Precipitin reaction, 185
Proceedings of the Zoologica Society of London, 119, 122
Progeny, sterile, 34
Progeny-testing, 85
Progress, 327
— biological, definition, 315
— of life, 332
Pseudo-hybrids, 75, 76
Psilotaceæ, 116, 117
Psychoanalysis, 221, 222
Psychoanalysis as Science, 223, 363
Psychoneurosis, 220, 221
Psychosomatic disorder, 220, 226
— state, 221
Pterobranchia, 116

Qualitative differences, 287
Quantitatively varying characters, 287
Quercus, 52

Rabbit, 87, 125
Race, 20, 29, 30, 51, 327
— of man, 27, 28
— pure, 23, 328
Radiation, 13
Radio-active rocks, 143
Rana, 34, 49
"Random" process, 14
Reaction-formation, 222, 223
Recapitulation
— theory, 158, 199
Reasons, 349, 350, 351
Recombination, 62, 287
Redirection activity, 226
Reef
— barrier, 245 *e.s.*
— fringing, 252 *e.s.*
Regression, 222, 223
Releasers, social, 213
Releasing stimuli, 214
Reproduction
— asexual, 23, 55
— isolating mechanisms, 30, 33, 39, 40, 42
— "parasexual", 42
— of a population, 319
— sexual, 22–24, 41, 55, 322, 325
Reptilia, 110, 111, 125, 328
Rendel, J. M., 239, 364
Rhythms, autonomic, 291
Richens, R. H., 361
Riley, H. P., 53
Robertson, A., 33
Robins, 233
Rodents, 114, 125
— of South America, 131
Rotifers, 65, 66
Roux, W., 159
Royal Society, 264
Rubus, 43
Rudimentary organs, 102
Rudiments, 202 *e.s.*
Rye, 76, 77

Salamander, 47, 48
Scelidotherium, 136
Schultz, A. H., 198, 201, 363
Scott, J., 268, 278
Sedgwick, A., 148, 152, 161
Segregation, 286, 287
Selection, 346
— artificial, 87, 324, 325
— natural, xi, xii, xiii, 4 *e.s.*, 33, 34, 51, 54, 335 *e.s.*
— sexual, 31, Ch. 10 *passim*
Self-replicating particles, 65, 66, 67, 69
Serres, P. M. T. de, 154
Seward, A. C., 362, 366
Sex, 288

Sex, cell, 22
Sexual
— isolation, 30, 44
— maturity, 30
— selection, 31, Ch. 10 *passim*
Seymouria, 146
Sheep, 31, 87, 91, 92, 95, 98
Shorthorn, 87, 88
Sibley, C. J., 53
Silurian Age, 114, 117
Simpson, G. G., 182, 192, 360, 363
Skull
— gorilla, 188, 191
— Java man, 193
— man, 188, 193
— man-ape, 191
Smith, A., 299, 336, 337, 341
Snake, 110, 111
Social releasers, 213, 215
"Social signals", 213, 215
Sociology, developmental, 306 *e. s.*
Somatic cell, 59
Sonneborn, T. M., 44, 78, 361
South African "man ape", 167, 190, 195
— skull, 191
South America, 130 *e.s.*
— fossil mammals, 131
Space, conquest, 332
Spath, L. F., 160
Species, 2, 20, 30, 102, 103, 115 *e.s.*
— adaptation, 333
— distinction, 32
— diversity, 332
— as dynamic entity, 37
— existence, 32
— group, 125
— Linnæus' review, 19
— monogamous, 232, 233, 234
— naming, 45
— number, 35
— origin, 326
— polygamous, 232, 234
— problem, 117
— sibling, 43 *e.s.*
Speciation, Ch II *passim*, 280, 327
Spencer, H., xi, 226, 300, 306, 316, 345, 353
Spender, M., 365
Spiders, 215
— Liphistiomorph, 116
Sprengel, C., 281
Spuhler, N. J., 166
Squamata, 111
Stability, genetical, 321 *e.s.*
"Standards of value", 344–348, 353, 354, 357
Statements of fact contrasted with value-judgement, 349–351
Stebbins, L., 54
Stebbins, R. C., 48
Stephenson, A., 365
Stephenson, T. A., 365
Sterility, self-, 284, 285
Stockard, C. R., 62, 361
Stone, W. S., 36
"Strickland Code", 126, 127
Structural intergradations, 279
Structure and Distribution of Coral Reefs, 245, 249–251, 257, 258
"Struggle for existence", 335 *e.s.*
Studbooks, 86
Subclass, 124, 125
Subsidence, theory, 260–265
Survival, 344
— probability, 317 *e.s.*
"Survival of the fittest", xi, 24, 316, 344
Sweet-pea, 284
Sympathy, 336, 338–342, 344
Syndrome displacement, 221, 226
Systema Naturæ, 46
Systematist
— formal, 105
— professional, 120

Tadpole, 107
Tahiti, 246, 252
Tail, 199
Tancred, 306
Tandy, G., 365
Tarsier, 183
Taxon, 110
Taxonomists, 39
Taxonomy, 21, 103, 123, 181
— taxonomic category, 115
— "taxonomic inflation", 117, 123
Teeth, 195, 203
"Telegony", 56, 66
Teleological reasoning, 295
Teleostei, 111
Territory of birds, 232, 234, 238
Tertiary Age, 114, 143, 147
— beds, 136, 140
Thiselton-Dyer, W., 292, 365
Thoday, J. M., 366
Thompson, G., 366
Thomson, J. A., 366
Thorpe, W. H., 364
Tiedemann, F., 154
Tinbergen, N., 214, 235, 236, 238, 364
Tit, Great, 238
Tolerance, 72
Tomato, 81, 82
— grafting, 80
Tortoise, 111
Toxodon, 132, 136, 137
Tradition, 331
"Transduction", 78, 79 ,80
— phase, 82

Transplantation
— antigen, 71, 72
— nuclear, 72, 74
— of tissue, 71
Trend evolution, xii
Triassic Age, 148
Tribe, 113
Trilasmus, 127
Trochophore, 156
Tucker, J. M., 52
Turtle, 111
Twins, identical, 23
Tylor, E. B., 306, 336

Ungulates
— odd toed, 142
— primitive, 132
— progressive, 132
Uniformity principle, 153
Use or disuse, 7, 8, 13, 14
Ussher, J., 142, 173

Value, 344
Variability, 320
Variation, 6, 15
— continuous, 10
— genetical, xiii
— hereditary, 6, 11, 13
Variety, 20, 102, 117
— intermediate, 141
Verges, J. B., 92, 98, 362
Versatility
— genetical, 320 *e.s.* 326
— of individuals, 321
Vertebrata, 116
Vertebrata, classification, 110
Vico, G., 299
Virus, 67, 79
Volpe, E. P., 53
Voltaire, F. M. A. de, 299

Waddington, C. H., xii, 33, 347, 348, 349, 352, 353, 360, 336
Wallace, A. R., xi, xiv, 19, 119, 296
Wallace, L. R., 362
Weber, M., 311
Weismann, A., 56, 57, 59, 362
— diagram of heredity, 67
— " one-way theory", 59, 62
Westermarck, E. A., 308
Wheat, 76, 77
Whitehorn, J. C., 364
Will, 14
— concept, 7
Willow warbler, 119
Wilson, E. G., 361
Wolf, Tasmanian, 181, 182
Wollman, E. L., 362
Wright, S., 25, 38
Würtemberger, L., 159

X-rays, 13, 14

Yeates, N. T. M., 362
Yonge, C. M., 365
Young, J. Z., 108, 362

Zinder, N. D., 42